AF389434

Evolutionary Dynamics of Wild Populations

Evolutionary Dynamics of Wild Populations

Editors

Delphine Legrand
Simon Blanchet

MDPI • Basel • Beijing • Wuhan • Barcelona • Belgrade • Manchester • Tokyo • Cluj • Tianjin

Editors
Delphine Legrand
Station d'Ecologie Théorique et
Expérimentale (SETE)
Centre National pour la
Recherche Scientifique
Moulis
France

Simon Blanchet
Station d'Ecologie Théorique et
Expérimentale (SETE)
Centre National pour la
Recherche Scientifique
Moulis
France

Editorial Office
MDPI
St. Alban-Anlage 66
4052 Basel, Switzerland

This is a reprint of articles from the Special Issue published online in the open access journal *Genes* (ISSN 2073-4425) (available at: www.mdpi.com/journal/genes/special_issues/evolutionary_dynamics).

For citation purposes, cite each article independently as indicated on the article page online and as indicated below:

LastName, A.A.; LastName, B.B.; LastName, C.C. Article Title. *Journal Name* **Year**, *Volume Number*, Page Range.

ISBN 978-3-0365-2011-7 (Hbk)
ISBN 978-3-0365-2010-0 (PDF)

Contents

About the Editors

Delphine Legrand

Delphine Legrand is an evolutionary ecologist studying organisms' response to environmental changes. She accomplished her PhD in Gif-sur-Yvette (France) during which she studied the evolutionary history of Drosophila species belonging to the melanogaster complex using both genetic and ecological data. She then worked on metapopulation dynamics in both experimental and natural populations of butterflies during her post-doctorates. She especially studied how phenotypic and environmental drivers, and their interaction, modulate the dispersal decision. She was recruited as a CNRS researcher in Moulis (France) where she is now studying the evolutionary response of ciliates to environmental stressors in experimental (spatially-explicit) microcosms. She combines experimental evolution, whole-genome sequencing and phenome screening to study adaptation of populations exposed to combination of stressors, in relation with global change. She has special interest for interdisciplinary research projects at the interface between evolution, ecology and genomics in which it is possible to decipher the (molecular) mechanisms behind evolutionary patterns.

Simon Blanchet

Simon Blanchet is an evolutionary ecologist tackling various questions about biodiversity in freshwaters. He accomplished his PhD in Quebec city (Canada) during which he investigated the consequences of competition between native and non-native salmonids, through the lens of ecological and evolutionary theories. He pursued his career in France as a post-doctorate where he enlarges his research scopes by elucidating the evolutionary impacts of non-genetic sources of inheritance in animals. He was recruited as a CNRS researcher in Moulis (France) where he is developing researches focusing on the ecological and evolutionary dynamics of host-parasite interactions, and on the evolutionary causes and ecological consequences of intraspecific (genetic) diversity in river ecosystems. He is also strongly collaborating with local managers and practitioners to develop operational tools for quantifying the impacts of humans on freshwaters. He has always been fascinated by rivers and organisms living in these ecosystems, and most (if not all) his researchers are dedicated to these ecosystems.

Editorial

Evolutionary Dynamics of Wild Populations

Delphine Legrand * and Simon Blanchet

Station d'Ecologie Théorique et Expérimentale, UPR5321, CNRS, 09200 Moulis, France; simon.blanchet@lilo.org
* Correspondence: delphine.legrand@sete.cnrs.fr

Citation: Legrand, D.; Blanchet, S. Evolutionary Dynamics of Wild Populations. *Genes* **2021**, *12*, 778. https://doi.org/10.3390/genes12050778

Received: 11 May 2021
Accepted: 12 May 2021
Published: 19 May 2021

Publisher's Note: MDPI stays neutral with regard to jurisdictional claims in published maps and institutional affiliations.

Wild populations are facing rapid and sometimes extreme environmental changes that are currently exacerbated by pressing human activities. A major scientific endeavor is to reveal the evolutionary processes allowing wild populations to generate adaptive responses to these rapid and drastic environmental changes. In the recent decades, the accumulation of empirical data as well as the development of new theories and molecular tools have largely improved our ability to tackle such a major question. In particular, there is now growing evidence that evolutionary processes (gene flow, drift, mutation, and natural selection) interact in sometimes complex ways to shape the rapid responses of organisms to changing environments, and this can lead to unexpected feedback between evolutionary and ecological dynamics. These rapid responses are sustained by genetic determinants in addition to alternative inheritance systems, including those that are epigenetically controlled. Revealing these underlying molecular mechanisms of adaptation may change the way wild populations are managed and conserved.

This volume "Evolutionary dynamics of wild populations" synthesizes these novel and fascinating studies and provide a rare opportunity to generate a general overview of the ongoing projects tackling the difficult task of studying evolutionary dynamics in natural settings. Evaluating the adaptive potential of populations has long been a matter of interest in evolutionary biology, which recently became an important conservation criterion. By investigating the genome-wide molecular bases of insects' adaptation to toxic host plants, Ferreira et al. [1] reveal footprint of selection in a key chemosensory gene family. They illustrate the importance of standing genetic variation for the emergence of different adaptive behavioral strategies. Working on nearly panmictic sea-water fish populations at a large spatial scale, Baltazar-Soares et al. [2] use a candidate gene approach to show that haplotypic frequencies correlate with thermal and oxygen conditions, suggesting adaptation to local environmental conditions despite a high degree of connectivity in this species. This fish might thus be able to follow environmental optima induced by global change. The fate of populations facing new environments might be influenced by many processes, including those linked to the landscape in which they live. For instance, Han et al. [3] suggest that the conservation status of the evergreen broad-leaved oak in East Asia should be evaluated based on the geographic localization of each population along their core-edge situation. Edge populations can sometimes harbor a similar global level of genetic diversity, but a unique allelic composition compared to that of core populations, suggesting that edge populations must be considered as independent conservation units. At the regional scale, Bal et al. [4] show that patterns of adaptive divergence in two sympatric stickleback species result from the interactive effects between species-specific characteristics and landscape features. They highlight similar levels of genetic diversity and neutral genetic differentiation between the two species, but different levels of morphological and adaptive genomic divergence. One of the two sympatric species systematically displayed a higher level of adaptive divergence, demonstrating the difficulty of extrapolating evolutionary dynamics from one species to another, even if they share a similar environment and a shared ancestry. At a lower spatial scale, Legrand et al. [5] show that the functioning of a natural metapopulation of butterflies does not rely on local dynamics of extinction/recolonization, but rather on a long adaptive history of the species to its local conditions. Especially, the fine-tuning of dispersal rates between populations according to local conditions favors the metapopulation equilibrium. However, other processes than those related to the landscape can strongly influence the fate of natural populations. Using reciprocal transplants between two recently founded populations of brown trout, Labonne et al. [6] show that sexual selection can have important effects on rapid evolutionary dynamics, even more than local adaptation or genetic rescue. By revealing the mechanisms sustaining patterns of genomic introgression in Brook Charr populations supplemented with the same domestic strain, Leitwein et al. [7] show little evidence of shared patterns of domestic ancestry between recipient populations. Patterns of introgression of recipient populations are rather dependent upon their initial genetic diversity, patterns of recombination and the stocking intensity. With the increasing availability of detailed genomic data at the chromosome level, conservation practices should increasingly benefit from the integration of such precise mechanisms in the management of populations, as they should benefit from the use of new tool to evaluate the identification of proper conservation units. Accordingly, Fargeot et al. [8] show that methylation profiles quantified in two sympatric freshwater fish populations better discriminate populations than do neutral genetic markers as those commonly used in conservation genetics studies. Although epigenetic profiles are expected to be strongly associated to environmental variation, they did not find evidence for this pattern in any of the two fish species, suggesting that higher mutation rate in epigenetic than genetic markers and neutral processes (drift) may sustain the higher discriminant ability of epigenetic markers. Methylation marks, as a molecular pathway for the regulation of gene expression, can be an important source of phenotypic variation in natural populations, notably sustaining rapid adaptation by phenotypic plasticity. Mouginot et al. [9] tested this hypothesis in the Snapdragon plant by investigating the link between DNA genome-wide patterns of methylation and degree of phenotypic plasticity in vegetative traits. They show strong epigenetic and phenotypic responses to change in light treatment. However, they surprisingly failed to detect a causal link between the epigenetic and phenotypic responses, pleading for further study on the proximal (molecular) mechanisms sustaining phenotypic variation commonly observed in wild populations. The microbiome can be another important source of phenotypic variation that can be adaptive and sometimes heritable. Bulteel et al. [10] propose to test the original hypothesis that tolerance to parasite in Daphnia could partly be explained by the microbiome carried by each individual. Although they failed to detect an important role of this proximal mechanism on their tolerance to parasites, they demonstrate a substantial role of the genetic background of hosts to both parasite tolerance and microbiome composition. Further studies are required to isolate the fitness benefits of carrying particular microbiomes.

This volume highlights the richness of studies focusing on the evolutionary dynamics of wild populations. It shows the diversity of organisms and approaches that can be used to reveal and understand empirical patterns, with—often but not always—the goal of improving the long-term conservation of wild populations. This diversity reflects the diversity of questions that occupy evolutionary biologists working in wild populations, which go from revealing their global (epi)genetic and phenotypic structure at different spatial and temporal scales to the search of the inherited bases of ecologically relevant phenotypic traits. This volume should be an important contribution to the field because firstly, papers selected in this issue provide answers to timely questions in evolutionary biology. Secondly, it proves that much has to be explored to understand the causes and consequences of evolutionary dynamics of wild populations, and hence that scientists still have to put effort into the study of wild populations.

Author Contributions: Conceptualization, D.L. and S.B.; Writing—original draft preparation, D.L.; writing—review and editing, D.L. and S.B. All authors have read and agreed to the published version of the manuscript.

Funding: This work was carried out at the Station d'Ecologie Théorique et Expérimentale (Centre National de la Recherche Scientifique; CNRS), which is supported by the Laboratoires d'Excellence TULIP (ANR-10-LABX-41).

Data Availability Statement: Not applicable.

Conflicts of Interest: The authors declare no conflict of interest.

References

1. Ferreira, E.A.; Lambert, S.; Verrier, T.; Marion-Poll, F.; Yassin, A. Soft Selective Sweep on Chemosensory Genes Correlates with Ancestral Preference for Toxic Noni in a Specialist Drosophila Population. *Genes* **2020**, *12*, 32. [CrossRef] [PubMed]
2. Baltazar-Soares, M.; de Araújo Lima, A.R.; Silva, G. Targeted Sequencing of Mitochondrial Genes Reveals Signatures of Molecular Adaptation in a Nearly Panmictic Small Pelagic Fish Species. *Genes* **2021**, *12*, 91. [CrossRef] [PubMed]
3. Han, E.-K.; Cho, W.-B.; Park, J.-S.; Choi, I.-S.; Kwak, M.; Kim, B.-Y.; Lee, J.-H. A Disjunctive Marginal Edge of Evergreen Broad-Leaved Oak (Quercus Gilva) in East Asia: The High Genetic Distinctiveness and Unusual Diversity of Jeju Island Populations and Insight into a Massive, Independent Postglacial Colonization. *Genes* **2020**, *11*, 1114. [CrossRef] [PubMed]
4. Bal, T.M.P.; Llanos-Garrido, A.; Chaturvedi, A.; Verdonck, I.; Hellemans, B.; Raeymaekers, J.A.M. Adaptive Divergence under Gene Flow along an Environmental Gradient in Two Coexisting Stickleback Species. *Genes* **2021**, *12*, 435. [CrossRef]
5. Legrand, D.; Baguette, M.; Prunier, J.G.; Dubois, Q.; Turlure, C.; Schtickzelle, N. Congruent Genetic and Demographic Dispersal Rates in a Natural Metapopulation at Equilibrium. *Genes* **2021**, *12*, 362. [CrossRef]
6. Labonne, J.; Manicki, A.; Chevalier, L.; Tétillon, M.; Guéraud, F.; Hendry, A.P. Using Reciprocal Transplants to Assess Local Adaptation, Genetic Rescue, and Sexual Selection in Newly Established Populations. *Genes* **2020**, *12*, 5. [CrossRef]
7. Leitwein, M.; Cayuela, H.; Bernatchez, L. Associative Overdominance and Negative Epistasis Shape Genome-Wide Ancestry Landscape in Supplemented Fish Populations. *Genes* **2021**, *12*, 524. [CrossRef] [PubMed]
8. Fargeot, L.; Loot, G.; Prunier, J.G.; Rey, O.; Veyssière, C.; Blanchet, S. Patterns of Epigenetic Diversity in Two Sympatric Fish Species: Genetic vs. Environmental Determinants. *Genes* **2021**, *12*, 107. [CrossRef] [PubMed]
9. Mouginot, P.; Luviano Aparicio, N.; Gourcilleau, D.; Latutrie, M.; Marin, S.; Hemptinne, J.-L.; Grunau, C.; Pujol, B. Phenotypic Response to Light Versus Shade Associated with DNA Methylation Changes in Snapdragon Plants (Antirrhinum Majus). *Genes* **2021**, *12*, 227. [CrossRef] [PubMed]
10. Bulteel, L.; Houwenhuyse, S.; Declerck, S.A.J.; Decaestecker, E. The Role of Microbiome and Genotype in Daphnia Magna upon Parasite Re-Exposure. *Genes* **2021**, *12*, 70. [CrossRef] [PubMed]

genes

MDPI

Article

Soft Selective Sweep on Chemosensory Genes Correlates with Ancestral Preference for Toxic Noni in a Specialist *Drosophila* Population

Erina A. Ferreira [1,2], Sophia Lambert [2], Thibault Verrier [2], Frédéric Marion-Poll [1,3] and Amir Yassin [1,2,*]

[1] Laboratoire Évolution, Génomes, Comportement et Écologie, CNRS, IRD, Université Paris-Saclay, 91198 Gif-sur-Yvette, France; ferreira@egce.cnrs-gif.fr (E.A.F.); frederic.marion-poll@egce.cnrs-gif.fr (F.M.-P.)
[2] Institut Systématique Evolution Biodiversité (ISYEB) Centre National de la Recherche Scientifique, MNHN, Sorbonne Université, EPHE 57 rue Cuvier, CP 50, 75005 Paris, France; slambert@bio.ens.psl.eu (S.L.); thibault.verrier@free.fr (T.V.)
[3] AgroParisTech, Université Paris-Saclay, 75231 Paris, France
* Correspondence: yassin@egce.cnrs-gif.fr

Abstract: Understanding how organisms adapt to environmental changes is a major question in evolution and ecology. In particular, the role of ancestral variation in rapid adaptation remains unclear because its trace on genetic variation, known as soft selective sweep, is often hardly recognizable from genome-wide selection scans. Here, we investigate the evolution of chemosensory genes in *Drosophila yakuba mayottensis*, a specialist subspecies on toxic noni (*Morinda citrifolia*) fruits on the island of Mayotte. We combine population genomics analyses and behavioral assays to evaluate the level of divergence in chemosensory genes and perception of noni chemicals between specialist and generalist subspecies of *D. yakuba*. We identify a signal of soft selective sweep on a handful of genes, with the most diverging ones involving a cluster of gustatory receptors expressed in bitter-sensing neurons. Our results highlight the potential role of ancestral genetic variation in promoting host plant specialization in herbivorous insects and identify a number of candidate genes underlying behavioral adaptation.

Keywords: insect-plant interactions; standing genetic variation; genome-wide selection scan; gene family evolution; feeding behavior

Citation: Ferreira, E.A.; Lambert, S.; Verrier, T.; Marion-Poll, F.; Yassin, A. Soft Selective Sweep on Chemosensory Genes Correlates with Ancestral Preference for Toxic Noni in a Specialist *Drosophila* Population. *Genes* **2021**, *12*, 32. https://doi.org/10.3390/genes12010032

Received: 30 November 2020
Accepted: 22 December 2020
Published: 29 December 2020

Publisher's Note: MDPI stays neutral with regard to jurisdictional claims in published maps and institutional affiliations.

1. Introduction

Host plant specialization by herbivorous insects is a complex phenomenon requiring the simultaneous evolution of multiple adaptive phenotypes on the same genome. Traditionally, these phenotypes are classified under two broad categories: preference phenotypes inducing the choice of the particular host by the insect, and performance phenotypes improving the survival of the insect on the host [1]. Preference phenotypes could rely on visual, chemical, anatomical or phenological attributes of the host plant. The signals of each of these attributes need to be transmitted by the peripheral nervous system of the insect to the central nervous system, which following processing of the information would elicit attraction or repulsion behaviors. How the insect perceives the attributes of its host plant is a question of intense evolutionary and neurogenetic research [2–4].

Much of our knowledge on the genetic basis of perception of environmental cues comes from studies on the model fly *Drosophila melanogaster*. The family Drosophilidae contains a wide spectrum of fly-plant associations going from generalist detritivorous species such as *D. melanogaster* to strict herbivorous such as species of the genus *Scaptomyza* [5]. Comparative genomics of the chemosensory gene families between generalist and specialist taxa have provided significant insights on how host plant shift and specialization can be driven by these genes [6–8]. There are multiple gene families that encode transmembrane receptors localized on the peripheral nervous system, of which three have attracted much attention due to their high diversification rate, namely olfactory (ORs), gustatory (GRs) and ionotropic (IRs) receptors. A fourth family, odorant-binding-proteins (OBPs), encodes for proteins that bind with both volatile odors and soluble tastants to help them migrate within the intercellular hemolymph up to the chemosensory neuron.

Between-species comparative genomics revealed the dynamic evolution of each of chemosensory gene families in terms of gene gains through duplication and conversion, gene losses through deletions and pseudogenizations, and protein sequence evolution through extensive non-synonymous mutations [9,10]. Specialist species are often characterized by a higher rate of gene losses and protein changes. At the intra-specific level, which should correspond to the early stages of host shift, only very few cases have been identified, e.g., the many geographical races of the cactophilic species *Drosophila mojavensis* in North America [11,12]. Of particular interest is the parallel specialization of the species *Drosophila sechellia* and the subspecies *Drosophila yakuba mayottensis* on toxic noni (*Morinda citrifolia*) fruits on separate islands of the Indian Ocean, i.e., the Seychelles and Mayotte, respectively [13,14]. Noni toxicity is due to its high content of medium-chained carboxylic acids (hexanoic and octanoic acids) in its fruit [15]. Remarkably, *D. sechellia* has evolved a strong olfactory attraction to those acids [16,17], as well as to methyl hexanoate, the major ester characteristic of the rotten, non-toxic fruit of noni [18]. Genetic studies revealed that evolutionary changes at the ionotropic receptor *Ir75b* and at the olfactory receptor *Or22a* underlie *D. sechellia* preference to noni hexanoic acid and methyl hexanoate, respectively [19,20]. Matsuo et al. [21] also suggested that the odorant-binding-protein OBP57e/d may also play a role in the gustatory preference of this species to noni acids. However, in spite of evidence for rapid evolution of gustatory receptors in *D. sechellia* [6,7], the role of those receptors in noni specialization remains unknown. This may be partly because both sugar and bitter sensing neurons may be involved in dose-dependent acid sensing, as it has recently been demonstrated from functional studies in *D. melanogaster* [22], therefore complicating the dissection of this character. For *D. y. mayottensis*, an olfactory preference for noni fruits in adult flies exists [14], but we still do not understand of the chemosensory evolution underlying this preference, as well as possible gustatory preference, in this subspecies.

Intra-specific adaptive changes related to host shift could be detected through genome-wide selection scans (GWSS) comparing two populations or races with different hosts [2,4]. However, our ability to detect such changes through GWSS depends on multiple factors, most importantly, the frequency of the selected allele(s) in the ancestral population before encountering the new host. If the advantageous allele was present at a very low frequency in the ancestral population or was introduced by a new mutation, selection driving it to fixation or near fixation on the new host will leave a strong signal on the genome known as a 'hard selective sweep'. Depending on the rate of recombination, neutral alleles linked to the selected one will also increase in frequency facilitating the detection of the whole region through GWSS using large genomic windows. However, if the advantageous allele was already at intermediate frequency in the ancestral population, its detection through GWSS becomes more complicated, because fewer neutrally-linked loci will be associated to its fixation, a phenomenon known as 'soft selective sweep' [23,24].

In a previous GWSS study, we have used differentiation at 10-kb windows in highly-recombining regions to detect 'hard selective sweeps' associated with specialization on toxic noni in *D. yakuba* [14]. Although, we have identified multiple regions, surprisingly, none contained any member of the four chemosensory gene families, despite a significant difference in olfactory preference between generalist and specialist populations. In this paper, we test the hypothesis that chemosensory genes in *D. y. mayottensis* might be under 'soft selective sweeps' and that an ancestral, yet substantial, preference for noni chemicals may be present in populations from the ancestral range. Using a combination of population genomics and behavioral analyses, we confirmed both hypotheses and identified some candidate genes potentially involved on the specialization on noni in both *D. yakuba* and

D. sechellia. We then discuss the relevance of our results within the broader context of plant-insect interaction and adaptation.

2. Materials and Methods

2.1. Population Genomics Analyses of Chemosensory Genes

We used genomic data from Yassin et al. [14] produced from two pools of 22 isofemale lines of *D. y. mayottensis*. Each pool consisted of 33 F_1 females from 11 lines (i.e., 3 females per line). As in Yassin et al. [14], we used sequences produced by Rogers et al. [25] for 10 inbred lines of *D. y. yakuba* from Kenya and Cameroon. All reads were mapped to the *D. yakuba* reference genome v.1.05 obtained from Flybase (https://flybase.org/; [26]) using Minimap2 software package [27]. Minimap2-generated SAM files were converted to BAM format using samtools 1.9 software [28]. The BAM files were then cleaned and sorted using Picard v.2.0.1 (http://broadinstitute.github.io/picard/). However, instead of directly merging both *D. y. mayottensis* pools as in Yassin et al. [14], we first generated using Popoolation 2 [29] a synchronized mpileup file for the two pools. We then used a customized Perl script to extrapolate allele frequencies to 22 diploid counts for each pool, and by excluding sites with less than 3 reads. The two genotyped pools were then merged for subsequent analyses. We also generated synchronized files for the 20 *D. y. yakuba* lines using Popoolation 2. However, to account for residual heterozygosity in those inbred lines which was not considered in Yassin et al. [14], we genotyped each line to obtain its diploid alleles, after excluding sites with less than 3 reads and alleles with frequencies less than 25% for the total counts using a customized Perl script. We also excluded tri-allelic sites for those inbred lines. For each geographical population, the diploid genotypes were then pooled and a synchronized file using the two *D. y. yakuba* populations and the *D. y. mayottensis* population was then generated. Sites with less than 10 counts at any of the three populations were excluded. This file was used to estimate nucleotide diversity (π) and pair-wise F_{ST} estimates at each site using HSM [30] formula introduced in a Perl script. We obtained a list of chemosensory genes, their coordinates and their orthology to the *D. melanogaster* genome from FlyBase. For genes with multiple orthologs due to paralogy, we used BLAST software [31] to choose the ortholog with the highest hit. For each gene, we averaged π and F_{ST}, after including up- and downstream 2-kb regions to account for possible regulatory sequences relevant to the gene function. Perl scripts are provided in Supplementary File S1.

In order to estimate deviation of each gene from neutral expectations, we first inferred a demographic model from presumably-neutral short autosomal short introns between *D. y. mayottensis* and the Kenyan population of *D. y. yakuba* using the "prior_onegrow_mig" model implemented in the DADI ver. 1.7. software package [32] as in Yassin et al. [14]. Based on the most optimal model parameters and theta estimates inferred from DADI, we conducted 10,000 simulations of a 5 kb region (i.e., amounting to the average length of a chemosensory gene $\pm$ 2 kb) using the *msms* software package (https://www.mabs.at/ewing/msms/index.shtml, [33]. After considering a recombination rate of 2.5×10^{-8} corresponding to the average rate in *D. melanogaster* [34], the *msms* command was: ms 54 10,000 -t 67 -r 341.9534019 5001 -I 2 10 44 0 -n 1 1.345 -n 2 0.223 -eg 0 2 34.23591156 -ma x 0.205 0.205 x -ej 0.053 2 1 -en 0.43275868 1 1.

For each run, we estimated π in the two populations and F_{ST} between them, both averaged over all segregating loci using a Perl script. We also estimated the highest F_{ST} value at a site for each run (hereafter F_{ST_max}). The distribution of F_{ST_max} was then plotted to identify the 0.95 quantile, i.e., the value above which the hypothesis of a neutral differentiation at a given site could be rejected at $p < 0.05$, as in Bastide et al. [35]. We also reanalyzed the simulation outcome to count the number of sites, exceeding the neutral F_{ST_max} threshold.

2.2. Behavioral Feeding Preference Analyses

Our previous behavioral analysis in *D. yakuba* [14] involved testing long-range olfactory preference through releasing and recapturing flies using a choice between traps including banana or noni fruits. However, because the most differentiating chemosensory genes were found to be those encoding gustatory receptors (see below), we decided to evaluate the gustatory preferences using multiple capillary feeders (MultiCAFE) technique [36]. We tested the same two *D. yakuba* populations used in Yassin et al. [14], i.e., a strain of *D. y. mayottensis* collected from the Bay of Soulou in Mayotte in 2013 and a strain of *D. y. yakuba* collected from Kunden in Cameroon in 1967. We also used a strain of *D. sechellia* collected from the Seychelles in 1985 as in Yassin et al. [14]. In addition, we used a strain of *D. melanogaster* collected from Mt. Oku in Cameroon in 2016 (courtesy of Jean R. David), as well as a laboratory strain iso-1, which carried mutations in some of the genes of interest identified in this study (courtesy of Jean-Luc Da Lage). All these strains were kept in a constant size (N = ~2000 flies) on a standard *Drosophila* medium at 18 °C.

For the MultiCAFE test, 10 > 5-days-old females were sorted the day before the experiment and placed for starvation in tubes with a filter paper humidified with 2 mL of distilled water. The tubes were kept for 24 h in a dark incubator at 25 °C. Flies were then aspirated and placed in group in custom feeding chambers (5.5 × 0.6 × 0.5 cm) and six 5 µL capillaries were introduced through small holes in the ceiling of the chamber, alternating between those containing the control solutions with 30 g·L^{-1} sucrose and test solutions containing 30 g·L^{-1} sucrose mixed with methyl hexanoate (MH; Sigma, St Quentin Fallavier, France; CAS number 106-70-7), hexanoic acid (HA; Sigma; CAS number 142-62-1) or octanoic acid (OA; Sigma; CAS number 124-07-2). HA and OA are the main noni toxins that are abundant at the toxic ripe stage, whereas MH is the characteristic ester of the non-toxic overripe stage [15,18]. All substances were tested at 0.5% concentration (4.6 g·L^{-1}), and four replicates per strain were tested (except for *D. melanogaster* iso-1 mutant strain which was not tested for this concentration). HA was also tested at 1% concentration (9.3 g·L^{-1}) due to its higher solubility and lesser toxicity compared to OA. For this analysis, capillaries were introduced as pairs into individual feeding chambers (2.3 × 0.6 × 0.5 cm), and multiple individuals per strain were quantified: 27, 33, 70, 74 and 30 for *D. y. mayottensis*, *D. y. yakuba*, *D. sechellia*, *D. melanogaster* Oku and *D. melanogaster* iso-1, respectively. All solutions were colored with Brilliant Blue R (Sigma; CAS Number 6104-59-2) to facilitate consumption quantification.

Flies were videotaped in MultiCAFE chambers at 25 °C for 2 h, using Logitech HC920 webcams, by capturing one image per minute using an open source camera security software, Ispy (https://www.ispyconnect.com/). The stacks of images were analyzed using a custom plugin developed in Java to work with Icy [37] to calculate consumption of each solution. The evaporation was estimated by measuring changes in the liquid level of 2–4 capillaries placed in the same cages, but out of reach for the flies. At the end of the experiment (i.e., 2-h), a preference index (PI) was calculated for each substance as follows:

$$PI = \frac{(C_S - C_C)}{(C_S + C_C)} \tag{1}$$

where C_S and C_C are consumption from the substance and control capillaries, respectively, and computed as the actual volume drop minus drop observed in the evaporation capillaries in µL. For flies tested in groups, C_S and C_C were respectively averaged per experiment. Statistical analyses were conducted using the R package [38]. For analyses at 0.5% concentrations, we used pairwise Wilcoxon test, since only four values per strain per substance, i.e., the average consumption by a group of 10 flies, were compared. For analyses at 1%, since consumptions were measured individually, we used Student's *t* test for pairwise comparisons, in addition to the Wilcoxon test.

3. Results

3.1. A Handful of Chemosensory Genes, Mostly Encoding Gustatory Receptors, Deviate from Neutral Expectations

On average, the level of genetic variation for chemosensory genes differed between Mayotte and Kenya *D. yakuba* populations (π = 0.0095 and 0.0107, respectively; Kruskal-

Wallis test $p < 0.01$) but not between gene families in Mayotte (Kruskal-Wallis test $p = 0.51$). Mean genetic differentiation between the two populations also did not significantly differ among the four chemosensory gene families (averaged $F_{ST} = 0.08$; Kruskal-Wallis test $p = 0.46$; Figure 1a). Remarkably, these values were far lower than the average neutral F_{ST} expected from the msms simulations ($F_{ST} = 0.136$; Figure 1b). Interestingly, F_{ST} for a single gene exceeded the 95% quantile estimate ($F_{ST} = 0.161$) from the *msms* simulation. This gene was the gustatory receptor *Gr22c* ($F_{ST} = 0.168$). It was followed by its two adjacent paralogs *Gr22b* ($F_{ST} = 0.143$) and *Gr22d* ($F_{ST} = 0.147$). The three genes belong to a cluster of closely-related GR paralogs (hereafter the Gr22a clade) falling on the distal part of chromosome arm 2L. This clade contains six genes that are expressed in bitter tasting organs [39]. *Gr22c* was pseudogenized in noni-specialist *D. sechellia* and lost in Pandanus-specialist *D. erecta* [6]. *Gr22b* showed a particularly rapid non-synonymous substitution in *D. sechellia* ($\omega > 1$, [7]). Moreover, two other genes *Gr22d* and *Gr22f* have also been pseudogenized in *D. sechellia* [6]. For the other families, a single exception was noticed for the odorant-binding-protein gene *Obp46a* ($F_{ST} = 0.151$).

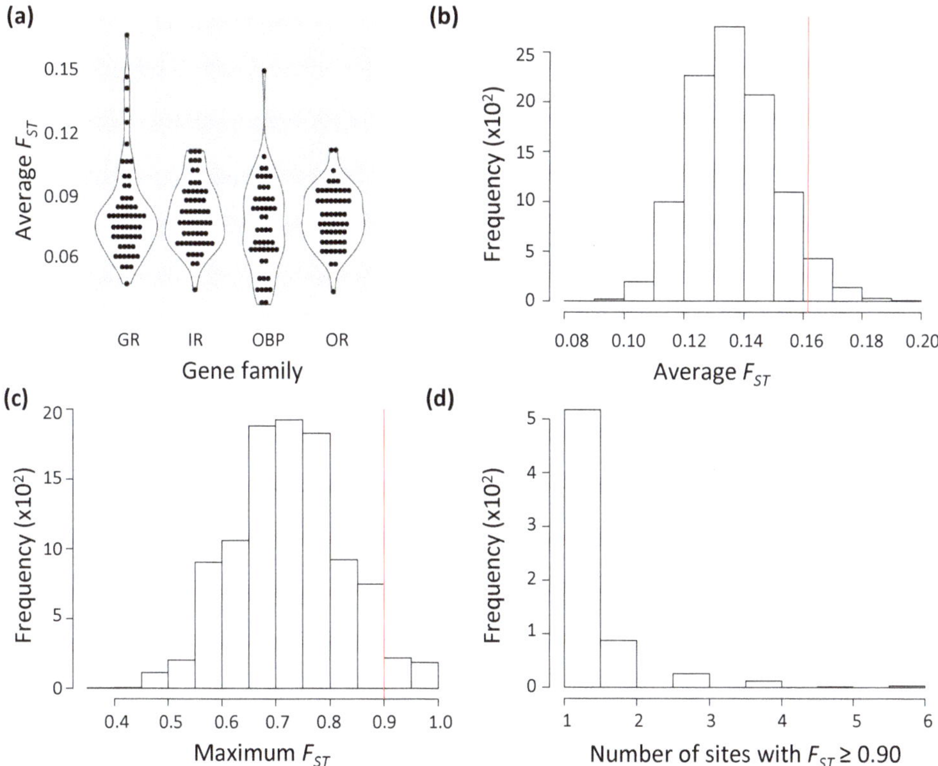

Figure 1. (**a**) Kernel density distributions of genetic differentiation (F_{ST}) estimates at chemosensory genes (±2 kb) between specialist *Drosophila yakuba mayottensis* from Mayotte and generalist *D. y. yakuba* from Kenya. Gene families are abbreviated as: GR = gustatory receptors, IR = ionotropic receptors, OBP = odorant-binding-proteins and OR = olfactory receptors. (**b–d**) Histograms of the outcome of the *msms* simulations based on DADI-inferred demographic model based on presumably neutral autosomal short intronic sequences for (**b**) average F_{ST}, (**c**) maximum F_{ST}, and (**d**) number of sites with $F_{ST} \geq 0.9$ per run. Vertical red lines indicate the 0.95 quantile F_{ST} value.

msms simulations based on the demographic model inferred from presumably neutral short autosomal intronic sites indicated that the probability to obtain a SNP with $F_{ST\_max} \geq 0.90$ in the absence of selection was $p \leq 0.05$ (Figure 1c). We found 13 sites crossing this threshold. Of which, 2, 2 and 9 belonged to GR, IR, and OR families, respectively (Table 1). Those 13 sites fell within 5 genes, with some genes having more than one presumably adaptively evolving site. The probability of having more than one site crossing the 0.90 threshold for one gene is strongly skewed (Figure 1d). Indeed, three genes had only a single site with $F_{ST} \geq 0.90$, i.e., *Gr22b*, *Gr93c* and *Or13a*. One gene, *Ir7a*, had two deviating sites ($p < 9 \times 10^{-3}$). However, the most interesting finding was the olfactory receptor *Or22a*, which underlies *D. sechellia* response to noni major ester ([20], where 8 sites deviated from neutral expectation. According to the *msms* simulations, the probability to obtain eight sites with $F_{ST} \geq 0.90$ in a single gene under neutral conditions is very small ($p < 3 \times 10^{-4}$). When we considered a less stringent threshold of $F_{ST} \geq 0.86$, which corresponded to the 0.90 quantile of the *msms* simulations, three additional sites, all falling in gustatory receptor genes, namely *Gr22d*, *Gr59a* and *Gr93b*, were found (Table 1). No site at OBP genes was found.

Table 1. List of most differentiating sites in chemosensory genes with F_{ST} values ≥ 0.86 between Kenya and Mayotte populations, corresponding to the 0.90 quantile value estimated from the *msms* simulation.

Position	Alleles (ref./alt.)	Gene	Effect	Allele Frequencies			F_{ST}
				Cameroon	Kenya	Mayotte	
X:10060167	G/A	Or13a	Downstream	8/4	1/11	22/0	0.917
X:13654476	T/A	Ir7a	Downstream	5/9	10/0	4/40	0.909
X:13654477	C/A	Ir7a	Downstream	5/9	10/0	4/40	0.909
2L:1497713	G/T	Or22a	Synonymous	6/10	15/1	0/44	0.938
2L:1497721	C/A	Or22a	Synonymous	6/10	15/1	0/44	0.938
2L:1497723	T/G	Or22a	Nonsynonymous (V/G)	6/10	15/1	0/44	0.938
2L:1497724	A/G	Or22a	Nonsynonymous (V/G)	6/10	15/1	0/44	0.938
2L:1497730	C/T	Or22a	Synonymous	6/14	15/1	0/44	0.938
2L:1497751	A/G	Or22a	Nonsynonymous (I/M)	9/5	16/0	0/44	1.000
2L:1497754	T/C	Or22a	Synonymous	9/5	16/0	0/44	1.000
2L:1497757	T/C	Or22a	Synonymous	5/13	16/0	0/44	1.000
2L:1764188	A/C	Gr22d	Downstream	4/14	0/20	19/3	0.864
2L:1768039	T/A	Gr22b	Synonymous	20/0	20/0	2/20	0.909
2R:18929879	G/A	Gr59a	Upstream	16/4	20/0	5/39	0.886
3R:18490921	G/A	Gr93b	Downstream	16/2	20/0	3/19	0.864
3R:18494463	A/G	Gr93c	Nonsynonymous (F/S)	12/8	18/0	4/40	0.909

Our F_{ST} values correspond to differentiation between a pair of populations, but they cannot tell, on their own, which of the two populations has experienced most differentiation since splitting from the common ancestor. Consequently, we checked the frequency of differentiating alleles between Kenyan *D. y. yakuba* and Mayotte *D. y. mayottensis* in another *D. y. yakuba* population from the presumably the ancestral region of the species, i.e., Cameroon [40]. We found that only sites at *Gr22b*, *Gr22d*, *Gr59a* and *Gr93b* had a differentiation between *D. y. mayottensis* and *D. y. yakuba* from Cameroon ≥ 0.50, i.e., the differentiating alleles at those genes have specifically increased in frequency only in *D. y. mayottensis*. For *Or22a* and the other genes, all *D. y. mayottensis* fixed alleles were more frequent in Cameroon than Kenyan alleles (Figure 2a,b). This suggests that those alleles did not likely originate in Mayotte.

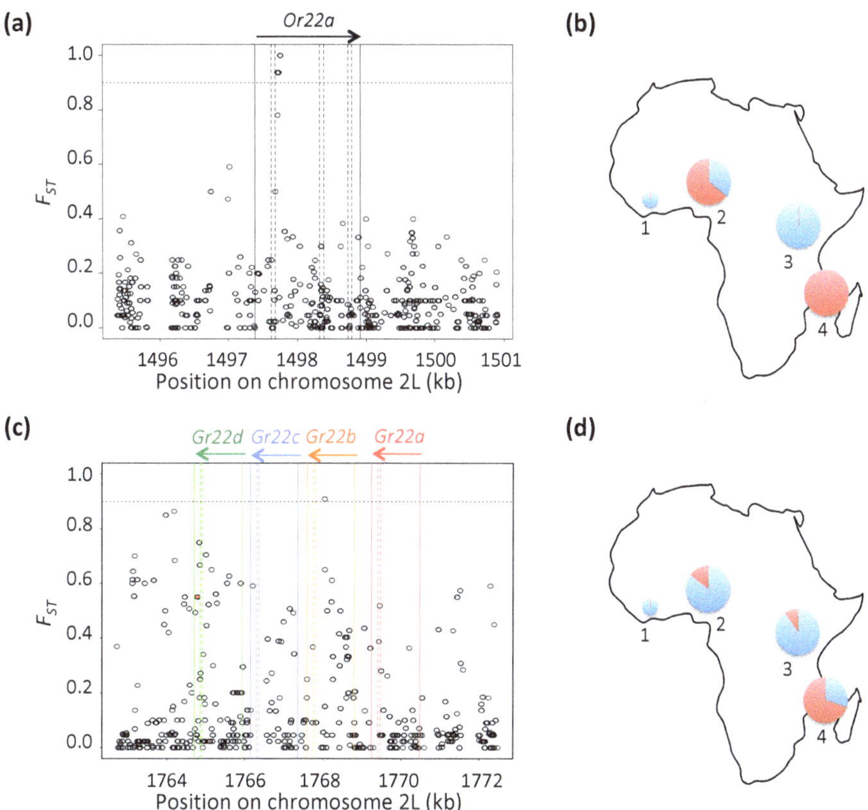

Figure 2. Genetic differentiation (F_{ST}) profile between *D. y. mayottensis* and *D. y. yakuba* (Kenya) at (**a**) the olfactory receptor *Or22a* gene, with (**b**) the geographical distribution of its most differentiating site in Mayotte, Kenya, Cameroon and the reference genome strain from Côte d'Ivoire, and at (**c**) the gustatory receptor genes of the Gr22a clade showing the non-sense mutation at *Gr22d* (red dot) and (**d**) its geographical distribution. Solid and dashed vertical lines indicate gene and exon boundaries, and arrows indicate the gene sense. Dotted horizontal line indicates the F_{ST} value corresponding to the 0.95 quantile of the *msms* simulations.

The overlap between genes of the Gr22a clade in our gene-level and population-specific site-level analyses, as well as the rapid evolution of this clade in *D. sechellia*, motivated us to investigate more thoroughly its polymorphism in *D. y. mayottensis* (Figure 2c). In *Gr22b*, the most differentiating site is a synonymous mutation that does not affect protein sequence. For *Gr22d*, the most differentiating site falls ~0.5 kb downstream the gene. Interestingly, it is close to a nonsense mutation in the second exon of *Gr22d* at 2L:1,764,799. This non-sense mutation has a high intermediate frequency in *D. y. mayottensis* (~0.70%) but does not reach fixation. In *D. y. yakuba*, the mutation is present in both Kenyan and Cameroonian populations but at very low (0.10%) to low (0.15%) frequencies, respectively (Figure 2d). The genetic differentiation pattern observed at *Gr22d* therefore could be explained by selection on the most-differentiating sites, if they play a role in the regulation of *Gr22d* or other genes of the Gr22a clade, on the nonsense mutation itself, or on the epistatic interactions between the most-differentiating sites and the nonsense mutation. Because our gene-level analyses also included 2kb up- and downstream sequences for each gene, the high average F_{ST} values for *Gr22c* that we found above is mostly due to the high differentiation sites in *Gr22b* and *Gr22d*, since the transcribed sequence of this gene does not include any particularly differentiating sites.

3.2. Generalist D. yakuba Has a Substantial Dose-Dependent Gustatory Preference for Noni Toxins

Among the three noni chemicals tested at 0.5% concentration (4.6 g·L^{-1}), no feeding preference was found for MH, the characteristic ester that increases in the rotting stage (Figure 3a). However, significant pairwise differences, especially between the generalist *D. melanogaster* and the specialist *D. sechellia*, were observed for the two toxins OA and HA (Wilcoxon's test $p = 0.029$; Figure 3b,c), characteristic of the toxic ripe stage. No significant difference between *D. y. yakuba* and *D. y. mayottensis* was found for either toxin. However, *D. y. mayottensis* did not significantly differ from specialist *D. sechellia* for either toxin, whereas *D. y. yakuba* suggestively differed from D. sechellia in preference for OA (Wilcoxon's test $p = 0.057$), the most lethal toxin, but not for HA (Figure 3b,c).

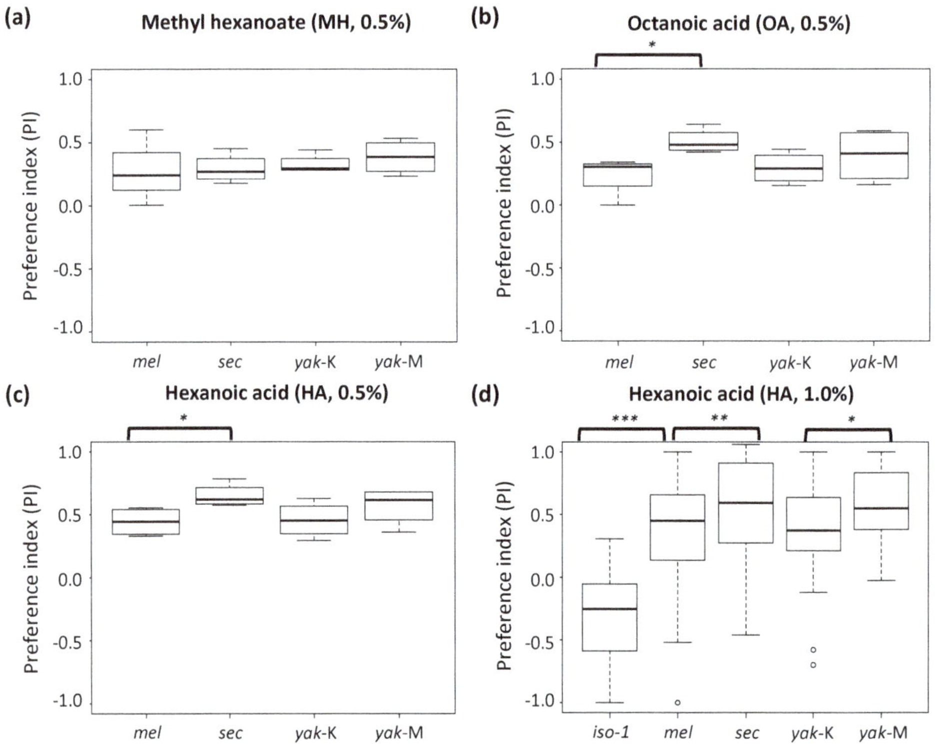

Figure 3. Preference indices of the feeding behavior assays for (**a**) methyl hexanoate at 0.5%, (**b**) octanoic acid at 0.5%, and (**c**) hexanoic acid at 0.5% and (**d**) 1.0%. Strain names are abbreviated as: *mel* = *D. melanogaster* from Mt Oku, iso-1 = *D. melanogaster* strain mutant for *Gr22b* and *Gr22d*, *sec* = *D. sechellia*, *yak*-K = *D. yakuba yakuba* from Kunden and *yak*-M from *D. y. mayottensis*. *p*-values are indicated as: * < 0.05, ** < 0.01 and *** < 0.001.

We also conducted an experiment using a higher concentration (9.3 g·L^{-1}) of HA. Here, both *D. sechellia* and *D. y. mayottensis* showed a significant preference for the acid solution compared to their respective, generalist relatives, i.e., *D. melanogaster* ($t = -2.708$, d.f. = 141.42, $p = 0.008$; Wilcoxon's test $p < 0.010$) and *D. y. yakuba* ($t = -2.133$, d.f. = 57.76, $p = 0.037$; Wilcoxon's test $p = 0.052$), respectively, although the difference was more pronounced between *D. sechellia* and *D. melanogaster* (Figure 3d). Interestingly, whereas no significant difference was found between *D. sechellia* and *D. y. mayottensis* ($t = -0.060$, d.f. = 61.21, $p = 0.952$; Wilcoxon's test $p = 0.670$), *D. sechellia*'s preference was significantly higher than that of *D. y. yakuba* (t = 2.251, d.f. = 62.43, p = 0.028; Wilcoxon's test p = 0.019) We also compared wild-type *D. melanogaster* with the strain iso-1, which carries non-sense mutations at both *Gr22b* and *Gr22d*. Here, a strongly significant difference was observed, with the mutant strain showing almost no preference for the acidic solution (t = −7.9002, d.f. = 63.76, p = 4.9 × 10^{-11}; Wilcoxon's test p = 9.5 × 10^{-10}; Figure 3d).

4. Discussion

4.1. Contrast between Olfactory and Gustatory Evolution in D. yakuba

Our results support the hypothesis that soft selective sweep on ancestral genetic variation in chemosensory genes might have promoted the evolution of *D. y. mayottensis* preference for noni chemicals, but they also unravel a major contrast in olfactory and gustatory evolution between molecular and phenotypic levels. On the one hand, specialist *D. yakuba* significantly differ in olfactory preference for noni from generalist flies [14] but there is weak signal of selection on olfactory receptors. On the other hand, gustatory receptors are significantly present in the pool of non-neutrally evolving SNPs although only suggestive, not significant, gustatory differences exist between specialist and generalist *D. yakuba*. The contrast at the phenotypic level may be explained by differences in the experimental settings we used for the two behaviors.

First, we measured olfactory preference by releasing flies from a distance from two traps each with a different fruits (i.e., banana vs. noni) [14]. Evolved repulsion against banana, for example, could have coupled with preference for noni to increase the difference between Mayotte and the mainland. No choice between two fruits was given to the flies in the gustatory experiment, and it could be that medium-chained carboxylic acids at the tested concentration were general appetizers. For example, even the generalist *D. melanogaster* had always a positive preference index for capillaries with noni chemicals (including the toxins) in agreement with previous reports in this species [22,41].

Second, flies were "trapped" by the choice they made in the olfactory experiments leaving no room for learning or plasticity (e.g., after consumption of a certain amount) [42]. This was not the case for the gustatory experiment where flies were confined in one chamber with capillaries containing alternative solutions placed only a few millimeters apart.

Third, the olfactory experiment used whole noni fruits with mixtures of chemicals, whereas only a single chemical was used per gustatory experiment. For oviposition site choice behavior, which is partly determined by gustatory perception, it has been noticed that mash of noni fruits, but not hexanoic or octanoic acids, elicits oviposition in *D. melanogaster* and *D. mauritiana* [17,43].

Fourth, we tested a single concentration for the two acids (4.6 g·L^{-1}) which discriminates oviposition site choice behavior between the four species of the melanogaster complex including *D. sechellia* and *D. melanogaster* [17]. We confirmed that these two species show different feeding behavior at this concentration but generalist and specialist *D. yakuba* did not seem to discriminate them. However, at a higher concentration of hexanoic acid (HA), a significant difference between generalist and specialist *D. yakuba* was found. In *D. melanogaster*, the concentration-dependent gustatory response to hexanoic acid is complex and depends on interactions between sweet- and bitter-tasting neurons [22]. Whereas, at concentration < 9.3 g·L^{-1} sweet-tasting neurons elicit an appetitive behavior, at higher concentrations bitter-tasting neurons induce a repulsive behavior. Given that most of the selection signal we identified here for *D. yakuba* was on bitter-tasting gustatory receptor genes, our behavioral assays suggest that those genes might have played a role in a reduced acid-repulsion in *D. y. mayottensis*.

Despite these experimental considerations, we were still able to recover a positive correlation between the results from the olfactory experiment and assays including the two carboxylic acids characteristic of the ripe stage of noni.

4.2. Apparent Lack of Selection on Acid-Sensing Ionotropic Receptor Genes

Octanoic and hexanoic acids constitute ~60% and 20% of the total volatiles of ripe noni fruits, respectively [15]. As rotting proceeds, these acids degrade and the concentration of their respective octanoate and hexanoate esters (e.g., methyl hexanoate) increases [44]. Generalist *Drosophila* are usually attracted to acetic acid produced by bacteria developed on rotten, fermented fruits, as well as by its derived acetate esters produced by the fermenting yeasts [45]. Specialization on noni in *D. sechellia* and *D. y. mayottensis* should have involved increasing responsiveness to medium-chained carboxylic acids and their ester derivatives in couple with decreasing responsiveness to short-chained carboxylic acids and their ester derivatives.

Prieto-Godino et al. [19] have recently shown that increased responsiveness to hexanoic (but not octanoic) acid in *D. sechellia* has involved coding and *cis-* and *trans*-regulatory changes in the ionotropic receptor *Ir75b*, which is responsive to butyric acid in its generalist relatives. They have also suggested that *cis*-regulatory changes expanding *Ir75b* have evolved in the ancestor of *D. sechellia* and *D. simulans* but have probably remained silent until trans-regulatory and coding changes occurred in the *D. sechellia* lineage. The increase of *Ir75b* responsiveness was accompanied by a decrease in responsiveness of the tandem paralog *Ir75a* to acetic acid [46]. Remarkably, we did not detect any signal of selection on these two genes in *D. yakuba*. Moreover, we found that, unlike *D. melanogaster*, both generalist and specialist *D. yakuba* did not significantly differ from *D. sechellia* in hexanoic acid preference, further suggesting that this acid-sensing pathway might not have played an important role in *D. yakuba* specialization.

4.3. Variation in the Ester-Sensing Olfactory Receptor Or22a in Populations from the Ancestral Range

Dekker et al. [18] showed that *D. sechellia* antennae responded most strongly to noni esters (methyl hexanoate) than to the two acids. They found that *D. sechellia* sensitivity to methyl hexanoate was also present in its closest relative *D. simulans* and correlated with an expansion of ab3 sensilla that are more sensitive to ethyl hexanoate in *D. melanogaster* and *D. yakuba*. However, both coding sequence and transcription level comparisons of *Or22a*, the main receptor of ab3 sensilla, suggest that hypersensitivity to methyl hexanoate might have evolved in the ancestor of *D. sechellia* and *D. simulans* [18,46,47]. Using elegant transgenic experiments in *D. sechellia*, Auer et al. [20] have recently confirmed the role of a coding change in *Or22a* in the olfactory response to noni fruits and provided evidence for the ancestry of that change in the simulans species complex.

Remarkably, we found strong allelic differentiation between *D. y. mayottensis* and *D. y. yakuba* from Kenya at *Or22a*, mostly including coding changes. However, closer examination of sequence evolution of this gene in *Or22a*, indicated that variants in both populations are also found at intermediate frequencies in the Cameroon population of *D. y. yakuba*, which likely represents the ancestral range of the species [40]. Indeed, Auer et al. [20] examined four geographical lines of *D. yakuba* including the three populations studied here as well as the reference genome line from Côte d'Ivoire. They found that the *D. y. mayottensis* line did not show significant differences in number of Or22a-expressing olfactory sensory neurons and their physiological responses to ethyl and methyl esters in spite of differences in their protein sequences. Given the level of polymorphism at this gene within and between *D. yakuba* populations that we found here, additional lines from Cameroon and other populations from West Africa should be tested in the future to fully understand the potential role of ancestral *Or22a* variation in driving the specialization of *D. y. mayottensis* on noni.

4.4. Selection on Bitter-Sensing Gustatory Receptors

Unlike olfactory perception, the role of gustatory receptor evolution in noni specialization remains unclear. Indeed, previous comparative genomics studies have revealed faster protein evolution and turnover for gustatory receptor genes than for olfactory receptors [6,7], but a functional link remains to be identified. Our phenotypic assay found no significant difference in feeding preference between *D. melanogaster* and *D. sechellia* for the methyl hexanoate ester. This may suggest that acids, not esters, are the most important constituents for the gustatory specialization on noni. In *D. melanogaster*, appetitive response to low concentration hexanoic and octanoic acids is induced by the gustatory ionotropic receptors *Ir25a*, *Ir76b* and *Ir56d* in sweet-tasting sensilla [22]. However, at high concentrations of hexanoic acids (>9.3 g·L^{-1}), this behavior is suppressed by the activation of bitter-sensing sensilla independent of the three ionotropic receptors [22]. We did not detect any deviation from neutral evolution on these ionotropic receptor genes. Instead, we detected a signal of selection on four bitter-tasting gustatory receptors *Gr22b*, *Gr22d*, *Gr59a* and *Gr93c*. Because generalist *D. y. yakuba* significantly differs in its preference for hexanoic acid (at 1%) from the specialist *D. sechellia*, whereas specialist *D. y. mayottensis* does not, the four gustatory receptors might have evolved to reduce *D. y. mayottensis* avoidance of high-concentration hexanoic acid. Unlike olfactory receptors whose evolution is strongly modular, gustatory receptors act in an integrative way so that changes in one receptor unpredictably change the response of an entire set of receptors to the same component [39]. For example, and most relevant to our findings, Sung et al. [47] showed epistatic interactions between loss-of-function mutations of the gustatory receptors *Gr22e* and *Gr59c*, two paralogs of our detected receptors, that depended on the tested bitter substance. In *D. sechellia*, three out of the five genes of the Gr22a clade were pseudogenized, namely *Gr22f*, *Gr22c* and *Gr22d* [6]. Although none of the most differentiating sites at the Gr22a clade in *D. yakuba* affects the protein sequence, we found a nonsense mutation at *Gr22d* that was nearly fixed in *D. y. mayottensis* whereas segregating at low frequencies in *D. y. yakuba*. Because nonsense mutations could be post-transcriptionnally corrected in neural cells [46], the role of this mutation remains to be tested. Future functional genetic analyses of the genes identified here would definitively shed light on their potential effects. Those future gustatory analyses should also include either whole noni fruits, or other noni chemical components that could be perceived as bitter by *D. y. mayottensis*. Nonetheless, the significant difference found in *D. melanogaster* iso-1 strain, which carries nonsense mutations at both *Gr22b* and *Gr22d*, to hexanoic acid indeed suggests the involvement of those genes in noni toxins perception.

In summary, our results support a role for standing genetic variation in promoting host plant preference evolution in herbivorous insects. Because such variation may be abundant in several insect-plant systems [48], its probable maintenance by various forms of balancing selection in ancestral populations may be an initial prerequisite for rapid host shifts, facilitating the gradual built-up of adaptations in simultaneous host use traits. Future analyses should therefore focus on how standing variation in preference alleles accumulate and interact with host performance traits at the onset of ecological specialization and adaptation.

Supplementary Materials: Supplementary materials are available online at https://www.mdpi.com/2073-4425/12/1/32/s1.

Author Contributions: Conceptualization, A.Y.; methodology, A.Y.; software, F.M.-P.; formal analysis, E.A.F., S.L. and T.V.; investigation, E.A.F., S.L. and T.V.; resources, F.M.-P. and A.Y.; writing—original draft preparation, E.A.F. and A.Y.; writing—review and editing, F.M.-P. and A.Y.; visualization, E.A.F. and A.Y.; supervision, F.M.-P. and A.Y.; funding acquisition, F.M.-P. and A.Y. All authors have read and agreed to the published version of the manuscript.

Funding: This research was funded by the French Agence Nationale de la Recherche (ANR) grant number ANR-18-CE02-0008 to A.Y., and grant number ANR-19-CE34-011 to F.M.-P. and A.Y., as well as by the Lounsbery Foundation and ATM-MNHN grants to A.Y.

Institutional Review Board Statement: Not applicable.

Informed Consent Statement: Not applicable.

Data Availability Statement: Experimental data could be obtained from the authors upon request.

Acknowledgments: We would like to thank two anonymous reviewers for their constructive comments, Perrine Colombi and Béatrice Denis for technical advices during the MultiCAFE assays, and Jean David and Jean-Luc Da Lage for supplying *D. melanogaster* strains used in this analysis.

Conflicts of Interest: The authors declare no conflict of interest. The funders had no role in the design of the study; in the collection, analyses, or interpretation of data; in the writing of the manuscript, or in the decision to publish the results.

References

1. Futuyma, D.J.; Moreno, G. The Evolution of Ecological Specialization. *Annu. Rev. Ecol. Syst.* **1988**, *19*, 207–233. [CrossRef]
2. Simon, J.-C.; d'Alençon, E.; Guy, E.; Jacquin-Joly, E.; Jacquiéry, J.; Nouhaud, P.; Peccoud, J.; Sugio, A.; Streiff, R. Genomics of Adaptation to Host-Plants in Herbivorous Insects. *Brief. Funct. Genom.* **2015**, *14*, 413–423. [CrossRef] [PubMed]
3. Gloss, A.D.; Abbot, P.; Whiteman, N.K. How Interactions with Plant Chemicals Shape Insect Genomes. *Curr. Opin. Insect Sci.* **2019**, *36*, 149–156. [CrossRef] [PubMed]
4. Vertacnik, K.L.; Linnen, C.R. Evolutionary Genetics of Host Shifts in Herbivorous Insects: Insights from the Age of Genomics. *Ann. N. Y. Acad. Sci.* **2017**, *1389*, 186–212. [CrossRef]
5. Etges, W.J. Evolutionary Genomics of Host Plant Adaptation: Insights from Drosophila. *Curr. Opin. Insect Sci.* **2019**, *36*, 96–102. [CrossRef]
6. McBride, C.S.; Arguello, J.R. Five Drosophila Genomes Reveal Nonneutral Evolution and the Signature of Host Specialization in the Chemoreceptor Superfamily. *Genetics* **2007**, *177*, 1395–1416. [CrossRef]
7. McBride, C.S. Rapid Evolution of Smell and Taste Receptor Genes during Host Specialization in *Drosophila Sechellia*. *Proc. Natl. Acad. Sci. USA* **2007**, *104*, 4996–5001. [CrossRef]
8. Almeida, F.C.; Sánchez-Gracia, A.; Campos, J.L.; Rozas, J. Family Size Evolution in Drosophila Chemosensory Gene Families: A Comparative Analysis with a Critical Appraisal of Methods. *Genome Biol. Evol.* **2014**, *6*, 1669–1682. [CrossRef]
9. Kulmuni, J.; Wurm, Y.; Pamilo, P. Comparative Genomics of Chemosensory Protein Genes Reveals Rapid Evolution and Positive Selection in Ant-Specific Duplicates. *Heredity* **2013**, *110*, 538–547. [CrossRef]
10. Sánchez-Gracia, A.; Vieira, F.G.; Almeida, F.C.; Rozas, J. Comparative Genomics of the Major Chemosensory Gene Families in Arthropods. *Encycl. Life Sci.* **2011**. [CrossRef]
11. Matzkin, L.M. Ecological Genomics of Host Shifts in Drosophila Mojavensis. *Adv. Exp. Med. Biol.* **2014**, *781*, 233–247. [CrossRef] [PubMed]
12. Matzkin, L.M.; Watts, T.D.; Bitler, B.G.; Machado, C.A.; Markow, T.A. Functional Genomics of Cactus Host Shifts in Drosophila Mojavensis. *Mol. Ecol.* **2006**, *15*, 4635–4643. [CrossRef] [PubMed]
13. R'Kha, S.; Capy, P.; David, J.R. Host-Plant Specialization in the *Drosophila Melanogaster* Species Complex: A Physiological, Behavioral, and Genetical Analysis. *Proc. Natl. Acad. Sci. USA* **1991**, *88*, 1835–1839. [CrossRef] [PubMed]
14. Yassin, A.; Debat, V.; Bastide, H.; Gidaszewski, N.; David, J.R.; Pool, J.E. Recurrent Specialization on a Toxic Fruit in an Island Drosophila Population. *Proc. Natl. Acad. Sci. USA* **2016**, *113*, 4771–4776. [CrossRef]
15. Farine, J.-P.; Legal, L.; Moreteau, B.; Le Quere, J.-L. Volatile Components of Ripe Fruits of Morinda Citrifolia and Their Effects on Drosophila. *Phytochemistry* **1996**, *41*, 433–438. [CrossRef]
16. Higa, I.; Fuyama, Y. Genetics of Food Preference in *Drosophila Sechellia*. *Genetica* **1993**, *88*, 129–136. [CrossRef]
17. Amlou, M.; Moreteau, B.; David, J.R. Genetic Analysis of *Drosophila Sechellia* Specialization: Oviposition Behavior Toward the Major Aliphatic Acids of Its Host Plant. *Behav Genet* **1998**, *28*, 455–464. [CrossRef]
18. Dekker, T.; Ibba, I.; Siju, K.P.; Stensmyr, M.C.; Hansson, B.S. Olfactory Shifts Parallel Superspecialism for Toxic Fruit in *Drosophila Melanogaster* Sibling, *D. Sechellia*. *Curr. Biol.* **2006**, *16*, 101–109. [CrossRef]
19. Prieto-Godino, L.L.; Rytz, R.; Cruchet, S.; Bargeton, B.; Abuin, L.; Silbering, A.F.; Ruta, V.; Peraro, M.D.; Benton, R. Evolution of Acid-Sensing Olfactory Circuits in Drosophilids. *Neuron* **2017**, *93*, 661–676.e6. [CrossRef]
20. Auer, T.O.; Khallaf, M.A.; Silbering, A.F.; Zappia, G.; Ellis, K.; Álvarez-Ocaña, R.; Arguello, J.R.; Hansson, B.S.; Jefferis, G.S.X.E.; Caron, S.J.C.; et al. Olfactory Receptor and Circuit Evolution Promote Host Specialization. *Nature* **2020**, *579*, 402–408. [CrossRef]
21. Matsuo, T.; Sugaya, S.; Yasukawa, J.; Aigaki, T.; Fuyama, Y. Odorant-Binding Proteins OBP57d and OBP57e Affect Taste Perception and Host-Plant Preference in *Drosophila Sechellia*. *PLoS Biol.* **2007**, *5*, e118. [CrossRef] [PubMed]
22. Ahn, J.-E.; Chen, Y.; Amrein, H. Molecular Basis of Fatty Acid Taste in Drosophila. *Elife Sci.* **2017**, *6*, e30115. [CrossRef] [PubMed]
23. Messer, P.W.; Petrov, D.A. Population Genomics of Rapid Adaptation by Soft Selective Sweeps. *Trends Ecol. Evol.* **2013**, *28*, 659–669. [CrossRef] [PubMed]
24. Hermisson, J.; Pennings, P.S. Soft Sweeps and beyond: Understanding the Patterns and Probabilities of Selection Footprints under Rapid Adaptation. *Methods Ecol. Evol.* **2017**, *8*, 700–716. [CrossRef]
25. Rogers, R.L.; Cridland, J.M.; Shao, L.; Hu, T.T.; Andolfatto, P.; Thornton, K.R. Landscape of Standing Variation for Tandem Duplications in Drosophila Yakuba and Drosophila Simulans. *Mol. Biol. Evol.* **2014**, *31*, 1750–1766. [CrossRef]
26. Thurmond, J.; Goodman, J.L.; Strelets, V.B.; Attrill, H.; Gramates, L.S.; Marygold, S.J.; Matthews, B.B.; Millburn, G.; Antonazzo, G.; Trovisco, V.; et al. FlyBase 2.0: The next Generation. *Nucleic Acids Res.* **2019**, *47*, D759–D765. [CrossRef]
27. Li, H. Minimap2: Pairwise Alignment for Nucleotide Sequences. *Bioinformatics* **2018**, *34*, 3094–3100. [CrossRef]

28. Li, H.; Handsaker, B.; Wysoker, A.; Fennell, T.; Ruan, J.; Homer, N.; Marth, G.; Abecasis, G.; Durbin, R. The Sequence Alignment/Map Format and SAMtools. *Bioinformatics* **2009**, *25*, 2078–2079. [CrossRef]
29. Kofler, R.; Pandey, R.V.; Schlötterer, C. PoPoolation2: Identifying Differentiation between Populations Using Sequencing of Pooled DNA Samples (Pool-Seq). *Bioinformatics* **2011**, *27*, 3435–3436. [CrossRef]
30. Hudson, R.R.; Slatkin, M.; Maddison, W.P. Estimation of Levels of Gene Flow from DNA Sequence Data. *Genetics* **1992**, *132*, 583–589.
31. Altschul, S.F.; Madden, T.L.; Schäffer, A.A.; Zhang, J.; Zhang, Z.; Miller, W.; Lipman, D.J. Gapped BLAST and PSI-BLAST: A New Generation of Protein Database Search Programs. *Nucl Acids Res* **1997**, *25*, 3389–3402. [CrossRef] [PubMed]
32. Gutenkunst, R.N.; Hernandez, R.D.; Williamson, S.H.; Bustamante, C.D. Inferring the Joint Demographic History of Multiple Populations from Multidimensional SNP Frequency Data. *PLoS Genet.* **2009**, *5*. [CrossRef] [PubMed]
33. Ewing, G.; Hermisson, J. MSMS: A Coalescent Simulation Program Including Recombination, Demographic Structure and Selection at a Single Locus. *Bioinformatics* **2010**, *26*, 2064–2065. [CrossRef] [PubMed]
34. Comeron, J.M.; Ratnappan, R.; Bailin, S. The Many Landscapes of Recombination in *Drosophila Melanogaster*. *PLoS Genet.* **2012**, *8*, 33–35. [CrossRef] [PubMed]
35. Bastide, H.; Lange, J.D.; Lack, J.B.; Yassin, A.; Pool, J.E. A Variable Genetic Architecture of Melanic Evolution in *Drosophila Melanogaster*. *Genetics* **2016**, *204*, 1307–1319. [CrossRef]
36. Sellier, M.-J.; Reeb, P.; Marion-Poll, F. Consumption of Bitter Alkaloids in *Drosophila Melanogaster* in Multiple-Choice Test Conditions. *Chem. Senses* **2011**, *36*, 323–334. [CrossRef]
37. De Chaumont, F.; Dallongeville, S.; Chenouard, N.; Hervé, N.; Pop, S.; Provoost, T.; Meas-Yedid, V.; Pankajakshan, P.; Lecomte, T.; Le Montagner, Y.; et al. Icy: An Open Bioimage Informatics Platform for Extended Reproducible Research. *Nat. Methods* **2012**, *9*, 690–696. [CrossRef]
38. *R Core Team R: A Language and Environment for Statistical Computing*; R Foundation for Statistical Computing: Vienna, Austria, 2016.
39. Delventhal, R.; Carlson, J.R. Bitter Taste Receptors Confer Diverse Functions to Neurons. *eLife* **2016**, *5*, 1–23. [CrossRef]
40. Lachaise, D.; Harry, M.; Solignac, M.; Lemeunier, F.; Bénassi, V.; Cariou, M.-L. Evolutionary Novelties in Islands: Drosophila Santomea, a New Melanogaster Sister Species from São Tomé. *Proc. R. Soc. Lond. B Biol. Sci.* **2000**, *267*, 1487–1495. [CrossRef]
41. Masek, P.; Keene, A.C. Drosophila Fatty Acid Taste Signals through the PLC Pathway in Sugar-Sensing Neurons. *PLoS Genet.* **2013**, *9*, e1003710. [CrossRef]
42. Gerber, B.; Stocker, R.F.; Tanimura, T.; Thum, A.S. Smelling, Tasting, Learning: Drosophila as a Study Case. *Results Probl. Cell Differ.* **2009**, *47*, 139–185. [CrossRef] [PubMed]
43. Moreteau, B.; R'Kha, S.; David, J.R. Genetics of a Nonoptimal Behavior: Oviposition Preference ofDrosophila Mauritiana for a Toxic Resource. *Behav. Genet.* **1994**, *24*, 433–441. [CrossRef]
44. Legal, L.; Chappe, B.; Jallon, J.M. Molecular Basis ofMorinda Citrifolia (L.): Toxicity on Drosophila. *J. Chem. Ecol.* **1994**, *20*, 1931–1943. [CrossRef] [PubMed]
45. Mansourian, S.; Stensmyr, M.C. The Chemical Ecology of the Fly. *Curr. Opin. Neurobiol.* **2015**, *34*, 95–102. [CrossRef]
46. Prieto-Godino, L.L.; Rytz, R.; Bargeton, B.; Abuin, L.; Arguello, J.R.; Peraro, M.D.; Benton, R. Olfactory Receptor Pseudo-Pseudogenes. *Nature* **2016**, *539*, 93–97. [CrossRef]
47. Sung, H.Y.; Jeong, Y.T.; Lim, J.Y.; Kim, H.; Oh, S.M.; Hwang, S.W.; Kwon, J.Y.; Moon, S.J. Heterogeneity in the Drosophila Gustatory Receptor Complexes That Detect Aversive Compounds. *Nat. Commun.* **2017**, *8*, 1–10. [CrossRef] [PubMed]
48. Gloss, A.D.; Groen, S.C.; Whiteman, N.K. A Genomic Perspective on the Generation and Maintenance of Genetic Diversity in Herbivorous Insects. *Annu. Rev. Ecol. Evol. Syst.* **2016**, *47*, 165–187. [CrossRef]

Article

Targeted Sequencing of Mitochondrial Genes Reveals Signatures of Molecular Adaptation in a Nearly Panmictic Small Pelagic Fish Species

Miguel Baltazar-Soares *, André Ricardo de Araújo Lima and Gonçalo Silva

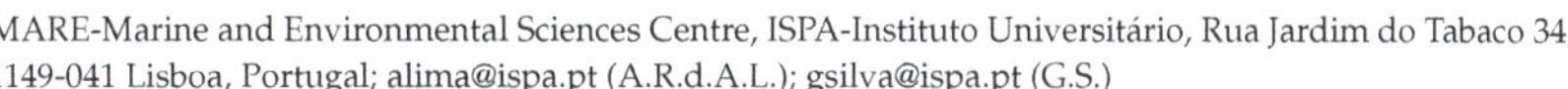

MARE-Marine and Environmental Sciences Centre, ISPA-Instituto Universitário, Rua Jardim do Tabaco 34, 1149-041 Lisboa, Portugal; alima@ispa.pt (A.R.d.A.L.); gsilva@ispa.pt (G.S.)
* Correspondence: miguelalexsoares@gmail.com

Abstract: Ongoing climatic changes, with predictable impacts on marine environmental conditions, are expected to trigger organismal responses. Recent evidence shows that, in some marine species, variation in mitochondrial genes involved in the aerobic conversion of oxygen into ATP at the cellular level correlate with gradients of sea surface temperature and gradients of dissolved oxygen. Here, we investigated the adaptive potential of the European sardine *Sardina pilchardus* populations offshore the Iberian Peninsula. We performed a seascape genetics approach that consisted of the high throughput sequencing of mitochondria's *ATP6*, *COI*, *CYTB* and *ND5* and five microsatellite loci on 96 individuals coupled with environmental information on sea surface temperature and dissolved oxygen across five sampling locations. Results show that, despite sardines forming a nearly panmictic population around Iberian Peninsula, haplotype frequency distribution can be explained by gradients of minimum sea surface temperature and dissolved oxygen. We further identified that the frequencies of the most common *CYTB* and *ATP6* haplotypes negatively correlate with minimum sea surface temperature across the sampled area, suggestive of a signature of selection. With signatures of selection superimposed on highly connected populations, sardines may be able to follow environmental optima and shift their distribution northwards as a response to the increasing sea surface temperatures.

Keywords: small pelagic fishes; OXPHOS complex; adaptive potential; climate change

Citation: Baltazar-Soares, M.; de Araújo Lima, A.R.; Silva, G. Targeted Sequencing of Mitochondrial Genes Reveals Signatures of Molecular Adaptation in a Nearly Panmictic Small Pelagic Fish Species. *Genes* **2021**, *12*, 91. https://doi.org/10.3390/genes12010091

Received: 4 December 2020
Accepted: 11 January 2021
Published: 13 January 2021

Publisher's Note: MDPI stays neutral with regard to jurisdictional claims in published maps and institutional affiliations.

1. Introduction

Ongoing climatic change translates into a series of environmental fluctuations with the potential to transform the current distribution of global biodiversity. As biological responses fall in the spectrum of "adapt, move or go extinct", the range shifts, biological invasions and re-shaping of entire ecosystems, are thus expected to increase in frequency [1]. The marine environment is particularly susceptible to climatic shifts [1], and phenomena such as rising sea surface temperatures, increases in CO_2 concentrations (ocean acidification) and depletion of O_2 in deep oceanic waters, translate into tropicalization of fish assemblages in temperate zones [2,3] and hypoxia in coastal regions and estuaries [4,5]. Investigating the adaptive potential of species to cope with environmental shifts relies on the identification of phenotypic or molecular changes, expected to emerge as evolutionary responses [6]. Advances in sequencing technologies have greatly improved research on the molecular variation across non-model organisms [7]. Coupled with the extensively available environmental data that exists in marine regions, seascape genetic approaches are frequently being used to identify signatures of evolutionary processes [8,9].

On top of the environmental pressures imposed by climate change, marine species of commercial relevance are also subject to overexploitation. Unsustainable harvesting compromises the viability of stocks in multiple ways. For example, underestimating stock sizes leads to a logical reduction in the critical mass of spawners necessary to maintain population dynamics [10]; capturing larger individuals imposes an early age at maturation, reducing the overall quality and quantity of brood stock [11,12]; disproportionally harvesting of unequal stocks may eliminate unique genetic diversity, thus reducing the species with adaptive potential, and therefore limiting the breath of possible responses to environmental pressures [13]. Coastal species with commercial value are heavily exposed to a synergistic effect of environmental and anthropogenic pressures. In this context, stocks of small pelagic fishes such as sardines, anchovies or herrings are among those that are particularly vulnerable [14]. "Small pelagic fishes" defines a group of ray-finned and fusiform short size fishes that tend to form large shoals that inhabit the water column near coastal locations. They mostly aggregate at the edge of continental shelves in order to benefit from the oxygen and nutrient-rich waters that emerge during upwelling events [15,16]. Due to their abundance, biomass, and trophic level occupied, those fish are an important component of an ecosystem that expands beyond the coastal limitations of their distribution [17–19].

The European sardine *Sardina pilchardus* is a small epipelagic fish inhabiting the north eastern Atlantic coast, the Mediterranean, and the Black Sea. It is a key component of the Mediterranean blue economy, supported by centuries-old tradition of consumption, and more recently, production of canned products [20]. Both the Mediterranean and Eastern Atlantic stocks along a costal line from Senegal to Galicia, in Spain, are in a particularly fragile state [21–23]. Consequently, sardine fisheries have a recent, yet long record of fishing restrictions, quota limitations, and premature closures of fishing seasons.

In this work, we aim to investigate the adaptive potential of sardine populations surrounding the temperate zone of the Iberia Peninsula to fluctuations on sea surface temperature and dissolved oxygen. We focused on genetic variation at the mitochondrial level, as the mitochondrial genome encodes for genes involved on the oxidative phosphorylation (OXPHOS) pathway [24,25]. The central role of mitochondria in organismal metabolism questions longstanding assumptions of the organelle's evolutionary neutrality [26]. It is thus not surprising that signatures of selection have been identified among the mitochondrial-encoded OXPHOS genes in several fish species, including Atlantic and Pacific salmons, anchovies, or sardines in the Indian Ocean, and associated with thermal gradients or dissolved oxygen concentration [26–28]. We followed a target-gene approach to screen the molecular diversity of sardine's mitochondrial ATP6, CYTB, ND5 and COI, which code for complexes I,III–V of the OXPHOS pathway, and specifically sequenced key functional regions where signatures of selection have been detected in other marine fishes [26,27]. We also sampled a broad geographic area, enveloping the coastal waters of all the Iberian Peninsula, and consequently a gradient of coastal environments and putative barriers to dispersal. To explore the hypothesis that the frequency of mitochondrial variants among mitochondrial-encoded OXPHOS genes is related to the spatial variation of temperature and dissolved oxygen concentration, we collated environmental information from sampling locations. Finally, we complemented the survey of the species' genetic diversity with five microsatellite markers, which allowed us to inspect the connectivity among sampled sites.

2. Methods

2.1. Sample Collection and DNA Extraction

A total of 96 individuals from 5 locations, Bay of Biscay (Spain), Gulf of Lion (France), Sesimbra (Portugal), Olhão (Portugal) and Tarragona (Spain) were collected, spanning the Atlantic-Mediterranean region surrounding the Iberian Peninsula (Figure 1). DNA was extracted with EZ-DNA Genomic DNA Isolation kit© (Zymo Research, California, CA, USA) following manufacturer instructions. Quantity and quality of extracted DNA were first assessed with Nanodrop© and later with Qubit© fluorometer.

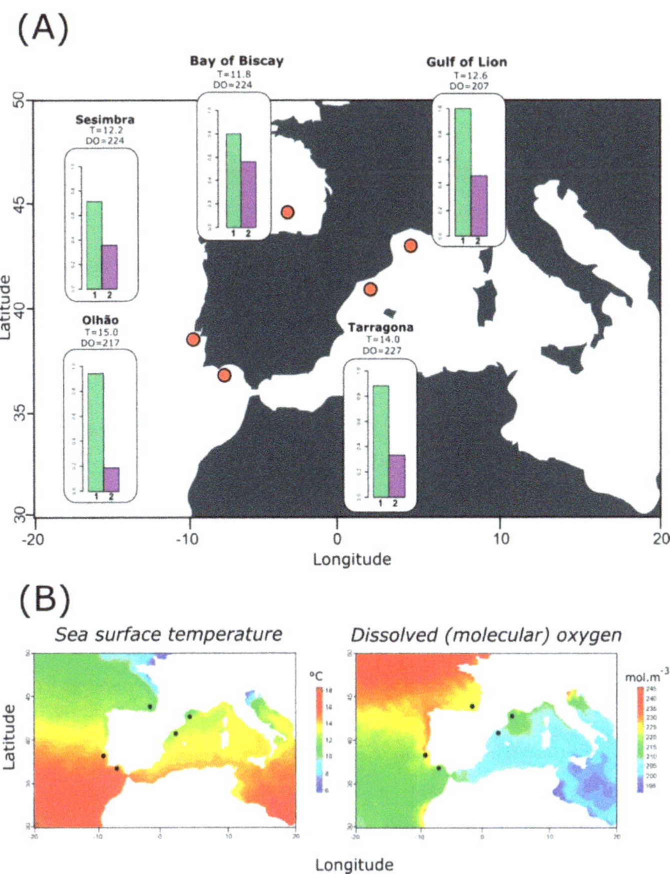

Figure 1. Sampling locations and associated environmental variation. (**A**) Sampling locations and plots of the frequencies (*y*-axis = frequency of the most common haplotype among ATP6, ATP6-Hap1 (**1**) and CYTB, CYTB-Hap3 (**2**). The minimum sea surface temperature (T °C) and minimum dissolved oxygen (mol·m^{-3}) obtained from each location is also shown. (**B**) Heatmap of minimum sea surface temperature and minimum dissolved oxygen distribution in the marine area surrounding the Iberian Peninsula. Black dots represent the sampling locations.

2.2. Mitochondrial DNA: Defining Target-Regions for Amplicon Sequencing

To explore whether the mitochondrial regions identified as putative under selection on other small pelagic fishes would exhibit similar signatures on sardines, 4 mitochondrial genes, each encoding different subunits involved in specific complexes of the mitochondrial respiratory chain were used: ND5 (NADH: ubiquinone oxireductase core subunit 5) for respiratory complex I; CYTB (cytochrome b) for respiratory complex III; COI (cytochrome c oxidase I) for respiratory complex IV; ATP6 (ATP synthase membrane subunit 6) for respiratory complex V. To facilitate the identification which within-gene regions should target sardines, we collected sequences from NCBI of other small pelagic species on which signatures of selection have been reported on our target genes. Those were then aligned with BioEdit [29] against the sardine mitogenome (reference NCBI: NC_009592.1) to verify where matching occurs. The conservation of chosen regions was further checked against other fish species. Primers were designed on the sardine genome to flank putative regions of interest. Reference numbers of utilized sequences are available on Table S1.

2.2.1. Sequencing, Filtering and Variant Calling at Targeted Mitochondrial Regions

Extracted DNA was sent to CIBIO (Centro de Investigação em Biodiversidade e Recursos Genéticos) in Vairão, Portugal, to primer optimization, library preparation and amplicon sequencing. The primer-pairs specifically designed for this study are available on Table S2. All obtained sequences are available in Text S1. Detailed protocols on library preparation and amplicon sequencing are available in Appendix A.1.

Demultiplexing, barcode removal and QC control were performed in Illumina's Basespace online platform. Only reads with Phred quality score (Q) > 30 were utilized in this study. Demultiplexed reads were aligned against a reference, i.e., sardine's mitogenome (reference NCBI: NC_009592.1) with *bwa-mem* algorithm [30] implemented in bwa [31]. Only reads that were properly paired were kept, mapping qualities above 10 and coverage above 2 were kept for variant calling. Variant calling was performed with Stacks v 1.48 [32] through imported .bam files. Specifically, we ran ref_map.pl pipeline with the sardine mitogenome as reference and default parameters.

2.2.2. Diversity and Distribution of Genetic Variation at Targeted Mitochondrial Regions

Haplotype diversity (Hd), number of haplotypes (nH), nucleotide diversity (π) and segregation sites (S) were calculated both independently for each gene in DNAsp [33]. In total, 4 datasets, corresponding to each sequenced mitochondrial region were built. To assess the connectivity among sampled locations, the population structure in Arlequin for all datasets was estimated via F_{ST} calculation (10,000 permutations). The level of significance was set to $\alpha = 0.05$ [34]. Isolation by distance was inferred by using log-transformed distance-by-sea among sampling locations and F_{ST} pairwise values in the R package ecodist, with the level of significance set to $\alpha = 0.05$ [35]. The number of shared haplotypes among locations was calculated to inspect whether geographic distance could play a role in the distribution of haplotype diversity. Finally, median-joining networks were calculated and drawn to better understand the genealogy among identified haplotypes in POPART [36].

2.2.3. Relationship between Haplotype Composition and Environmental Variables at Each Location

All computations were performed in the R environment [37]. A Redundancy Analysis (RDA) was performed to investigate whether haplotype distribution could be explained by environmental conditions found at each collection site. Environmental parameters (annual monthly means) were extracted from the Bio-Oracle database [38] with scripts customized to collect data on the maximum, minimum, mean and the range of values of sea surface temperature (SST) and dissolved oxygen (DO) given the approximate GPS positions of exact sampling locations. Thus, a matrix of haplotype frequencies was utilized as a response variable in an RDA [39], while a matrix of values related to surface temperature and dissolved oxygen were utilized as putative explanatory variables. Haplotypes unique to single locations were excluded from the dataset for this analysis.

The overall statistical significance of the RDA model was assessed using an ANOVA-like permutation function, anova.cca, implemented in the in the R package *vegan* [40]. Significance of constrained axis and effects and significance of each variable were also assessed with anova.cca, defining the "axis" and "margin" to avoid the sequential test of terms [41]. Lastly, a putative association of specific haplotype frequencies and environmental variation was inferred by considering, as significant, those relationships that diverged 2.5 standard deviations from the mean distribution on the RDA plot (i.e., two-tailed *p*-value = 0.0125).

2.2.4. Inferring Historical Events of Selection across Phylogenies with dN/dS Ratios-Based Tests

As we were interested in screening for signatures of selection on each specific respiratory complex, a phylogenetic inference of events of selection was performed independently for each gene-specific dataset. As such, we utilized methods included in Hyphy via the Datamonkey web interface, which are based on the inference of nonsynonymous and synonymous substitution rates across phylogenies [42]. A single-likelihood ancestor counting (SLAC) [43], a fixed effect likelihood (FEL) [43] and fast unconstrained Bayesian approximation (FUBAR) [44] algorithms were applied. While sharing the same theoretical basis, i.e., selective pressure for each site is constant along the entire phylogeny; algorithms vary on the methodology to infer dN and dS substitution rates. Specifically, SLAC uses the maximum-likelihood and counting approach, FEL uses a maximum-likelihood approach and FUBAR relies on a Bayesian framework. Significance was assessed accordingly to the default value of each specific test, namely, with the level of significance set to $\alpha = 0.1$ (rounded-up to the decimal) in likelihood models SLAC and FEL, and with a posterior probabilities threshold of >0.9 (rounded-up to the decimal), for ($\alpha > \beta$) in FUBAR (default).

2.3. Microsatellite Amplification and Diversity Estimates

Nuclear genetic variation was investigated with five microsatellites previously described in other sardine studies–SAB07, SAR09, SR15, SAR112, SAR218, and SP17 [45,46]. Amplification and genotyping protocols can be found in Appendix A.1. Nei's unbiased heterozygosity (He), observed heterozygosity (Ho), Wright's inbreeding coefficient (FIS) and Garza-Williams' index were computed in Arlequin v 3.5.2.2 [34]. Rarefied allelic richness (Ar) and private rarefied allelic richness (pAr) were estimated for both each location and averaged over loci in HP-Rare 1.1 [47]. The presence of null alleles was inspected with Micro-Checker [48].

2.3.1. Estimates of Genetic Differentiation and Structure among Sampled Locations

We first assessed the likelihood of each locus to be in the Hardy-Weinberg equilibrium (HWE), as consistent deviations could compromise assessment of neutral differentiation. HWE was computed per loci and population in Arlequin v 3.5.2.2. The possibility of population structure was then investigated through multiple approaches. First, we utilized a Discriminant Analysis of Principal Components (DAPC) to explore the number of clusters (K) that most likely explains the observed distribution of nuclear diversity implemented in the R package adegenet [49]. This was achieved through the application of the function find.clusters with max.n .clusters = 5, identification of most likely K through the inflection point of the Bayesian Inference Criteria (BIC) curve and subsequent calculation of discriminant components. Lastly, we plotted the 95% confidence interval ellipses of the identified K groups overlaid on the original distribution. For that, we utilized the two major linear discriminants (LD1 and LD2) extracted from the discriminant analysis as graphical coordinates. We further estimated the pairwise F_{ST} among locations, derived from allelic frequencies in Arlequin v3.5.2.2 (10,000 bootstrap). Isolation by distance was also inspected and analyses were performed similarly to those described in the previous Section 2.2.2.

2.3.2. Inferences on Deviations from Evolutionary Neutrality

Additional locus-specific analyses were performed because one of the utilized markers (SAR112) was reported to be the subject of clinal selection along the Mediterranean [46]. In this context, first, the frequency of HWE deviation across locations for each locus was explored. Then, the locus-specific allele frequency distribution among populations was analyzed by comparing the average F_{ST} obtained from respective pairwise comparisons. Frequencies were compared with ANOVA and Tukey's post-hoc tests being applied to investigate the significance of pairwise relationships.

3. Results

3.1. Mitochondrial DNA Sequencing and Variant Calling

The 96 individuals sequenced for the four targeted genes produced a total of 912,609 usable reads. On average, 2223.84 (SE $\pm$ 128.96) paired-end reads per individual were utilized for the ATP6, 2599.17 (SE $\pm$ 115.99) for the COI, 2693.29 (SE $\pm$ 140.73) for the CYTB and 1990.04 (SE $\pm$ 97.33) for the ND5. Average coverage per individual varied from 304.07 (SE $\pm$ 18.80) on ATP6 to 208.19 (SE $\pm$ 9.22) at CYTB. The position of variant sites identified with *Stacks* was inspected visually and utilized to construct haplotypes from the reference mitogenome. As genes were analyzed as fully independently, fragments were trimmed to remove uninformative invariant sites at the edges of the DNA string prior to population genomic analysis. Sizes of the fragments utilized varied across genes, specifically ATP6 (381 bp), COI (291 bp), CYTB (201 bp) and ND5 (330 bp).

3.2. Estimates of Mitochondrial Genetic Diversity and Population Structure

Genetic diversity indices expectedly varied among genes. Here, CYTB exhibited the higher indices of diversity (μ_S = 3.6, SE_S = 0.219; μ_{nHap} = 4.6, SE_{nHap} = 0.219; μ_{Hd} = 0.717, SE_{Hd} = 0.02; μ_π = 6 $\times$ 10^{-3}, and SE_π = 3 $\times$ 10^{-4}), while the COI showed a high conservation rate (μ_S = 1.4, SE_S = 0.219; μ_{nHap} = 2.4, SE_{nHap} = 0.219; μ_{Hd} = 0.164, SE_{Hd} = 0.02; and μ_π = 5 $\times$ 10^{-4}, SE_π = 8 $\times$ 10^{-5}). Among the sampled locations, genetic diversity was only pronouncedly different for the ATP6, as Tarragona and the Gulf of Lions reported a single haplotype in contrast to the Bay of Biscay and Sesimbra, where we observed three different haplotypes (Table 1). Estimates of pairwise F_{ST} only revealed two significant differentiation values, specifically for ATP6 haplotype frequencies between Sesimbra and the Gulf of Lions (F_{ST} = 0.045, p = 0.042) and ND5 between Tarragona and Olhão (F_{ST} = 0.114, p = 0.033). Investigating the overall number of haplotypes shared among locations revealed that closer sampled locations shared more haplotypes, suggesting that geographic distance might play a role in connectivity, despite the absence of clearly structured populations (Figure 2) and non-significant Mantel tests across genes (Table S3).

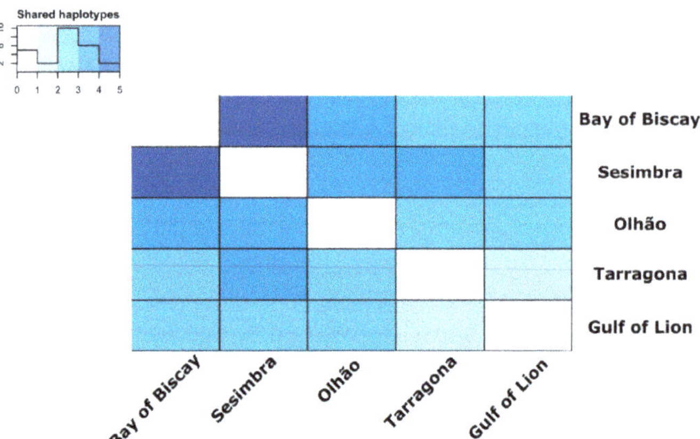

Figure 2. Heatmap of shared haplotypes. Quantification of the number of haplotypes shared by sampling locations. The legend refers to absolute numbers shared among locations, with the histogram line representing the frequency distribution.

Table 1. Diversity estimates obtained for each mitochondrial gene region coding and respective complex of the electron transport chain.

Gene	Population	*n*	S	nHap	Hd	π	Tajima's-D
ATP6 Respiratory Complex V	Bay of Biscay	15	3	4	0.371	0.00135	−1.31654
	Gulf of Lion	15	0	1	0	0	na
	Olhão	18	2	2	0.11	0.00058	−1.50776
	Sesimbra	14	3	4	0.396	0.0012	−1.67053
	Tarragona	17	0	1	0	0	na
COI Respiratory Complex IV	Bay of Biscay	17	2	3	0.228	0.00081	−1.50358
	Gulf of Lion	20	1	2	0.1	0.00034	−1.16439
	Olhão	15	1	2	0.133	0.00046	−1.15945
	Sesimbra	14	1	2	0.143	0.00049	−1.15524
	Tarragona	18	2	3	0.216	0.00076	−1.50776
CYTB Respiratory Complex III	Bay of Biscay	16	3	4	0.642	0.0051	0.38767
	Gulf of Lion	17	3	4	0.676	0.00563	0.78185
	Olhão	16	4	5	0.717	0.00589	−0.05743
	Sesimbra	14	4	5	0.769	0.00716	0.476
	Tarragona	15	4	5	0.781	0.00625	0.07027
ND5 Respiratory Complex I	Bay of Biscay	19	4	5	0.696	0.00284	−0.53717
	Gulf of Lion	20	1	2	0.521	0.00158	1.53133
	Olhão	19	2	3	0.526	0.0017	−0.04521
	Sesimbra	13	2	3	0.615	0.0021	0.2084
	Tarragona	18	4	5	0.66	0.00273	−0.67309

n = number of individuals per population; S = segregation sites; nHap = number of haplotypes; Hd = Haplotype diversity; π = nucleotide diversity.

3.3. Relating Haplotype Frequencies and Environmental Variation

The low number of different haplotypes in relation to explanatory variables prevented us building a single, full-scale model to incorporate maximum, minimum and averages of both dissolved oxygen and sea surface temperature. However, to preserve the fundamental objective of exploring and disentangling the role of those variables in sardine's distribution, we built four RDA models for each specific set of minima, maxima, mean and range. Only the model that incorporated minima was revealed to be statistically significant (ANOVA: F = 2.84, d.f. = 2, p = 0.05), and with an adjust R^2 = 0.48 (Figure 3A) (Table 2). CAP1 was revealed to significantly explain 43% of observed variability (ANOVA: F = 3.343, p = 0.050) and the marginal effect of each environmental variable showed minimum sea surface temperature (sst_min) to be significant (ANOVA: F = 3.109, p = 0.033) (Table 3).

We further identified haplotype 1 of ATP6 and haplotype 3 of CYTB to present a significant negative correlation with (minimum) sea surface temperature (Figure 3b). Models with average and maxima dissolved temperature and sea surface temperature did not significantly explain the haplotype frequencies distribution (Table 2).

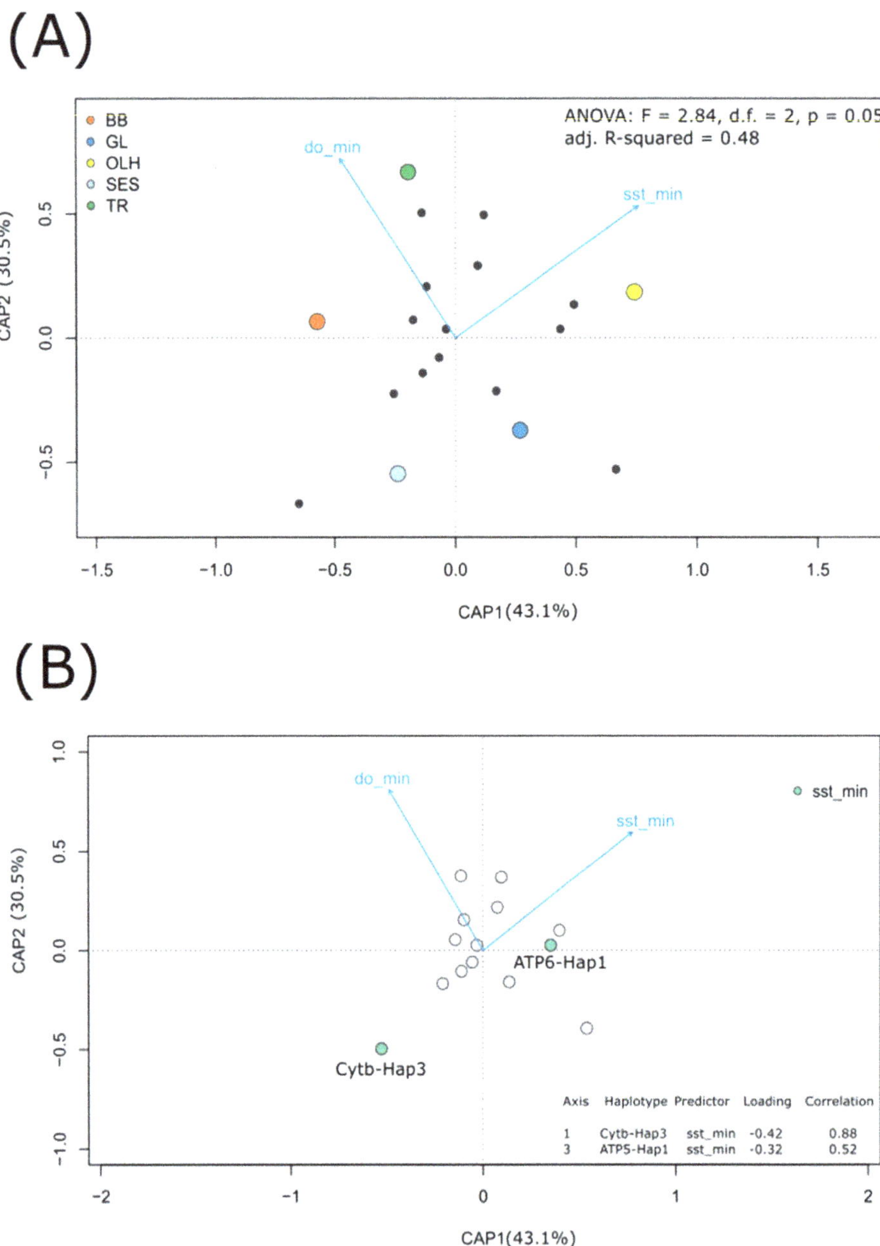

Figure 3. Presentation of redundancy analysis ordination with minimum values of environmental variables. (**A**) Ordination plot with haplotypes (response variables) as grey dots, populations in coloured dots and environmental variables (explanatory variables) as vectors. BB = Bay of Biscay, GL = Gulf of Lion, TR = Tarragona, OLH = Olhão, SES = Sesimbra; sst_min = minimum sea surface temperature, do_min = minimum dissolved oxygen. (**B**) Ordination plot highlighting haplotypes, Cytb-Hap3 and ATP6-Hap1, whose frequency is predicted to vary with minimum sea surface temperature (Cytb-Hap3: $r = 0.88$, $p < 0.05$; ATP6-Hap1: $r = 0.52$, $p < 0.05$), also showing respective axis and correlation coefficients.

Table 2. Model results obtained with the *anova.cca* function for each pair of parameter sea surface temperature (T °C) and dissolved oxygen (mol·m^{-3}). Significant models are highlighted in bold.

Maximum Temperature and Dissolved Molecular Oxygen				
	df	SS	F	*p* (>F)
model	2	0.366	1.858	0.117
Residuals	2	0.197	-	-
Minimum Temperature and Dissolved Molecular Oxygen				
	df	SS	F	*p* (>F)
model	2	0.416	**2.841**	**0.050**
Residuals	2	0.146	-	-
Range of Temperature and Dissolved Molecular Oxygen				
	df	SS	F	*p* (>F)
model	2	0.150	1.087	0.483
Residuals	2	0.413	-	-
Mean of Temperature and Dissolved Molecular Oxygen				
	df	SS	F	*p* (>F)
model	2	0.202	0.560	0.867
Residuals	2	0.361	-	-

Model formula: Haplotype frequencies ~ SST+ DO. In bold, values significant for $p \leq 0.05$.

Table 3. ANOVA on the minimum SST and DO model to test the marginal effect of terms with 500 permutations to access.

	df	**SS**	**F**	*p* (**>F**)
CAP1	1	0.244	3.343	**0.050**
CAP2	1	0.171	2.339	0.150
Residual	2	0.146		
	df	**SS**	**F**	*p* (**>F**)
sst_min	1	0.230	3.109	**0.033**
do_min	1	0.202	2.756	0.058
Residual	2	0.146		

In bold, values significant for $p \leq 0.05$.

3.4. Inferences of Episodic Selection across the Phylogeny of Sampled Specimens

Investigating the occurrence of pervasive selection across the entire gene phylogenies revealed sites deviating from a neutral evolution background on all our target genes. SLAC detected six sites on CYTB, three of which revealed signatures of positive or directional selection and three others on negative or purifying selection. FEL and FUBAR, whose statistical power is more robust than SLACs, both identified episodes of pervasive selection on genes ATP6 and ND5. For ATP6, FEL detected site 53 ($p < 0.1$), while FUBAR detected sites 53, 49, 24, 36, and 52 (posterior probability of $\alpha > \beta$, ≥ 0.9). For ND5, FEL detected site 58 ($p = 0.05$), while FUBAR detected sites 58, 64, and 90 (posterior probability of $\alpha > \beta$, ≥ 0.9). The Bayesian approach further identified four sites on the COI, namely, 8, 44, 53, and 57 (posterior probability of $\alpha > \beta$, ≥ 0.9) (Table 4). All these signatures relate to negative/purifying selection.

Table 4. Phylogenetic inferences on episodic or pervasive historical events of selection and adopting three different algorithms.

Method	Gene	Codon	Site	*p* Value	pp [α > β]
SLAC	*CYTB*	14	42	**0.1**	n/a
		15	45	**0.1**	n/a
		16	48	**0.1**	n/a
		26	78	**0.1**	n/a
		40	120	**0.1**	n/a
		48	144	**0.1**	n/a
		53	159	**0.1**	n/a
		57	171	**0.1**	n/a
FEL	*ATP6*	53	159	**0.1**	n/a
	ND5	58	174	**0.1**	n/a
FUBAR	*ATP6*	24	72	n/a	**0.9**
		49	147	n/a	**0.9**
		36	108	n/a	**0.9**
		52	156	n/a	**0.9**
		53	159	n/a	**0.9**
	ND5	58	174	n/a	**0.9**
		64	192	n/a	**0.9**
		90	270	n/a	**0.9**
	COI	8	24	n/a	**0.9**
		44	132	n/a	**0.9**
		53	159	n/a	**0.9**
		57	171	n/a	**0.9**

Only sites whose deviations from the null hypothesis from neutral evolution were reported and are shown. Significance was determined with threshold for $p \leq 0.1$, in likelihood models SLAC and FEL, and with posterior probabilities in FUBAR. Only sites denoting significant values (in bold) are reported. Note that the *p*-value on SLAC relates to P (dN/Ds < 1). All sites are reporting synonymous substitutions. n/a denotes statistic not available in the specific test, as defined in the Methods Section.

3.5. Microsatellite Amplification and Diversity Estimates

In general, we did not detect prominent variation regarding population specific diversity. Observed heterozygosity (Ho) values ranged between 0.65 (SD ± 0.27) in Sesimbra and 0.74 (SD ± 0.16) for the Gulf of Lion. Rarefied allelic richness over all loci varied between Sesimbra (12) and the Bay of Biscay (13.44). Private allelic richness, and indicator of population-specific alleles, was also similar from one location to another, being the extremes Tarragona (1.18) and the Bay of Biscay (2.24). Garza-Williams index, which estimates the effects of potential bottlenecks from observed heterozygosity, also indicated a striking similarity across populations (Table 5). Observed and expected heterozygosity (He) were revealed to significantly differ across sampled locations (ANOVA: $F_{1,4}$ = 6.02, p = 0.01) (Figure S1A).

Table 5. Diversity estimates regarding the five microsatellite markers.

Locations	*n*	rAR	pA	H_o	H_e
Bay of Biscay	20	13.44	2.24	0.71	0.86
Gulf of Lion	20	12.41	1.80	0.73	0.83
Olhão	20	12.89	2.10	0.74	0.86
Sesimbra	16	12.00	1.78	0.65	0.84
Tarragona	20	12.18	1.18	0.63	0.80

n = number of individuals; rAR = rarefied allelic richness; pA = private alleles; H_o = Observed heterozygosity and H_e = Expected heterozygosity.

3.6. Estimates of Microsatellite Genetic Differentiation and Structure among Sampled Locations

Inferences on the structure of sardine populations revealed a nearly panmictic scenario. Discriminant component analyses suggested a K = 2, though 95% CI ellipses largely overlap (Figure S2A). Pairwise F_{ST} comparisons further reinforced such a scenario as all but one showed a significant result, for $p = 0.02$ ($F_{ST\ Bay\ of\ Biscay\ vs\ Gulf\ Lion} = 0.05$). While those locations are at the maximum geographic distance possible within our sampled sites, correlating geographic and genetic distances obtained no significant value. Lastly, STRUCTURE analyses also revealed panmixia (Figure S2B).

Regarding locus-specific analyses, we first detected locus SAR218 to be constantly departing from Hardy-Weinberg equilibrium: it was the only locus whose *Ho* significantly deviated from *He* in all populations. This is probably due to the presence of null alleles at this locus, which was found to be consistent across all examined populations (Table S4).

Exploring the structure and explaining each locus, we did not observe significant departures from what would be expected from panmixia: CIs of the 95% ellipses largely overlapped in every locus (Figure S3). Pairwise F_{ST} analyses suggested a very similar picture, with only one significant pairwise comparison between Sesimbra and Olhão for the locus SAR112 ($F_{ST} = 0.013$, $p = 0.036$). However, we detected averaged pairwise F_{ST} among comparisons and this varied significantly across loci (ANOVA: $F_{1,5} = 21.385$, $p < 0.01$) (Figure 1B). Post-hoc comparisons further revealed locus SAR218 to be the one originating in significantly higher average F_{ST} than all the other loci, Tukey HSD ($Avg_{SAR218} = 0.023$, $SD_{SAR218} = 0.005$).

4. Discussion

Environmental alterations driven by the ongoing rapid climatic shifts impose intense selective pressures on populations of marine organisms, and particularly on those that are currently exploited. Understanding the response repertoire of those species is crucial to develop effective and sustainable exploitation plans to accommodate potential shifts in the distribution range and overall impacts on the viability of populations. By screening the variation on genes directly involved in metabolic performance, we aimed to infer the adaptive potential of sardine populations to shifts in temperature and dissolved oxygen. We observed a high degree of connectivity among populations and further revealed a putative molecular basis of adaptation to cold temperatures. Under the predicted increase in sea surface temperature, our results indicate that sardines might be able to follow environmental optima and extend their distribution northwards.

4.1. Population Structure Reveals High Connectivity around Iberian Peninsula

Genetic diversity estimates across investigated markers suggested the Bay of Biscay harboring higher diversity among investigated populations. At the mitochondrial level, three out of four genes (CYTB being the exception) exhibited a higher number of haplotypes and/or nucleotide diversity, and microsatellite markers consistently showed a higher rarefied number of alleles and private alleles among sardines collected on the Bay of Biscay. The expected a positive correlation between genetic diversity and effective population size and may simply suggest that the Bay of Biscay assembles a larger number of reproducing individuals in comparison to all others [50]. The results are coherent with observations reporting the Atlantic sardine stocks holding higher abundances than those in the Mediterranean [51–53]. Alternatively, differences among observed diversity estimates may be associated with variable fishing efforts in relation to stock size, as overfished stocks are predicted to exhibit lower genetic diversity than those that are sustainably managed [54]. Population structure analyses revealed a scenario of near panmixia, punctuated by deviations likely associated to geographic distance among sampled sites. Specifically, not only nearby locations shared a higher number of haplotypes, but also the F_{ST} pairwise estimate associated with microsatellite data between the Bay of Biscay and the Gulf of Lion. As Mantel tests did not support a general pattern of isolation by distance, we consider having identified a signature of limited dispersal among sampled locations, possibly due to the natural barrier imposed by the Alboran Sea, as suggested by Ramon and Castro (1997) [55].

Generally, our observations are consensual within the existing studies dedicated to sardine population structure in the region, though the majority suggests a sardine panmictic stock [22,51,56], and others point towards a soft structure on a specific region of the species distribution [55]. While our study did not aim to exhaustively characterize the population structure of the species, we believe that it reinforces the need to undertake genome-wide approaches and temporal sampling to convincingly resolve the stock structure of the European sardine.

4.2. Temperature-Dependent Distribution of Most Frequently Observed Haplotypes

Exploring the role of specific environmental factors on the distribution of sardine's mitochondrial diversity revealed that only minimum values of temperature and dissolved oxygen across sampled locations significantly explained the gradients of genetic variation. By showing that the distribution of the more frequently observed haplotypes of two of the targeted genes—ATP and CYTB—negatively correlated with minimum surface temperature, we can infer that the sardine stock in the screened geographical area is selected to tolerate colder sea surface temperatures. These results agree with previous works on mitochondrial DNA selection on other marine fishes, particularly on anchovies [27] and on Atlantic salmon [26], probably indicating a functional need to increase metabolic efficiency at lower temperatures [26]. North eastern Atlantic latitudes covered by our work are characterized by a strong seasonal upwelling, that creates oxygen- and nutrient-rich environments nearby coastal costal locations but originates from lower sea surface temperatures [57]. One of the anticipated effects of ongoing climate change on coastal upwelling systems is an increase in intensity of spring and summer upwelling events [57]. As the Iberian Peninsula is particularly sensitive to increases in upwelling intensity [58], we interpret the relationship identified in our work as a putative genomic signature of the species' response. Interestingly, historical signatures of purifying selection identified amongst the phylogenies of targeted genes were associated with low frequency haplotypes. As all variants are exposed to selection in haploid systems, such observation suggests the evolutionary dynamics of the sardine genetic pool purged specific haplotypes from the population due potentially deleterious effects [59,60]. While synonymous mutations do not alter the sequencing of encoded protein, they are known to impact secondary structures affecting thermodynamics and stability, as observed on mRNA molecules [61], misfolding of proteins [62] or translational selection and codon use bias in both proto and metazoans [63,64]. In addition, and specific to small pelagic fishes, the often large effective population sizes would reduce the efficiency of genetic drift on removing rare variants, but see Cvijović et al. (2018) for alternative theoretical expectations on site-frequency distribution under puryfying selection [65]. Lastly, genetic signatures on mitochondrial genes of the OXPHOS complexes are also worth exploring at the light of cyto-nuclear co-evolution [66] or having their functional impact on organismal fitness effectively assessed given the existence of compensatory mechanisms to buffer the expression of deleterious alleles [67].

5. Conclusions and Future

Under and intense fishing pressure, the capacity for the European sardines to positively respond to ongoing climatic shifts must be resolved before developing practices of sustainable exploitation. Here, we focused on the European sardine stocks around the Iberian Peninsula and provided additional evidence for the existence of an apparent near panmictic population with a high degree of connectivity. This connectivity might be important to maintain frequency distribution of key mitochondrial haplotypes across the species distribution range, ensuring not only the species adaptive potential, but also an overall capacity to migrate to follow environmental optima: eggs and larvae of European sardine have been reported as North as the Baltic Sea [68]. Strong northward shifts are nevertheless expected for temperate-cold water species such as European sardines, according to species distribution models under future climatic conditions [69].

Shifts in thermal conditions are particularly stressful for ectotherms. Additionally, there are certainly other mechanisms involved in a putative adaptive response to thermal variation. For instances, adaptive and non-adaptive phenotypic plasticity are known to play a decisive role in assisting organisms to cope with shifting temperatures [70,71]. Epigenetic variation is also another possible route towards adaptation, and evidence of such keeps on accumulating among both natural populations and under controlled experiments [72–74]. Though we interpreted the relationship between haplotype frequency and minimum sea surface temperature as a signature of selection, we cannot disregard the potential effect of small sample sizes or the fact that we did not screen the edges of specie's distribution. Still, with the seascape genetics approach we undertook, we are confident that our study provides a robust glimpse of sardines' adaptive potential and is an important contribution to the growing body of evidence supporting adaptation among populations of small pelagic fishes.

Supplementary Materials: The following are available online at https://www.mdpi.com/2073-4425/12/1/91/s1, Figure S1: Comparison of diversity and differentiation estimates obtained with microsatellite loci. (A) Expected (He) and observed heterozygosity (Ho); (B) Averaged F_{ST} per locus considering scores of the five populations, Figure S2: Population differentiation analyses produced with microsatellite loci. (A) DAPC plots considering all loci–ellipses are at 95% C.I.; (B) STRUCTURE assignment plots for K = 5, Figure S3: DAPC plots considering each locus separately-ellipses are at 95% C.I, Table S1: NCBI references sequences utilized to design mitochondrial primers, Table S2: Primers utilized to amplify mitochondrial genes, Table S3: Mantel tests for each mitochondrial gene and overall microsatellite loci, Table S4: Micro-checker reports for each loci, per population. In bold, loci where null alleles were detected. Text S1: sequences in *fasta* format of mitochondrial's amplified regions obtained in this work.

Author Contributions: Conceptualization, M.B.-S. and G.S.; methodology, M.B.-S. and G.S.; validation, M.B.-S., G.S. and A.R.d.A.L.; formal analysis, M.B.-S. and A.R.d.A.L.; investigation, M.B.-S., G.S. and A.R.d.A.L.; resources, G.S.; data curation, M.B.-S.; writing—original draft preparation, M.B.-S.; writing—review and editing, M.B.-S., G.S. and A.R.d.A.L.; visualization, M.B.-S. and A.R.d.A.L.; supervision, G.S.; project administration, G.S.; funding acquisition, G.S. All authors have read and agreed to the published version of the manuscript.

Funding: This study had the support of Fundação para a Ciência e Tecnologia (FCT), through the strategic project UIDB/04292/2020 granted to MARE, and the project PTDC/BIA-BMA/32209/2017, co-financed by FCT through national funds and by the European Union through FEDER (European Regional Development Fund).

Data Availability Statement: Sequence data is available as supplementary file. Sequences are deposited in NCBI under the accession numbersMW437948-MW438277.

Acknowledgments: The authors would like to acknowledge to Naiara Rodriguez-Ezpeleta, Jordi Viñas, Marta Frigola, Xènia Tepe, Joana Boavida, Emilie Villar, David Piló, Laurie Casalot, Didier Aurelle and Fauvelle Vincent for they availability, and their collaboration on collecting and shipping tissue samples. The authors also thank Flying Sharks for transporting samples collected on the Gulf of Lions and Susana Costa Lopes (CIBIO) for sequencing and library preparation.

Conflicts of Interest: The authors declare no conflict of interests.

Appendix A

Appendix A.1. Library Preparation and Amplicon Sequencing Protocol

Multiplexed Illumina MiSeq amplicon libraries for the four mitochondrial regions were amplified using a two-stage PCR approach [75]. First-stage PCR conditions differed among genes. The ATP6 fragment was amplified in a 10 ML reaction with 5 ML of MyTaq© Mix (BioLine, UK), 0.4 mM of each primer with overhangs as in McInnes et al. [76] and 10 ng of DNA template. Thermal cycling was performed in a T100 Thermal Cycler© (BioRad) consisting of an initial step at 95 °C for 10′, followed by 40 cycles at 95 °C for 30″, 58 °C for 30″ and 72 °C for 30″ and a final extension of 72 °C for 10′. Amplification of the remaining genes was performed under the following conditions: an initial step at 95 °C for 10′, followed by 95 °C for 30″, 68 °C −0.5 °C/cycle for 30″, 72 °C for 20″, followed by 31 cycles at 95 °C for 30″, 64 °C for 30″, 72 °C for 20″, and a final extension of 72 °C for 10″. A template-free PCR was included in each amplification to control for potential contamination. PCR products were purified using reversible immobilization (SPRI) paramagnetic beads© (Agencourt AMPure XP, Beckman Coulter, CA, USA) with 0.8× of beads per microliter of PCR product.

Individual tags from McInnes et al. (2017) [76] were added to each sample PCR products obtained from first-stage amplifications. Second-stage PCRs were prepared with 12.5 ML of KAPA HiFi HotStart ReadyMixPCR Kit© (Kapa Biosystems, Wilmington, MA, USA), 2.5 ML of 1 mM of each tag primer and 2.5 ML of 1:10 diluted PCR product in 25 ML final volume. Thermal cycler conditions were as following: 95 °C for 3′, followed by 10 cycles of 95 °C for 30″, 55 °C for 30″, 72 °C for 30″, and a final extension of 72 °C for 5′. PCR products were purified using reversible immobilization (SPRI) paramagnetic beads.

All samples were then pooled after normalizing amplified quantities to 10 nM. Amplicon's size distribution was analysed on a TapeStation 2200 using the HS D1000 kit (Agilent Technologies, Santa Clara, CA, USA). Libraries concentration was checked by quantitative PCR on a BIORAD C1000 Real Time Thermo cycler using KAPA Library Quantification kit (Kapa Biosystems, Wilmington, MA, USA). Amplicon libraries were sequenced on a MiSeq genome sequencer (Illumina, San Diego, CA, USA) using MiSeq Reagent Kit Nano v2 (500 cycles) with paired-end reads.

Appendix A.2. Amplification and Genotyping of Microsatellite Loci

Amplification was performed with a Qiagen Multiplex PCR kit (Hilden, Germany) in a single multiplex under the following conditions: initial denaturation step at 95 °C for 15 min, 17 cycles comprising an initial step of at 95 °C for 30 s, an annealing step at 56 °C for 1 min and 30 s (with an increase of 0.5 °C/cycle during the first 13 cycles) and an elongation at 72 °C for 30 s, followed by 27 cycles which included a denaturation at 95 °C for 30 s, annealing at 50 °C for 1 min and an elongation step at 72 °C for 30 s. Final elongation occurred at 60 °C for 30 min. PCR products were separated by capillary electrophoresis on an automatic sequencer ABI3130xl Genetic Analyzer (AB Applied Biosystems, Foster City, CA, USA). Fragments were scored against the GeneScan-500 LIZ Size Standard using the GENEMAPPER 4.1 (Applied Biosystems) and manually checked twice.

References

1. Calosi, P.; De Wit, P.; Thor, P.; Dupont, S. Will life find a way? Evolution of marine species under global change. *Evol. Appl.* **2016**, *9*, 1035–1042. [CrossRef] [PubMed]
2. Costa, B.H.E.; Assis, J.; Franco, G.; Erzini, K.; Henriques, M.; Gonçalves, E.; Caselle, J.; Gonçalves, E.J. Tropicalization of fish assemblages in temperate biogeographic transition zones. *Mar. Ecol. Prog. Ser.* **2014**, *504*, 241–252. [CrossRef]
3. Vergés, A.; Steinberg, P.D.; Hay, M.E.; Poore, A.G.B.; Campbell, A.H.; Ballesteros, E.; Heck, K.L.; Booth, D.J.; Coleman, M.A.; Feary, D.A.; et al. The tropicalization of temperate marine ecosystems: Climate-mediated changes in herbivory and community phase shifts. *Proc. R. Soc. B Biol. Sci.* **2014**, *281*, 20140846. [CrossRef] [PubMed]
4. Melzner, F.; Thomsen, J.; Koeve, W.; Oschlies, A.; Gutowska, M.A.; Bange, H.W.; Hansen, H.P.; Körtzinger, A. Future ocean acidification will be amplified by hypoxia in coastal habitats. *Mar. Biol.* **2012**, *160*, 1875–1888. [CrossRef]
5. Altieri, A.H.; Gedan, K.B. Climate change and dead zones. *Glob. Chang. Biol.* **2014**, *21*, 1395–1406. [CrossRef] [PubMed]
6. Eizaguirre, C.; Baltazar-Soares, M. Evolutionary conservation—Evaluating the adaptive potential of species. *Evol. Appl.* **2014**, *7*, 963–967. [CrossRef]
7. Nielsen, E.E.; Hemmer-Hansen, J.; Larsen, P.F.; Bekkevold, D. Population genomics of marine fishes: Identifying adaptive variation in space and time. *Mol. Ecol.* **2009**, *18*, 3128–3150. [CrossRef]
8. Grummer, J.A.; Beheregaray, L.B.; Bernatchez, L.; Hand, B.K.; Luikart, G.; Narum, S.R.; Taylor, E.B. Aquatic Landscape Genomics and Environmental Effects on Genetic Variation. *Trends Ecol. Evol.* **2019**, *34*, 641–654. [CrossRef]
9. Selkoe, K.A.; Henzler, C.M.; Gaines, S.D. Seascape genetics and the spatial ecology of marine populations. *Fish Fish.* **2008**, *9*, 363–377. [CrossRef]
10. Hutchinson, W.F. The dangers of ignoring stock complexity in fishery management: The case of the North Sea cod. *Biol. Lett.* **2008**, *4*, 693–695. [CrossRef]
11. Ernande, B.; Dieckmann, U.; Heino, M. Adaptive changes in harvested populations: Plasticity and evolution of age and size at maturation. *Proc. R. Soc. B Biol. Sci.* **2004**, *271*, 415–423. [CrossRef] [PubMed]

12. Dunlop, E.S.; Eikeset, A.M.; Stenseth, N.C. From genes to populations: How fisheries-induced evolution alters stock productivity. *Ecol. Appl.* **2015**, *25*, 1860–1868. [CrossRef] [PubMed]

13. Spielman, D.; Brook, B.W.; Frankham, R. Most species are not driven to extinction before genetic factors impact them. *Proc. Natl. Acad. Sci. USA* **2004**, *101*, 15261–15264. [CrossRef] [PubMed]

14. Deyle, E.R.; Fogarty, M.; Hsieh, C.-H.; Kaufman, L.; MacCall, A.D.; Munch, S.B.; Perretti, C.T.; Ye, H.; Sugihara, G. Predicting climate effects on Pacific sardine. *Proc. Natl. Acad. Sci. USA* **2013**, *110*, 6430–6435. [CrossRef]

15. Cury, P.; Bakun, A.; Crawford, R.J.M.; Jarre, A.; Quiñones, R.A.; Shannon, L.J.; Verheye, H.M. Small pelagics in upwelling systems: Patterns of interaction and structural changes in "wasp-waist" ecosystems. *ICES J. Mar. Sci.* **2000**, *57*, 603–618. [CrossRef]

16. Rykaczewski, R.R.; Checkley, D.M. Influence of ocean winds on the pelagic ecosystem in upwelling regions. *Proc. Natl. Acad. Sci. USA* **2008**, *105*, 1965–1970. [CrossRef]

17. Koslow, J.A.; Davison, P.; Lara-Lopez, A.; Ohman, M.D. Epipelagic and mesopelagic fishes in the southern California Current System: Ecological interactions and oceanographic influences on their abundance. *J. Mar. Syst.* **2014**, *138*, 20–28. [CrossRef]

18. Yamaguchi, A.; Matsuno, K.; Abe, Y.; Arima, D.; Imai, I. Latitudinal variations in the abundance, biomass, taxonomic composition and estimated production of epipelagic mesozooplankton along the 155° E longitude in the western North Pacific during spring. *Prog. Oceanogr.* **2017**, *150*, 13–19. [CrossRef]

19. Duarte, L.O.; Garcia, C. Trophic role of small pelagic fishes in a tropical upwelling ecosystem. *Ecol. Model.* **2004**, *172*, 323–338. [CrossRef]

20. González-García, S.; Villanueva-Rey, P.; Belo, S.; Vázquez-Rowe, I.; Moreira, M.T.; Feijoo, G.; Arroja, L. Cross-vessel eco-efficiency analysis. A case study for purse seining fishing from North Portugal targeting Euro-pean pilchard. *Int. J. Life Cycle Assess.* **2015**, *20*, 1019–1032. [CrossRef]

21. Vázquez-Rowe, I.; Villanueva-Rey, P.; Hospido, A.; Moreira, M.T.; Feijoo, G. Life cycle assessment of European pilchard (*Sardina pilchardus*) consumption. A case study for Galicia (NW Spain). *Sci. Total Environ.* **2014**, *475*, 48–60. [CrossRef] [PubMed]

22. Jemaa, S.; Bacha, M.; Khalaf, G.; Dessailly, D.; Rabhi, K.; Amara, R. What can otolith shape analysis tell us about population structure of the European sardine, *Sardina pilchardus*, from Atlantic and Mediterranean waters? *J. Sea Res.* **2015**, *96*, 11–17. [CrossRef]

23. Antonakakis, K.; Giannoulaki, M.; Machias, A.; Somarakis, S.; Sanchez, S.; Ibaibarriaga, L.; Uriarte, A. Assessment of the sardine (*Sardina pilchardus* Walbaum, 1792) fishery in the eastern Mediterranean basin (North Aegean Sea). *Mediterr. Mar. Sci.* **2011**, *12*, 333–357. [CrossRef]

24. Letts, J.A.; Fiedorczuk, K.; Sazanov, J.A. The architecture of respiratory supercomplexes. *Nat. Cell Biol.* **2016**, *537*, 644–648. [CrossRef] [PubMed]

25. Formosa, L.E.; Ryan, M.T. Mitochondrial OXPHOS complex assembly lines. *Nat. Cell Biol.* **2018**, *20*, 511–513. [CrossRef] [PubMed]

26. Consuegra, S.; John, E.; Verspoor, E.; de Leaniz, C.G. Patterns of natural selection acting on the mitochondrial genome of a locally adapted fish species. *Genet. Sel. Evol.* **2015**, *47*, 58. [CrossRef]

27. Silva, G.; Lima, F.P.; Martel, P.; Castilho, R. Thermal adaptation and clinal mitochondrial DNA variation of European anchovy. *Proc. R. Soc. B Biol. Sci.* **2014**, *281*, 20141093. [CrossRef]

28. Sebastian, W.; Sukumaran, S.; Zacharia, P.; Muraleedharan, K.; Kumar, P.D.; Gopalakrishnan, A. Signals of selection in the mitogenome provide insights into adaptation mechanisms in heterogeneous habitats in a widely distributed pelagic fish. *Sci. Rep.* **2020**, *10*, 1–14.

29. Hall, T.A. BioEdit: A User-Friendly Biological Sequence Alignment Editor and Analysis Program for Windows 95/98/NT. In *Nucleic Acids Symposium Series*; Information Retrieval Ltd.: London, UK, 1999; Volume 41, pp. 95–98.

30. Li, H. Aligning sequence reads, clone sequences and assembly contigs with BWA-MEM. *arXiv* **2013**, arXiv:1303.3997.

31. Li, H.; Durbin, R. Fast and accurate long-read alignment with Burrows–Wheeler transform. *Bioinformatics* **2010**, *26*, 589–595. [CrossRef]

32. Catchen, J.; Hohenlohe, P.A.; Bassham, S.; Amores, A.; Cresko, W.A. Stacks: An analysis tool set for population genomics. *Mol. Ecol.* **2013**, *22*, 3124–3140. [CrossRef] [PubMed]

33. Librado, P.; Rozas, J. DnaSP v5: A software for comprehensive analysis of DNA polymorphism data. *Bioinformatics* **2009**, *25*, 1451–1452. [CrossRef]

34. Excoffier, L.; Lischer, H.E.L. Arlequin suite ver 3.5: A new series of programs to perform population genetics analyses under Linux and Windows. *Mol. Ecol. Resour.* **2010**, *10*, 564–567. [CrossRef] [PubMed]

35. Goslee, S.C.; Urban, D.L. The ecodist package for dissimilarity-based analysis of ecological data. *J. Stat. Softw.* **2007**, *22*, 1–19. [CrossRef]

36. Leigh, J.W.; Bryant, D. POPART: Full-feature software for haplotype network construction. *Methods Ecol. Evol.* **2015**, *6*, 1110–1116. [CrossRef]

37. R Core Team. *R: A Language and Environment for Statistical Computing*; R Foundation for Statistical Computing: Vienna, Austria, 2013.

38. Assis, J.; Tyberghein, L.; Bosch, S.; Verbruggen, H.; Serrão, E.A.; De Clerck, O. Bio-ORACLE v2.0: Extending marine data layers for bioclimatic modelling. *Glob. Ecol. Biogeogr.* **2018**, *27*, 277–284. [CrossRef]

39. Rao, C.R. The use and interpretation of principal component analysis in applied research. *Sankhyā Indian J. Stat. Ser. A* **1964**, *26*, 329–358.

40. Wickham, H.; Francois, R.; Henry, L.; Müller, K. Vegan: Community Ecology Package. R package version 1.17-4. 2010. Available online: https://cran.r-project.org/web/packages/vegan/index.html (accessed on 15 July 2020).

41. Legendre, P.; Oksanen, J.; ter Braak, C.J.F. Testing the significance of canonical axes in redundancy analysis. *Methods Ecol. Evol.* **2011**, *2*, 269–277. [CrossRef]

42. Weaver, S.; Shank, S.D.; Spielman, S.J.; Li, M.; Muse, S.V.; Kosakovsky Pond, S.L. Datamonkey 2.0: A Mod-ern Web Application for Characterizing Selective and Other Evolutionary Processes. *Mol. Biol. Evol.* **2018**, *35*, 773–777. [CrossRef]

43. Pond, S.L.K.; Frost, S.D.W. Not So Different After All: A Comparison of Methods for Detecting Amino Acid Sites Under Selection. *Mol. Biol. Evol.* **2005**, *22*, 1208–1222. [CrossRef]

44. Murrell, B.; Moola, S.; Mabona, A.; Weighill, T.; Sheward, D.; Pond, S.L.K.; Scheffler, K. FUBAR: A Fast, Unconstrained Bayesian AppRoximation for Inferring Selection. *Mol. Biol. Evol.* **2013**, *30*, 1196–1205. [CrossRef] [PubMed]

45. Ruggeri, P.; Splendiani, A.; Bonanomi, S.; Arneri, E.; Cingolani, N.; Santojanni, A.; Belardinelli, A.; Giovannotti, M.; Barucchi, V.C. Temporal genetic variation as revealed by a microsatellite analysis of European sardine (*Sardina pilchardus*) archived samples. *Can. J. Fish. Aquat. Sci.* **2012**, *69*, 1698–1709. [CrossRef]

46. Kasapidis, P.; Silva, A.; Zampicinini, G.; Magoulas, A. Evidence for microsatellite hitchhiking selection in European sardine (*Sardina pilchardus*) and implications in inferring stock structure. *Sci. Mar.* **2011**, *76*, 123–132. [CrossRef]

47. Kalinowski, S.T. HP-RARE 1.0: A computer program for performing rarefaction on measures of allelic richness. *Mol. Ecol. Notes* **2005**, *5*, 187–189. [CrossRef]

48. Van Oosterhout, C.; Hutchinson, W.F.; Wills, D.P.M.; Shipley, P. MICRO-CHECKER: Software for identifying and correcting genotyping errors in microsatellite data. *Mol. Ecol. Notes* **2004**, *4*, 535–538. [CrossRef]

49. Jombart, T. adegenet: A R package for the multivariate analysis of genetic markers. *Bioinformatics* **2008**, *24*, 1403–1405. [CrossRef] [PubMed]

50. Frankham, R.; Briscoe, D.A.; Ballou, J.D. *Introduction to Conservation Genetics*; Cambridge University Press: Cambridge, UK, 2002.

51. Baibai, T.; Oukhattar, L.; Quinteiro, J.V.; Mesfioui, A.; Rey-Mendez, M. First global approach: Morphological and biological variability in a genetically homogeneous population of the European pilchard, *Sardina pilchardus* (Walbaum, 1792) in the North Atlantic coast. *Rev. Fish Biol. Fish.* **2012**, *22*, 63–80. [CrossRef]

52. Silva, A.; Carrera, P.; Masse, J.; Uriarte, A.; Santos, M.; Oliveira, P.B.; Soares, E.; Porteiro, C.; Stratoudakis, Y. Geographic variability of sardine growth across the northeastern Atlantic and the Mediterranean Sea. *Fish. Res.* **2008**, *90*, 56–69. [CrossRef]

53. Van Beveren, E.; Fromentin, J.-M.; Rouyer, T.; Bonhommeau, S.; Brosset, P.; Saraux, C. The fisheries history of small pelagics in the Northern Mediterranean. *ICES J. Mar. Sci.* **2016**, *73*, 1474–1484. [CrossRef]

54. Pinsky, M.L.; Palumbi, S.R. Meta-analysis reveals lower genetic diversity in overfished populations. *Mol. Ecol.* **2014**, *23*, 29–39. [CrossRef]

55. Ramon, M.; Castro, J. Genetic variation in natural stocks of *Sardina pilchardus* (sardines) from the western Mediterranean Sea. *Heredity* **1997**, *78*, 520–528. [CrossRef]

56. Correia, A.T.; Hamer, P.; Carocinho, B.; Silva, A. Evidence for meta-population structure of *Sardina pilchardus* in the Atlantic Iberian waters from otolith elemental signatures of a strong cohort. *Fish. Res.* **2014**, *149*, 76–85. [CrossRef]

57. Bakun, A.; Black, B.A.; Bograd, S.J.; Garcia-Reyes, M.; Miller, A.J.; Rykaczewski, R.R.; Sydeman, W.J. Anticipated effects of climate change on coastal upwelling ecosystems. *Curr. Clim. Chang. Rep.* **2015**, *1*, 85–93. [CrossRef]

58. Miranda, P.; Alves, J.; Serra, N. Climate change and upwelling: Response of Iberian upwelling to atmospheric forcing in a regional climate scenario. *Clim. Dyn.* **2013**, *40*, 2813–2824. [CrossRef]

59. Avery, P.J. The population genetics of haplo-diploids and X-linked genes. *Genet. Res.* **1984**, *44*, 321–341. [CrossRef]

60. Levin, L.; Zhidkov, I.; Gurman, Y.; Hawlena, H.; Mishmar, D. Functional Recurrent Mutations in the Human Mitochondrial Phylogeny: Dual Roles in Evolution and Disease. *Genome Biol. Evol.* **2013**, *5*, 876–890. [CrossRef] [PubMed]

61. Chamary, J.V.; Hurst, L.D. Evidence for selection on synonymous mutations affecting stability of mRNA secondary structure in mammals. *Genome Biol.* **2005**, *6*, R75. [CrossRef]

62. Drummond, D.A.; Wilke, C.O. Mistranslation-induced protein misfolding as a dominant constraint on coding-sequence evolution. *Cell* **2008**, *134*, 341–352. [CrossRef]

63. Castellana, S.; Vicario, S.; Saccone, C. Evolutionary Patterns of the Mitochondrial Genome in Metazoa: Exploring the Role of Mutation and Selection in Mitochondrial Protein–Coding Genes. *Genome Biol. Evol.* **2011**, *3*, 1067–1079. [CrossRef]

64. Uddin, A.; Chakraborty, S. Synonymous codon usage pattern in mitochondrial CYB gene in pisces, aves, and mammals. *Mitochondrial DNA Part A* **2015**, *28*, 187–196. [CrossRef]

65. Cvijović, I.; Good, B.H.; Desai, M.M. The Effect of Strong Purifying Selection on Genetic Diversity. *Genetics* **2018**, *209*, 1235–1278. [CrossRef] [PubMed]

66. Barreto, F.S.; Watson, E.T.; Lima, T.G.; Willett, C.S.; Edmands, S.; Li, W.; Burton, R.S. Genomic signatures of mitonuclear coevolution across populations of *Tigriopus californicus*. *Nat. Ecol. Evol.* **2018**, *2*, 1250–1257. [CrossRef] [PubMed]

67. Hill, G.E. Mitonuclear Compensatory Coevolution. *Trends Genet.* **2020**, *36*, 403–414. [CrossRef] [PubMed]

68. Alheit, J.; Pohlmann, T.; Casini, M.; Greve, W.; Hinrichs, R.; Mathis, M.; O'Driscoll, K.; Vorberg, R.; Wagner, C. Climate variability drives anchovies and sardines into the North and Baltic Seas. *Prog. Oceanogr.* **2012**, *96*, 128–139. [CrossRef]

69. Schickele, A.; Goberville, E.; Leroy, B.; Beaugrand, G.; Hattab, T.; Francour, P.; Raybaud, V. European small pelagic fish distribution under global change scenarios. *Fish Fish.* **2021**, *22*, 212–225. [CrossRef]

70. Schulte, P.M.; Healy, T.M.; Fangue, N.A. Thermal Performance Curves, Phenotypic Plasticity, and the Time Scales of Temperature Exposure. *Integr. Comp. Biol.* **2011**, *51*, 691–702. [CrossRef]
71. Yampolsky, L.Y.; Schaer, T.M.M.; Ebert, D. Adaptive phenotypic plasticity and local adaptation for temperature tolerance in freshwater zooplankton. *Proc. R. Soc. B Biol. Sci.* **2014**, *281*, 20132744. [CrossRef]
72. McCaw, B.A.; Stevenson, T.J.; Lancaster, L.T. Epigenetic Responses to Temperature and Climate. *Integr. Comp. Biol.* **2020**, *60*, 1469–1480. [CrossRef]
73. Best, C.; Ikert, H.; Kostyniuk, D.J.; Craig, P.M.; Navarro-Martin, L.; Marandel, L.; Mennigen, J.A. Epigenetics in teleost fish: From molecular mechanisms to physiological phenotypes. *Comp. Biochem. Physiol. Part B Biochem. Mol. Biol.* **2018**, *224*, 210–244. [CrossRef]
74. Heckwolf, M.J.; Meyer, B.S.; Häsler, R.; Höppner, M.P.; Eizaguirre, C.; Reusch, T.B. Two different epigenetic information channels in wild three-spined sticklebacks are involved in salinity adaptation. *Sci. Adv.* **2020**, *6*, eaaz1138. [CrossRef]
75. Binladen, J.; Gilbert, M.T.P.; Bollback, J.P.; Panitz, F.; Bendixen, C.; Nielsen, R.; Willerslev, E. The Use of Coded PCR Primers Enables High-Throughput Sequencing of Multiple Homolog Amplification Products by 454 Parallel Sequencing. *PLoS ONE* **2007**, *2*, e197. [CrossRef] [PubMed]
76. McInnes, J.C.; Jarman, S.N.; Lea, M.-A.; Raymond, B.; Deagle, B.E.; Phillips, R.A.; Catry, P.; Stanworth, A.; Weimerskirch, H.; Kusch, A. DNA metabarcoding as a marine conservation and management tool: A circumpolar examination of fishery discards in the diet of threatened albatrosses. *Front. Mar. Sci.* **2017**, *4*, 277. [CrossRef]

Article

A Disjunctive Marginal Edge of Evergreen Broad-Leaved Oak (*Quercus gilva*) in East Asia: The High Genetic Distinctiveness and Unusual Diversity of Jeju Island Populations and Insight into a Massive, Independent Postglacial Colonization

Eun-Kyeong Han [1,†], Won-Bum Cho [2,†], Jong-Soo Park [3], In-Su Choi [4], Myounghai Kwak [5], Bo-Yun Kim [6] and Jung-Hyun Lee [2,*]

[1] Department of Biological Sciences and Biotechnology, Chonnam National University, Gwangju 61186, Korea; urinara-han@hanmail.net
[2] Department of Biology Education, Chonnam National University, Gwangju 61186, Korea; rudis99@hanmail.net
[3] Department of Biological Sciences, Inha University, 100, Inha-ro, Michuhol-gu, Incheon 22212, Korea; beeul25@gmail.com
[4] School of Life Sciences, Arizona State University, Tempe, AZ 85287, USA; 86ischoi@gmail.com
[5] Biological and Genetic Resources Utilization Division, National Institute of Biological Resources, Incheon 22689, Korea; mhkwak1@korea.kr
[6] Plant Resources Division, National Institute of Biological Resources, Incheon 22689, Korea; bykim416@korea.kr
* Correspondence: quercus@jnu.ac.kr
† These authors contributed equally to this work.

Received: 13 August 2020; Accepted: 22 September 2020; Published: 23 September 2020

Abstract: Jeju Island is located at a marginal edge of the distributional range of East Asian evergreen broad-leaved forests. The low genetic diversity of such edge populations is predicted to have resulted from genetic drift and reduced gene flow when compared to core populations. To test this hypothesis, we examined the levels of genetic diversity of marginal-edge populations of *Quercus gilva*, restricted to a few habitats on Jeju Island, and compared them with the southern Kyushu populations. We also evaluated their evolutionary potential and conservation value. The genetic diversity and structure were analyzed using 40 polymorphic microsatellite markers developed in this study. Ecological Niche Modeling (ENM) has been employed to develop our insights, which can be inferred from historical distribution changes. Contrary to our expectations, we detected a similar level of genetic diversity in the Jeju populations, comparable to that of the southern Kyushu populations, which have been regarded as long-term glacial refugia with a high genetic variability of East Asian evergreen trees. We found no signatures of recent bottlenecks in the Jeju populations. The results of STRUCTURE, neighbor-joining phylogeny, and Principal Coordinate Analysis (PCoA) with a significant barrier clearly demonstrated that the Jeju and Kyushu regions are genetically distinct. However, ENM showed that the probability value for the distribution of the trees on Jeju Island during the Last Glacial Maximum (LGM) converge was zero. In consideration of these results, we hypothesize that independent massive postglacial colonization from a separate large genetic source, other than Kyushu, could have led to the current genetic diversity of Jeju Island. Therefore, we suggest that the Jeju populations deserve to be separately managed and designated as a level of management unit (MU). These findings improve our understanding of the paleovegetation of East Asian evergreen forests, and the microevolution of oaks.

Keywords: marginal edge; *Quercus gilva*; genetic diversity; massive colonization; Jeju Island; conservation

1. Introduction

It is well known that the population genetic structure in extant plants is affected by various factors, including the dispersal ability of pollinators, seed dispersal modes, reproductive systems, and historical migration patterns; the historical range change during Quaternary climatic oscillations is also considered a primary factor [1–3]. Although their relative importance may vary across time and space, the genetic features of populations in East Asian temperate regions likely reflect historical, rather than current, levels of gene flow [4–8]. East Asia has experienced complex and dynamic changes in land configurations during the Quaternary period, which led to a high richness and endemism of plant species in forests [9]. The range change of warm temperate evergreen forests was larger than that of temperate deciduous forests, especially in Korea and Japan.

Peripheral, especially marginal/edge, populations, might reflect genetic impoverishment as a result of genetic drift and reduced gene flow when compared to core populations [10–13]. Such genetic determinants have the potential to further expand species ranges through adaptation to the selection pressures of a marginal environment, assumed to furnish less fitness for their survival [14]. The populations have played decisive roles for species facing and responding to rapidly changing environmental conditions [15,16]. Many of these studies of local fitness have improved our knowledge of how a given species adapts to a changing environment [17,18]. Nonetheless, our understanding of the evolution of species in warm temperate evergreen forests in East Asia is still lacking.

The volcanic Jeju Island of South Korea is characterized by its high endemism, unique altitudinal zonation of vegetation, and untouched environments [19], and thus it was designated as a UNESCO Biosphere Reserve in 2002 and a World Heritage Site in 2007. The island is disjunctively located at the marginal edge of the distributional range of East Asian evergreen broad-leaved forests. The lowland zone of Jeju Island is covered with forests, dominated by warm temperate and subtropical evergreen broad-leaved plants [20]. These species commonly inhabit large ranges across East Asia, including South Korea, Japan, and China, but exhibit disjunctive distributions, with a heterogeneous boundary of habitat preferences [21]. Therefore, conserving the populations at the marginal edge of their range can be beneficial to the long-term survival of a species.

Quercus gilva Blume (Fagaceae) is a large, ecologically important tree of evergreen broad-leaved forests in East Asia [22]. Although *Q. gilva* is widely distributed in East Asian forests [23,24], its habitat is decreasing due to anthropogenic pressure. The main cause of habitat decrease is human-mediated disturbance, such as large-scale regional development [25] and logging [26]. The population of Jeju Island is extremely small (total: ca. 600 individuals; [27]) and distributed in a unique habitat, called Gotjawal, where the trees occur in an area made up of numerous fragmented rocks. This species is listed as Vulnerable (VU) in the Korea Red Data Book [28]. As of 2012, it has also been protected under the Endangered Species Act (ESA) within Korean law. Since tracking of intraspecific Conservation Units (CUs) is one of the most important tasks for the long-term conservation of a given species, a population genetic examination for *Q. gilva* was attempted using RAPD (Random Amplified Polymorphic DNA) [29] and ISSR (Inter Simple Sequence Repeat) [25] analysis. However, previous assessments have not been performed with other comparative populations, which could be a criterion for accurately recognizing their genetic status. From a recent conservation genetics point of view, providing information on the population's establishment history is becoming a fundamental step in long-term conservation [30–32].

Fossils and pollen grains could be utilized for unraveling evolutionary clues that contribute to present genetic diversity, but the situation is complicated by historically complex distribution changes [33,34]. However, such past data are largely absent in East Asia, because the area in which the evergreen forests appear to have been historically distributed is now in the sea. Given this, a technique such as Ecological Niche Modeling (ENM) is the only auxiliary way to reveal the historical distribution of a species [35]. ENM is useful in genetic studies to infer climate change-associated correlations between distribution shifts and genetic structure [36–38].

As has been observed in other warm temperate species in East Asia [5,6,39], the extant Jeju populations of *Q. gilva* are most closely related to those in Kyushu, Japan, which is geographically adjacent and has a similar establishment history. Therefore, we characterized the genetic compositions of the *Q. gilva* populations in Jeju Island, located at the disjunctive edge of their distribution range. The genetic diversity was compared to the Kyushu populations, regarded as long-term glacial refugia with a high genetic variability of East Asian warm temperate evergreen broad-leaved trees. The purposes of the present study are (1) to develop a high-resolution and cost-effective polymorphic microsatellite set so that researchers can continue periodic genetic monitoring, (2) to evaluate the evolutionary potential and conservation value of marginal-edge Jeju populations by inferring the history of population establishment, and (3) to provide conservation guidelines for the recovery and management of the threatened Jeju populations. The genetic diversity and structure were analyzed using 40 polymorphic microsatellite markers developed in this study through high-throughput sequencing data. ENM was also employed to examine historical distribution changes.

2. Materials and Methods

2.1. Plant Material Sampling and DNA Extraction

We collected a total of 158 leaf samples of *Q. gilva* from three populations in Jeju Island, Korea and three populations in Kyushu, Japan. Since *Q. gilva* is protected as an endangered species in Korea, we first requested permission from the Ministry of Environment and then proceeded with the material collection. We selected the trees with a diameter at breast height (DBH) of more than 20 cm while maintaining minimal intervals of more than 5 m between individuals. One leaf sample was collected per individual to minimize damage to the species. In Jeju populations, a total of 77 leaves, including 32 from Gueok-ri (k-GU), 27 from Jeoji-ri (k-JJ), and 18 from Seogwang-ri (k-SG), were obtained, with an average of 25.6. In the Kyushu populations, an average of 27 and total of 81 leaf samples were collected from Kitadake, Kumamoto Prefecture (j-GM; 23), and Aoidake (j-MY; 29) and Enodake (j-NB; 29) in Miyazaki Prefecture. Collected leaf samples were stored at −80 °C in a deep freezer at the lab of Biological Education, Chonnam National University until use. Total genomic DNA was extracted from dried leaf samples using the DNeasy Plant mini kit (Qiagen, Seoul, Korea) following the manufacturer's instructions. The concentration of extracted DNA was determined using Nano-300 (Allsheng, Hangzhou, China), and diluted to 15 ng/µL to obtain the same concentration of template DNA in each sample.

2.2. Loci Isolation for Microsatellite Markers' Development and Genotyping

In order to develop polymorphic microsatellite markers for *Q. gilva*, we produced high-throughput sequencing data in a fresh leaf collected from Gueok-ri, Seogwipo-si, JeJu Island, Korea. A voucher specimen was deposited in the herbarium of Chonnam National University (BEC) (Voucher no. LeeQg20180502). A shotgun library construction for DNA sequencing was generated using the Illumina MiSeq platform (LAS, Seoul, Korea). According to the method of Cho et al. [40], we detected di-, tri-, or tetranucleotide motifs with flanking regions >100 bp and at least 10, six, or four repeats, respectively, through SSR_pipeline v. 0.951 [41]. After acquiring reads containing microsatellites from this screening, we attempted a reference mapping of the total paired reads to each remaining sequence using Geneious R11.0.5 [42]. In the reference-mapped results, after discarding putative multicopy loci with exceptionally high coverage (>20 reads), we used the final reads, showing the variation in length at the repeating site, no substitution of the site to produce the primer, and no additional insertion/deletion in the flanking region. Based on the final selected reads, we designed 54 primer pairs using Primer3 version 0.4.0 software [43] in the Geneious program according to the following parameters: primer size 18–22 bp, Tm (melting temperature) of 53–60 °C, and GC content of 35–65%.

The forward primers added three sets of M13 tag sequences (5′-CACGACGTTGTAAAACGAC-3′, 5′-TGTGGAATT GTGAGCGG-3′, and 5′-CTATAGGGCACGCGTGGT-3′) with 6-FAM, VIC, and NED fluorescent dye, respectively.

To assess the polymorphisms for the designed microsatellite loci, we conducted a preliminary PCR analysis with 32 individuals from the Gueok-ri population. PCR amplification was performed with a Veriti 96-well thermal cycler (Applied Biosystems, Foster City, CA, USA) using 5 μL volumes that were composed of 15 ng of extracted DNA, 2.5 μL Multiplex PCR Master Mix (Qiagen, Valencia, CA, USA), 0.01 μM forward primer, 0.2 μM reverse primer, and 0.1 μM of the M13 primer (fluorescently labeled). PCR amplification was performed as follows: initial denaturation at 95 °C for 15 min; 35 cycles of denaturation at 95 °C for 30 s, annealing at 56 °C for 1.5 min, and extension at 72 °C for 1 min; a final extension at 72 °C for 10 min. The PCR products were diluted at 1:30, and 1 uL was analyzed on an ABI 3730XL sequencer with GeneScan^{TM}-500LIZ^{TM} Size Standard (Applied Biosystems). Allele sizes and peaks for each sample were determined three times via Peak Scanner software 2 to minimize genotyping errors. We selected 46 polymorphic microsatellite loci with clear, strong peaks for each individual. Then, we tested the remaining 126 individuals from five populations according to the DNA extraction and PCR protocols described above.

2.3. Statistical Data Analysis

Before inferring the genotyping data, we estimated the null allele frequency using INEst (inbreeding/null allele estimation) software based on the individual inbreeding model (IIM), which calculates the null allele frequency regardless of the effect of inbreeding [44]. This analysis showed that six loci (Qrg009, Qrg013, Qrg026, Qrg030, Qrg036, and Qrg048) showed a null allele frequency of more than 5%. Therefore, we used a total of 40 microsatellite markers, except for the six loci, for statistical analysis.

The summary genetics statistics were calculated at the population and pooled regional population levels. These included the number of alleles (N_A), the number of private alleles (P_A), the private allele rate (P_{riv}), the mean expected heterozygosity (H_E), the mean observed heterozygosity (H_O), and the fixation index (F_{IS}), calculated using GenAlEx 6.5 [45]. The allele richness (A_R) and genetic differentiation among populations (F_{ST}) were determined by calculating the overall F_{IS} according to the method of Weir and Cockerham [46], using FSTAT 1.2 [47]. The statistical significance of F_{ST} was tested using the log-likelihood (G)-based exact test in FSTAT. To test for departures from Hardy–Weinberg equilibrium (HWE) and linkage equilibrium, we conducted exact tests based on a Markov chain method (1000 permutations), using GENEPOP 4.0 [48]. The possibility of recent bottleneck for population was detected using BOTTLENECK 1.2.02 [49] (1000 iterations). We utilized two models for evolution—a two-phase model (TPM; the proportion of the stepwise mutation model (SMM) in TPM = 0.000, variance of the geometric distribution for TPM = 0.36), and a stepwise mutation model (SMM)—in a BOTTLENECK analysis that included the Bayesian Wilcoxon signed-rank test, to evaluate departures from a 1:1 deficiency/excess ratio [50]. The possibility of population bottleneck was also estimated by a mode-shift test, which detects disruptions in the distribution of allelic frequencies [50].

To analyze the population structure, we used a Bayesian clustering approach implemented in STRUCTURE 2.3, as calculated from microsatellite markers [51], using 1,000,000 Markov Chain Monte Carlo (MCMC) iterations (100,000 burn-in, with admixture). The simulation used 20 iterations, with $K = 1$ to $K = 7$ clusters. The optimal number of clusters, K, was found via the K method, using STRUCTURE HARVESTER [52]. CLUMPP v. 1.1.2 [53] with the Greedy algorithm was used to combine the membership coefficient matrices (Q-matrices) from 1000 iterations for $K = 2$, using random input orders.

To test for the presence of isolation-by-distance (IBD), we used Mantel tests in GenAlEx 6.5 [45] with 999 random permutations; this requires a correlation analysis between the pairwise F_{ST} values, and measurements of geographic distance between populations. To identify genetic boundaries between populations, we performed a barrier analysis [54] based on Monmonier's algorithm [55] with 1000 bootstrap matrices of pairwise D_A standard genetic distance [56] that were calculated by MICROSATELLITE ANALYZER (MSA) v. 4.05 [57]. The distance matrices were also used to construct a 50% consensus tree by the Neighbor-Joining (N-J) method, as implemented in PHYLIP v. 3.68 [58]. To find the genetic structure of *Q. gilva*, a principal coordinate analysis (PCoA) was conducted by the covariance standardized approach of pairwise Nei's genetic distances in GenAlEx 6.5.

2.4. Ecological Niche Modeling

We modeled the present and past (during LGM) potential distributions of *Quercus gilva* using Maxent 3.4.1 [59]. Occurrence data for this species included sample localities from our study as well as published data [27] and GBIF data with preserved specimens [60]. We obtained 242 occurrence data points and the occurrence data were spatially rarefied using SDMtoolbox 2.4 [61] to reduce bias in developing the distribution model. Two occurrence points of Korean Peninsula (inland) and Toyama Prefecture in Japan were excluded because they were estimated to be distributed in uncertain and inappropriate climate zones. A total of 97 occurrence data points were finally used in ENM. We obtained 19 bioclimatic variables (Online Resource 2) for the present and LGM from Climatologies at High Resolution for the Earth's Land Surface Areas (CHELSA, http://chelsa-climate.org/; [62]). We obtained elevation data for the present—the Global Multi-resolution Terrain Elevation Data (GMTED2010) dataset [63]—from the USGS EROS Archive (https://www.usgs.gov/land-resources/eros/coastal-changes-and-impacts/gmted2010), and for the LGM from CHELSA. To reconstruct the historical distributions, we utilized three past climate models for LGM: the Community Climate System Model (CCSM4; [64]), the Earth System Model based on the Model for Interdisciplinary Research on Climate (MIROC-ESM; [65]), and the Max Planck Institute for Meteorology Earth System model (MPI-ESM-P). We selected one of the climate variables and elevation data sharing a high Spearman correlation efficient (>0.7) by using SDMtoolbox 2.4 [61], in order to avoid multicollinearity problems. Therefore, 7 of 20 variables were used in ENM. To reduce the effects of uncertainty in the historical climate models, we averaged the historical distributions that were based on each of the three climate models. The climate data, for 20–37° N and 115–145° E (30 arcsecond resolution), were extracted using ArcGIS 10.5 (ESRI 2017). Maxent runs were performed in batch mode with these settings: create response curves, conduct jackknife tests, use 20 replicates, generate logistic output, select random seeds, and we used 10,000 background points and 1,000 iterations.

3. Results

3.1. Development of Polymorphic Microsatellite Markers

In total, 11,957,206 reads were generated by Illumina paired-end sequencing (Short Read Archive accession number: PRJNA649602). The total number of reads containing microsatellites identified through the SSR-pipeline was 100,849 reads, including 55,084 reads with dinucleotide motifs, 41,037 reads with trinucleotide motifs, and 4,728 reads with tetranucleotide motifs. Of these, the di-, tri-, and tetranucleotide motifs with planking areas of >100 bp and having repeating units of at least 12, 6, and 6, respectively, were 29,058 reads, 21,424 reads, and 2,562 reads.

As a result of applying 54 designed microsatellite loci to 32 individuals of *Q. gilva* from Gueok-ri populations in Korea, 46 polymorphic microsatellite markers with clear and strong peaks for each allele were selected (Table 1). Regarding the results of the genetic diversity analysis, a total of 385 alleles were detected in 46 microsatellite loci across all samples. The number of alleles (N_A) per locus ranged from 2 to 19, with an average of 8.370 alleles per locus. Values for observed heterozygosity (H_O) and expected heterozygosity (H_E) ranged from 0.044 to 0.918 (mean: 0.616) and from 0.067 to 0.899 (mean: 0.664), respectively (Table 2). The inbreeding coefficient (F_{IS}) for each locus ranged from −0.135 to 0.669. The null allele frequency identified by INEst software ranged from 0.0018 to 0.2774. Comparing the genetic diversity (N_A and H_E) by the locus of di-, tri-, and tetranucleotide motifs, the results were higher in loci with dinucleotide motifs (mean N_A = 11.125, H_E = 0.784) than with tri- (mean N_A = 5.571, H_E = 0.527) or tetranucleotide motifs (mean N_A = 5.000, H_E = 0.547) (Figure 1).

Table 1. Characterization of six multiplexes of 46 microsatellite loci for *Quercus gilva*.

Locus	Primer Sequence (5′–3′)	Repeat Motif	Number of Alleles	Size Range (bp)	Fluorescent Label	GenBank Accession No.
Multiplex mix A						
Qrg001	F: TCTGATGAGGTGCTGGAA R: TTGTTATCCAATTCTCTCCCT	(TC)$_{12}$	7	100–118	6-FAM	MT811115
Qrg002	F: TGAGCTTGTTGATTGGAGAA R: CTTCAAGACGTACTACAGCA	(CA)$_{12}$	6	158–172	6-FAM	MT811116
Qrg003	F: TTGGTGGAAGAGATTGTGAG R: CTCTTTGGGTTCTCTGTTGT	(CT)$_{14}$	7	213–225	6-FAM	MT811117
Qrg004	F: TGGCTTCCTGACCATACATA R: GACTAACCCTGCCCTCAA	(GAA)$_6$	6	107–122	VIC	MT811118
Qrg006	F: CTCAATGGCGAAATCATCAG R: TCTATAGAGGCAGCAAACAC	(TTAG)$_8$	5	220–236	VIC	MT811119
Qrg007	F: GTTGGATTGGATTCTGTTGC R: TTCCCTCCTTGTCACGTT	(AG)$_{12}$	15	103–135	NED	MT811120
Qrg008	F: ATCGGAGCAAGAAATCAAAT R: CCACCAACTCTAATGCTGTA	(AAG)$_8$	3	159–168	NED	MT811121
Qrg009	F: CACTCTCTTCGACCTTCTTT R: TTCTGGGTTCTTGCTTATCG	(TCA)$_9$	6	225–240	NED	MT811122
Multiplex mix B						
Qrg011	F: CGTTCAGATCAGGGTACAAA R: ATAAGCAAAGCACCCATGTA	(CA)$_{14}$	5	160–170	6-FAM	MT811123
Qrg012	F: ATTAATGGAGAACTGCCCTC R: AGGATCATGAACTTCGACTG	(CTT)$_{11}$	5	223–235	6-FAM	MT811124
Qrg013	F: TCTCAAACGGACCCATTTAA R: TCCTGTGATTACTGTCTATGC	(CT)$_{13}$	5	108–120	VIC	MT811125
Qrg014	F: GTCAGTATAGCATGTGGTGT R: TTGGTGAGTTGAGATTGCAA	(GA)$_{14}$	8	159–189	VIC	MT811126
Qrg015	F: TTCCCATTTCAGACAAGAGG R: GATTCGAACCCTCCTACAAA	(TAAC)$_7$	7	209–237	VIC	MT811127
Qrg016	F: CTCTACCATCAACATCCTGC R: AATTCCAGTTTTGCAGTCCA	(AGAC)$_6$	6	124–148	NED	MT811128
Qrg017	F: ACACCAAACAAAGCAAACAA R: TACGAACACAATCCAAACCT	(AACA)$_6$	3	163–171	NED	MT811129
Qrg018	F: CAACCACAATGTGTAAAGACA R: GCAAAGAGTGTATGTGCTC	(ACA)$_{10}$	4	218–236	NED	MT811130
Multiplex mix C						
Qrg019	F: AACTCTTGCTCCATTCATTT R: GGGTCTACAATTGAATTATGGC	(AG)$_{13}$	8	133–149	6-FAM	MT811131
Qrg020	F: AGGATTTGTAGCTGACCCTA R: GCCAAGTAATCAAATTGACTGA	(GTT)$_8$	4	166–178	6-FAM	MT811132
Qrg021	F: ACAAAGACTACGTGTGCATA R: TTTCTATGAAACGCAACAGC	(CT)$_{14}$	10	229–253	6-FAM	MT811133
Qrg022	F: GGATGACATGGCTGATCTTC R: ATAACTGGAATGGCATGGAG	(AAG)$_7$	3	123–135	VIC	MT811134
Qrg024	F: CCTAAGAAGCACAGGTAAGG R: AGAGCAAGTGAGAAAGAGTC	(CT)$_{14}$	11	237–263	VIC	MT811135
Qrg025	F: CATATAGCCGAGGAAGAAGT R: GAAGGCAGAGGTTGGTTAAA	(GAA)$_6$	2	134–137	NED	MT811136
Qrg026	F: GATGGGAATGCTCTTAGGTC R: TTGTGAAGTCGCCTACAATT	(ATAG)$_6$	3	180–188	NED	MT811137
Qrg027	F: TGGAAATGACATTGTTACCCT R: CCGATGACAAGAATCCCAAT	(GA)$_{14}$	12	235–271	NED	MT811138

Table 1. *Cont.*

Locus	Primer Sequence (5′–3′)	Repeat Motif	Number of Alleles	Size Range (bp)	Fluorescent Label	GenBank Accession No.
Multiplex mix D						
Qrg028	F: TAAAGGAGTGCATGGTGAAA R: AGTGAAGCCTCTTTCCTAGA	$(CT)_{13}$	9	127–147	6-FAM	MT811139
Qrg029	F: AAGATAACTGCACGCTTGTA R: TCAGAAATCGCTCATACCTG	$(TG)_{13}$	7	184–196	6-FAM	MT811140
Qrg030	F: CTATTCATGGACTCCTCTGT R: AATTGCAAGGCCTTAGAACT	$(AG)_{15}$	7	235–249	6-FAM	MT811141
Qrg031	F: GGTTAGGGCTCTTTCCAAAT R: CTCTCCCTTTCTTTCACTGT	$(GA)_{13}$	8	131–145	VIC	MT811142
Qrg033	F: TCTTGCCAATCTAAATCCCA R: TGCATGATACAGAAACACCA	$(AAGA)_{7}$	2	239–247	VIC	MT811143
Qrg034	F: GGACATCTACAGCCTACAAA R: CGCAGACCAAATATCATTCTC	$(CT)_{12}$	12	143–173	NED	MT811144
Qrg036	F: TAACTTTGTTCTCGCCTGA R: AATGTAGAGCCTGTTTGCAT	$(GA)_{13}$	7	239–259	NED	MT811145
Multiplex mix E						
Qrg037	F: TTCGAGATAGGACAGAGGAG R: TGTGTTTGATTAGCGGAGAA	$(AAGA)_{8}$	5	128–144	6-FAM	MT811146
Qrg038	F: TGGCTATGATAATTGTGGGT R: CTCAACCCTGTTATCTCACC	$(GA)_{17}$	8	182–204	6-FAM	MT811147
Qrg039	F: AAAGTGGATTTGCAGCCTAA R: GACAATGGAGAAGGGACAAT	$(TC)_{14}$	6	244–260	6-FAM	MT811148
Qrg040	F: GCATTTCTCTCTCTGGTTCA R: AAGTACCCTCCATCTACGTT	$(AAG)_{6}$	3	128–146	VIC	MT811149
Qrg041	F: CTTCCTCGTCAATAGTCCAC R: AGTGAGTTTGATACGCTTGT	$(AAG)_{12}$	9	186–228	VIC	MT811150
Qrg042	F: CCCACACATTATACCACGAA R: CTACTAACAACCGCAACTCT	$(AG)_{17}$	8	227–253	VIC	MT811151
Qrg043	F: CATACATCCTAGTGCAGCAG R: GGTAGCTCAAGTTCACAGTT	$(CAA)_{6}$	2	149–155	NED	MT811152
Multiplex mix F						
Qrg046	F: CTGCCCCTAACTAATCTGTT R: GTAGATGATGAGGTTGTGGG	$(TGT)_{6}$	2	149–152	6-FAM	MT811153
Qrg047	F: AGACCAGTAGATGCTTCAAA R: ATTCATGACCCTCCTTCTCA	$(AAG)_{9}$	3	208–217	6-FAM	MT811154
Qrg048	F: TCCATCGTCAACAAAGGATT R: AACCAGTTCTCACTCTCTCT	$(AG)_{17}$	7	235–269	6-FAM	MT811155
Qrg049	F: CAACTACTGTAGCCTTGTGT R: TATGCCTCCAGTGTACTACA	$(CA)_{12}$	7	146–166	VIC	MT811156
Qrg050	F: GGGACCATAGCAGTGTTAAT R: AGCCCTCCCTTATTTATTCC	$(TC)_{21}$	8	192–216	VIC	MT811157
Qrg051	F: CTCCTCTTGGCTATGACATC R: TCTTGTTTGAGGAAGTTGACA	$(TTC)_{14}$	10	235–259	VIC	MT811158
Qrg052	F: ACTTGTAACTAACCTGGCTC R: CTAGGAGGATGAAATGGCAA	$(CTAA)_{8}$	4	150–162	NED	MT811159
Qrg053	F: TGACAGTACATGGTAAAGCT R: TTCTTGGTCTTGAATGAGGA	$(CT)_{14}$	7	204–228	NED	MT811160

Table 2. Genetic parameters for 46 microsatellite loci across all samples developed for *Quercus gilva*.

Locus	N_A	H_O	H_E	Null	Locus	N_A	H_O	H_E	Null
Qrg001	8	0.703	0.619	0.0020	Qrg027	16	0.829	0.865	0.0031
Qrg002	10	0.741	0.755	0.0053	Qrg028	10	0.842	0.820	0.0018
Qrg003	7	0.741	0.794	0.0112	Qrg029	8	0.677	0.743	0.0275
Qrg004	7	0.778	0.786	0.0057	Qrg030	9	0.253	0.765	0.2774 *
Qrg006	6	0.772	0.763	0.0033	Qrg031	10	0.810	0.841	0.0091
Qrg007	18	0.918	0.899	0.0012	Qrg033	3	0.222	0.209	0.0048
Qrg008	4	0.361	0.366	0.0058	Qrg034	16	0.867	0.885	0.0038

Table 2. *Cont.*

Locus	N_A	H_O	H_E	Null	Locus	N_A	H_O	H_E	Null
Qrg009	8	0.570	0.748	0.0868 *	Qrg036	9	0.627	0.800	0.0910 *
Qrg011	6	0.684	0.745	0.0071	Qrg037	5	0.696	0.738	0.0118
Qrg012	7	0.759	0.756	0.0026	Qrg038	13	0.759	0.794	0.0076
Qrg013	8	0.513	0.657	0.0728 *	Qrg039	6	0.753	0.731	0.0035
Qrg014	11	0.861	0.835	0.0016	Qrg040	3	0.468	0.539	0.0218
Qrg015	7	0.620	0.671	0.0073	Qrg041	13	0.823	0.808	0.0025
Qrg016	6	0.608	0.622	0.0044	Qrg042	14	0.734	0.805	0.0204
Qrg017	5	0.589	0.597	0.0117	Qrg043	3	0.133	0.131	0.0058
Qrg018	6	0.532	0.631	0.0464	Qrg046	3	0.348	0.384	0.0192
Qrg019	10	0.759	0.824	0.0121	Qrg047	4	0.475	0.496	0.0067
Qrg020	5	0.633	0.674	0.0050	Qrg048	12	0.595	0.808	0.0977 *
Qrg021	19	0.861	0.886	0.0062	Qrg049	10	0.430	0.505	0.0353
Qrg022	3	0.165	0.179	0.0099	Qrg050	14	0.791	0.819	0.0044
Qrg024	13	0.741	0.857	0.0355	Qrg051	10	0.810	0.814	0.0036
Qrg025	2	0.044	0.067	0.0316	Qrg052	4	0.551	0.572	0.0106
Qrg026	4	0.139	0.203	0.0607 *	Qrg053	10	0.741	0.756	0.0029

N_A, number of alleles; H_O, observed heterozygosity number of alleles; H_E, expected heterozygosity; Null, null allele frequency estimate. * indicates that the frequency of the null allele exceeds 5%.

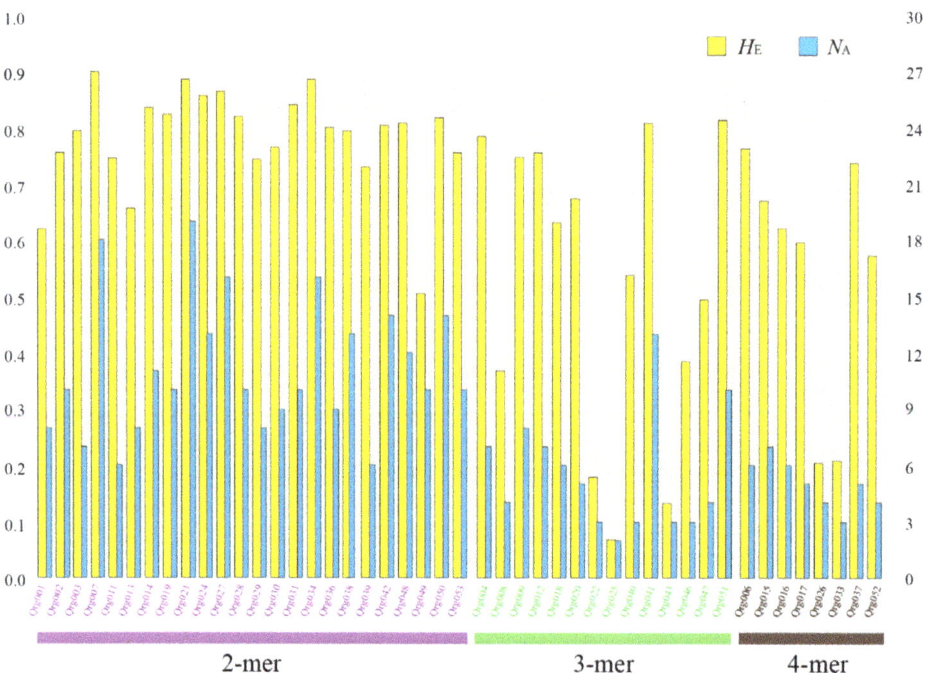

Figure 1. Comparing genetic diversity (N_A and H_E) by locus with di-, tri-, and tetranucleotide motifs. Genetic diversity is based on the allele frequency of six populations of *Quercus gilva* using 46 microsatellite loci.

All the developed markers were deposited in in the National Center for Biotechnology Information's GenBank database (Table 1). As a result, a cost-effective set of 46 polymorphic microsatellite markers with high resolutions has been successfully developed. These markers will be useful for conserving genetic resources through periodical monitoring management, creating a seed genogram, cloning detection, and postrestoration assessments in the endangered Jeju populations of *Q. gilva*.

3.2. Genetic Diversity

Genetic diversity parameters, evaluated at the population and pooled regional population levels for all 158 individuals of *Q. gilva*, are shown in Table 3. In total, 335 alleles were amplified from 40 microsatellite loci, with an average of 8.4 alleles per locus. They ranged from a minimum of two (Qrg025) to a maximum of 19 (Qrg021). Among the six *Q. gilva* populations, the levels of genetic diversity showed no noticeable difference; the number of alleles ranged from 224 to 261 (mean of 244.7); H_E ranged from 0.615 (j-NB) to 0.651 (k-SG) (mean of 0.639); A_R ranged from 5.201 (j-NB) to 5.940 (j-MY) (mean of 5.726); P_A ranged from 4 (k-JJ and j-NB) to 11 (k-GU) (mean of 7); F_{IS} ranged from −0.024 (k-GU) to 0.043 (k-SG) (mean of 0.007). The k-SG population had the highest genetic diversity values, while the j-NB population had the lowest (H_E, A_R). In the comparison between the Jeju and Kyushu populations, the levels of genetic diversity were almost equivalent, showing only a slight difference in degree depending on the parameters (Table 3). The BOTTLENECK analysis (Wilcoxon tests) showed no significant bottleneck effects across all populations under a TPM and SMM ($p > 0.05$), as well as a mode shift (Table 4).

Table 3. Summary statistics of genetic diversity for six populations based on 40 microsatellite loci of *Quercus gilva*.

ID	Location	Coordinates	E_V	N	N_A	A_R	P_A	P_{riv}	H_O (SE)	H_E (SE)	F_{IS}
Jeju Island											
k-GU	Gueok-ri, Daejeong-eup, Jeju	33°18′8.21″ N, 126°16′36.59″ E	136	32	254	5.729	11	0.043	0.660 (0.037)	0.645 (0.033)	– 0.024
k-JJ	Jeoji-ri, Hangyeong-myeon, Jeju	33°18′45.36″ N, 126°17′3.76″ E	168	27	254	5.905	4	0.016	0.656 (0.040)	0.640 (0.035)	0.003
k-SG	Seogwang-ri, Andeok-myeon, Jeju	33°17′57.47″ N, 126°18′59.97″ E	201	18	237	5.925	8	0.034	0.639 (0.039)	0.651 (0.033)	0.043
Mean			168.3	25.6	248.3	5.853	7.7	0.031	0.652	0.645	0.007
Pooled populations				77	301	7.525	33	0.110	0.641	0.657	0.018
Kyushu											
j-GM	Kuma-gun, Kumamoto Prefecture	32°17′39.5″ N, 130°52′17.6″ E	485	23	238	5.656	5	0.021	0.638 (0.039)	0.634 (0.036)	−0.005
j-MY	Miyakonojo-shi, Miyazaki Prefecture	31°50′55.7″ N, 131°13′30.4″ E	230	29	261	5.940	10	0.038	0.636 (0.034)	0.647 (0.035)	0.003
j-NB	Nobeoka-shi, Miyazaki Prefecture	32°39′15.8″ N, 131°41′14.3″ E	38	29	224	5.201	4	0.018	0.613 (0.043)	0.615 (0.038)	0.023
Mean			251	27	241	5.600	6.3	0.026	0.629	0.632	0.007
Pooled populations				81	302	7.503	34	0.113	0.628	0.648	0.015

E_V, elevation of sampling site (meter); N, number of individuals; N_A, number of alleles; A_R, allelic richness; P_A, number of private alleles; P_{riv}, private allelic rate; H_O, observed heterozygosity number of alleles; H_E, expected heterozygosity; SE, standard error; F_{IS}, inbreeding coefficient.

Table 4. Probability of a bottleneck estimated using the program BOTTLENECK for six populations of *Quercus gilva*, based on the two-phase model (TPM) or stepwise mutation model (SMM).

Population	Wilcoxon Test		Mode Shift
	TPM	SMM	
Jeju Island			
k-GU	0.476153	0.988818	No
k-JJ	0.465576	0.991747	No
k-SG	0.294988	0.925049	No
Kyushu			
j-GM	0.383349	0.974584	No
j-MY	0.292696	0.982035	No
j-NB	0.135276	0.765987	No

3.3. Population Structure

To infer the population structure of *Q. gilva*, we performed STRUCTURE, N-J phylogeny, and PCoA analysis with 40 microsatellite loci. The results clearly demonstrated that the Jeju and Kyushu regions are genetically distinct. The STRUCTURE analysis showed that the optimal *K*-value was 2 for $\Delta K = 199.678$ and the second fit value was 4 for $\Delta K = 6.319$. At $K = 2$, a strong genetic structure was found among populations, divided clearly into two regions (Figures 2 and 3). In terms of neighbor-joining criteria, the sampled populations of *Q. gilva* were clearly divided into two clusters (Jeju and Kyushu), in concordance with the clustering results obtained by STRUCTURE (Figure 4). The principal coordinate analysis (PCoA) results revealed a population structure that was in accordance with the STRUCTURE and N-J phylogeny analysis (Figure 5). The first two coordinates explained 7.32% (4.12% for axis 1 and 3.20% for axis 2) of the total genetic variation. Based on pairwise F_{ST}, the barrier analyses identified a strong barrier between the Jeju and Kyushu regions (Figure 2). Although the F_{ST} values of pairwise comparisons among the six populations showed a numerically low overall genetic differentiation, with a F_{ST} of 0.029 with 95 and 99% confidence intervals of 0.024–0.034 and 0.022–0.036, respectively, it was significant at all loci ($p < 0.001$). This differentiation was also seen between populations within Jeju Island (mean 0.010, with 95 and 99% confidence intervals of 0.005–0.017 and 0.003–0.019, $p < 0.001$), and within populations of Kyushu (mean 0.021, with 95 and 99% confidence intervals of 0.015–0.027 and 0.013–0.029, $p < 0.001$). Therefore, the fact that the highest value of genetic differentiation is from the overall population means that it results from a difference between Jeju Island and Kyushu (Figure 6b). Additionally, isolation by distance (IBD) analysis, as determined by Mantel test, showed a significant correlation between genetic and geographic distance among populations ($R^2 = 0.7611$, $p = 0.025$) (Figure 6a).

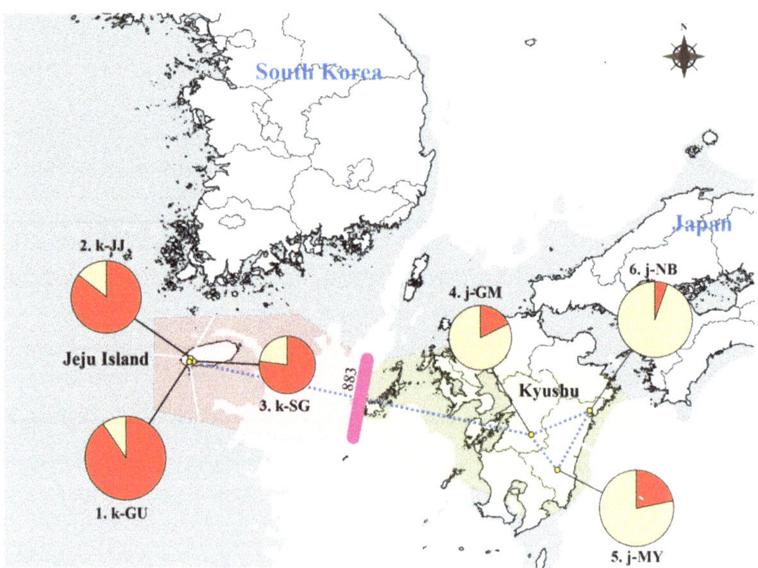

Figure 2. Genetic composition and a genetic barrier of *Quercus gilva* geographic populations using 40 microsatellite loci. Genetic composition is based on STRUCTURE clustering results ($K = 2$). The genetic barrier is marked with a thick purple line, estimated by BARRIER. The gray shading represents exposed coastal areas and sea basins during times of glacially induced alterations in sea levels during the Late Pleistocene.

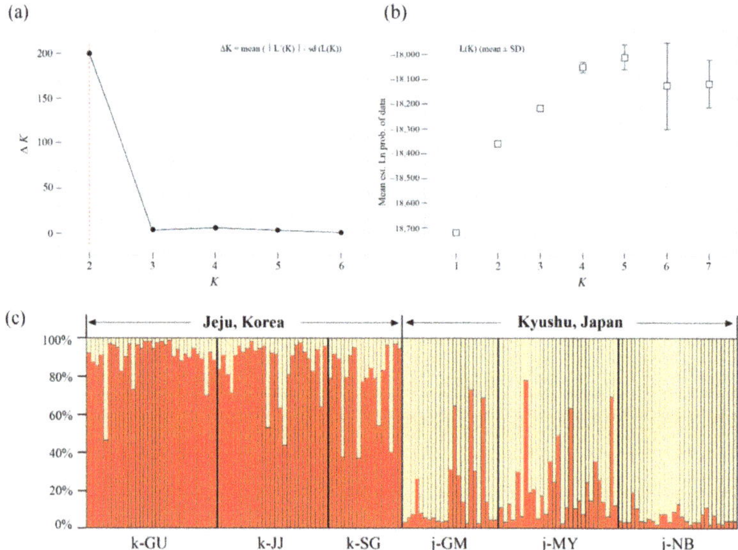

Figure 3. Plots generated in STRUCTURE Harvester (**a**) Evanno's delta K statistic; (**b**) the mean log-likelihood of the data L(K). Genetic structure of *Quercus gilva* populations based on Bayesian assignment tests performed in STRUCTURE. (**c**) Genetic structural plot of *Q. gilva* populations at $K = 2$. Each individual is represented by a single vertical line that represents the individual's estimated membership fractions in these two clusters.

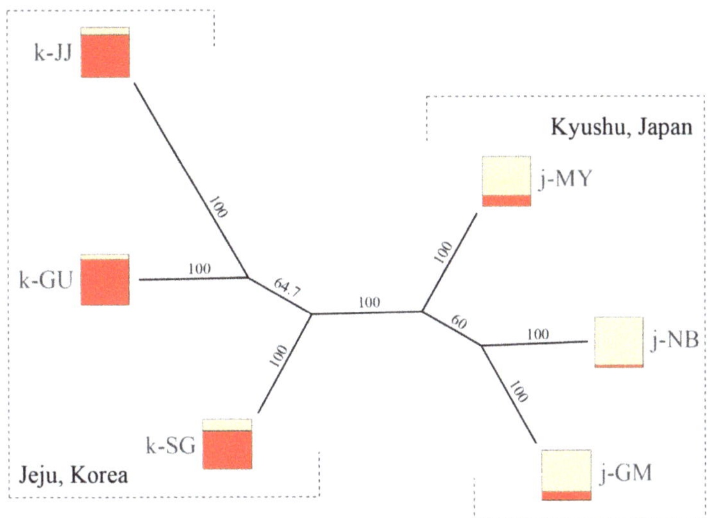

Figure 4. Neighbor-joining (N-J) tree based on F_{st} genetic distance among populations. Figures in tree branches are percentage bootstrap values estimated from 1000 reiterations. The square marks indicate the overall genotype assignment for each population to particular genetic clusters based on STRUCTURE analysis.

Principal Coordinates (PCoA)

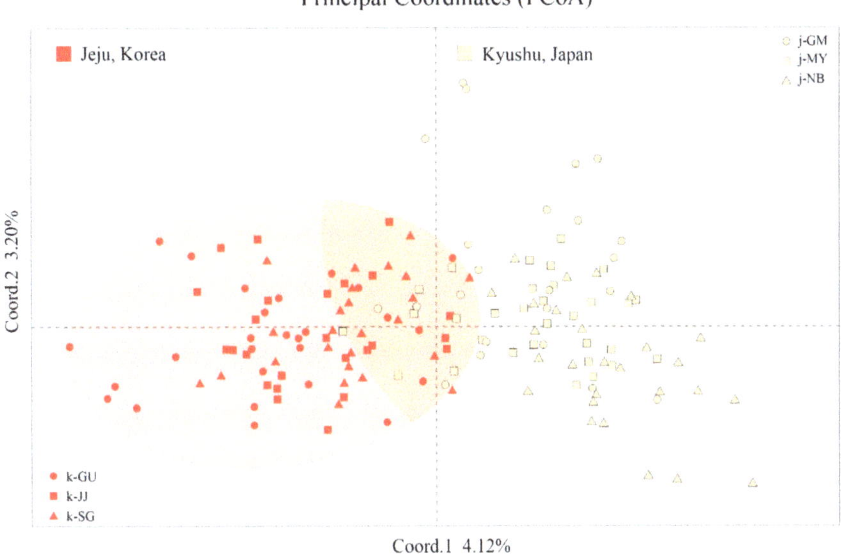

Figure 5. Principal coordinates analysis based on Nei's genetic distance calculated from the allele frequencies of the 158 individuals for *Quercus gilva*. The orange symbols indicate individuals of Jeju region, and the yellow ones indicate Kyushu region. Six subgroups indicate each populations of *Q. gilva*.

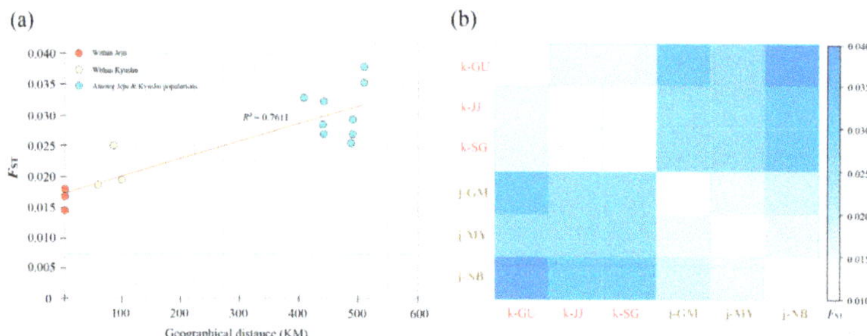

Figure 6. The genetic differentiation for the six populations of *Quercus gilva*. (**a**) Mantel tests between F_{ST} values and geographical distance among populations; (**b**) distance matrix of pairwise F_{ST} between populations.

3.4. Ecological Niche Modeling

The ENM of *Q. gilva* (Figure 7) had a high average AUC (area under the curve) (0.899), supporting its predictive power. The most important variable was bio_02 (mean diurnal range; 55.8%), followed by bio_12 (annual precipitation; 14.6%) and bio_15 (precipitation seasonality; 10.9%). The estimated LGM distribution was near the paleo-coastline with no inland potential distribution (Figure 7b). The potential value of more than 0.500 were shown in southern Kyushu, the central East China Sea, southeastern Taiwan, and the Ryukyu archipelago (Figure 7b).

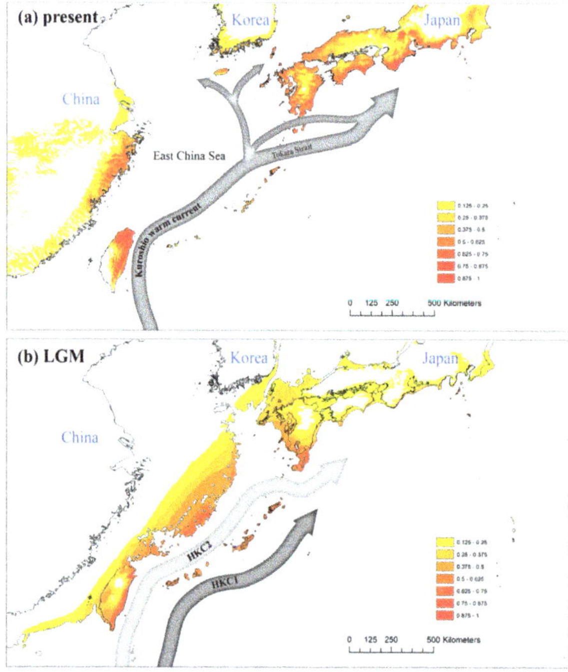

Figure 7. Potential distributions of *Quercus gilva* during (**a**) the present; (**b**) the Last Glacial Maximum, LGM. Distributions predicted by ecological niche modeling; potential distribution during the LGM was averaged from three general circulation models. HKC 1 indicates the main track of the Kuroshio Current during the LGM proposed by Ujiie et al. [66], Kao et al. [67], and Zheng et al. [68]. HKC 2 is the hypothetical Kuroshio Current during the LGM suggested by Vogt-Vincent and Mitarai [69].

4. Discussion

One of the most notable results of this study is that the geographically adjacent Jeju and Kyushu populations are genetically divergent. Comparison with other evergreen trees belonging to the same forest biomes may help to explain such a genetic pattern. Although the available data for genetic examination in this region are still minimal, consistent results point toward the fact that the warm temperate evergreen broad-leaved trees of South Korea, including the Jeju populations, have been affected by postglacial migration from those of Kyushu, Japan (*Neolitsea sericea*: [5]; *Machilus thunbergii*: [39]; *Quercus acuta*: [6]). Previous studies have shown that the Korean populations are homogeneous, with a genetic structure that is not very distinct from those of Kyushu, Japan. Furthermore, *Q. gilva* has an almost identical life history to *Q. acuta*; it is wind-pollinated, and the nuts contain a one-seeded fruit with a hard wall that is usually dispersed by small rodents such as squirrels and jays or animal-cached [70,71]. Therefore, the contrasting pattern of genetic structure suggest that historical factors are the most relevant. Therefore, we suggest that *Q. gilva* has a unique and separate evolutionary history.

Regardless of the latitude, the distribution of warm temperate evergreen broad-leaved forests in East Asia, such as in Korea, China, and Japan, is clearly related to the flow of the Kuroshio warm current (KC), showing a unique distribution structure [34,72,73]. Therefore, despite being located at relatively high latitudes, the southwestern portions of the Japanese main islands (i.e., southern areas of Kyushu, Shikoku, and Honshu) have long been regarded as crucial refugia with high genetic variation for evergreen broad-leaved trees that resulted from southward range shifts that paralleled glacial cycles [74–76]. ENM also revealed that *Q. gilva* populations are distributed like shadows along the flow of the present KC as well as the LGM (Figure 7). Therefore, considering the geographical location and population size of Jeju Island, it is worth noting that the levels of genetic diversity between the Jeju and Kyushu regions are almost equally similar in both populations and pooled regional populations. We found that the Jeju populations harbor a level of allele richness (A_R) and private allele rate (P_{riv}) comparable to those of Kyushu populations. Although, such indicators provide a strong possibility for the existence of refugia in this area [77–79], we are convinced that Jeju Island did not serve as the glacial refuge of *Q. gilva*. In agreement with our assumption, ENM showed that the probability values for the distribution of the trees on Jeju Island during the LGM converged at zero. East Asian evergreen oaks prefer different habitats depending on the elevation gradient along the mountain slopes. A previous study suggested that Jeju Island was the refugia of *Q. acuta* [6], which occurs at the highest elevation range [80]. By comparison, *Q. gilva* shows a bias towards low-elevation stands, generally forming a community with *Q. glauca* [80]. In fact, *Q. acuta* occupies the middle of Mt. Halla on Jeju Island (approximately 600 m a.s.l.), while *Q. gilva* is distributed very close to the southern coastline (average: 168 m a.s.l.). It should be taken into account that the Quaternary climate oscillations caused not only latitudinal changes in the distribution of a given species but also vertical elevation migration [38]. During the glacial periods, in addition to the cold climate, competition for vertical migration with other oak species would have prompted the Jeju populations to retreat to low latitudes. The ENM revealed that, due to an apparent range contraction southward along the paleo KC, with the exception of the Ryukyu Archipelago, the high distribution potentials have been narrowed to three places—southeastern Taiwan, central East China Sea, and southern Kyushu in LGM (Figure 7b). Given that the Jeju Island populations are genetically different from Kyushu's, the glacial refugia (i.e., genetic source) of the current Jeju populations would have most likely existed somewhere around the central East China Sea, which was land during the LGM.

One plausible scenario might be hypothesized for the unusual level of genetic diversity in Jeju Island, i.e., a massive postglacial immigration, demographically independent of Kyushu. Most oak species have common ecological traits, but their distinct adaptability also facilitates domination in such areas. In fact, evergreen *Quercus* species frequently dominate the landscape in extensive regions of East Asia, because they are better able to form dense and crowded stands [81,82]. In general, the larger the population is, the more reduced the effect of genetic drift is, which promotes a reduction in genetic diversity. We found no indication of significant recent bottlenecks, implying that the founding populations of *Q. gilva* might have been large enough to weaken the effect of genetic drift. Therefore, we suggest that a massive postglacial colonization, which maintained a high genetic diversity from separately stable and large genetic sources other than Kyushu, could have led to the current genetic diversity of Jeju Island. Further work using additional samples, including from broad areas such as the Ryukyu Islands and Eastern China, and the markers developed in this study might provide a better understanding of the historical migration of the warm temperate evergreen tree *Q. gilva* in Jeju Island. In particular, how the Jeju populations relate to East Asian populations other than Kyushu populations should be tested.

There are several definitions for determining the levels of CUs, such as Evolutionarily Significant Units (ESUs) and Management Units (MUs), because "high divergence" is too vague a term for practical purposes [83–85]. However, the criteria of the MUs clearly represent demographically independent units that merit separate management [86,87]. Considering the high genetic distinctiveness with a significant barrier, monophyletic phylogeny, population size, and other evidence, we suggest that the Jeju populations should be separately managed as a MU. The notable and unique genetic diversity of Jeju populations represents a high value in terms of conservation as it can contribute to the species' genetic diversity. Such genetic determinants should be well preserved and returned when East Asian populations are reconnected in response to the climate fluctuation. The low degree of genetic differentiation among the populations within Jeju Island suggests that all populations should be integrated and managed together rather than focusing conservation efforts on any particular subset of the population. From a long-term conservation genetics perspective, it is especially important for *Q. gilva* that conservation efforts should be focused on prohibiting the large-scale industrial development of the habitat, because large trees are not vulnerable to personal interference, such as overcollection. Thus, first, we recommend that all known habitats be protected in situ by law to prevent further damage. To prepare for inevitable land development, we suggest that ex situ preservation should be preceded by efforts to store good-quality seeds. Finally, if the artificial restoration of habitats is required, note that the source for the Jeju populations is not Kyushu.

Data Available: Sequence data for microsatellite loci developed in this study are available on GenBank (accession numbers MT811115–MT811160).

Author Contributions: E.-K.H. and W.-B.C. conducted data analyses and wrote the manuscript; E.-K.H. performed DNA extraction and PCR for genotyping and sequences; E.-K.H., W.-B.C., and J.H.L. performed the fieldwork; J.-S.P. performed the data analyses for the ENM; J.H.L., I.-S.C., and M.K. improved the manuscript; J.-H.L., B.-Y.K., and M.K. conceived and oversaw the study. All authors have read and agreed to the published version of the manuscript.

Funding: This research was funded by National Institute of Biological Resources, Korea, grant number NIBR201905102.

Acknowledgments: The authors thank G.H. Nam at the National Institute of Biological Resources for letting us distribute information of Jeju Island's plant materials. We are also grateful to the following researchers for commenting on a previous version of this manuscript: S. So (Korea National Park Research Institute), D. P. Jin (Inha University) and S. Yang (Korea Institute of Oriental Medicine).

Conflicts of Interest: The authors have no conflict of interest to declare.

References

1. Hewitt, G.M. Some genetic consequences of ice ages, and their role in divergence and speciation. *Biol. J. Linn. Soc.* **1996**, *58*, 247–276. [CrossRef]
2. Hewitt, G.M. The genetic legacy of the Quaternary ice ages. *Nature* **2000**, *405*, 907–913. [CrossRef] [PubMed]
3. Petit, R.J.; Aguinagalde, I.; De Beaulieu, J.-L.; Bittkau, C.; Brewer, S.; Cheddadi, R.; Ennos, R.; Fineschi, S.; Grivet, D.; Lascoux, M.; et al. Glacial Refugia: Hotspots but Not Melting Pots of Genetic Diversity. *Science* **2003**, *300*, 1563–1565. [CrossRef] [PubMed]

4. Sakaguchi, S.; Qiu, Y.-X.; Liu, Y.-H.; Qi, X.-S.; Kim, S.-H.; Han, J.; Takeuchi, Y.; Worth, J.R.P.; Yamasaki, M.; Sakurai, S.; et al. Climate oscillation during the Quaternary associated with landscape heterogeneity promoted allopatric lineage divergence of a temperate tree *Kalopanax septemlobus* (Araliaceae) in East Asia. *Mol. Ecol.* **2012**, *21*, 3823–3838. [CrossRef] [PubMed]

5. Lee, J.H.; Lee, D.H.; Choi, B.H. Phylogeography and genetic diversity of East Asian *Neolitsea sericea* (Lauraceae) based on variations in chloroplast DNA sequences. *J. Plant Res.* **2013**, *126*, 193–202. [CrossRef]

6. Lee, J.H.; Lee, D.H.; Choi, I.S.; Choi, B.H. Genetic diversity and historical migration patterns of an endemic evergreen oak, *Quercus acuta*, across Korea and Japan, inferred from nuclear microsatellites. *Plant Syst. Evol.* **2014**, *300*, 1913–1923. [CrossRef]

7. Jin, D.P.; Lee, J.H.; Xu, B.; Choi, B.H. Phylogeography of East Asian *Lespedeza buergeri* (Fabaceae) based on chloroplast and nuclear ribosomal DNA sequence variations. *J. Plant Res.* **2016**, *129*, 793–805.

8. Tamaki, I.; Kawashima, N.; Setsuko, S.; Lee, J.-H.; Itaya, A.; Yukitoshi, K.; Tomaru, N. Population genetic structure and demography of *Magnolia kobus*: Variety borealis is not supported genetically. *J. Plant Res.* **2019**, *132*, 741–758. [CrossRef]

9. Harrison, S.P.; Yu, G.; Takahara, H.; Prentice, I.C. Palaeovegetation: Diversity of temperate plants in East Asia. *Nature* **2001**, *413*, 129–130. [CrossRef]

10. Alleaume-Benharira, M.; Pen, I.R.; Ronce, O. Geographical patterns of adaptation within a species' range: Interactions between drift and gene flow. *J. Evol. Biol.* **2006**, *19*, 203–215. [CrossRef]

11. Eckert, C.G.; Samis, K.E.; Lougheed, S.C. Genetic variation across species' geographical ranges: The central–marginal hypothesis and beyond. *Mol. Ecol.* **2008**, *17*, 1170–1188. [CrossRef] [PubMed]

12. Myking, T.; Vakkari, P.; Skrøppa, T. Genetic variation in northern marginal *Taxus baccata* L. populations. Implications for conservation. *Forestry* **2009**, *82*, 529–539. [CrossRef]

13. Villellas, J.; Ehrlén, J.; Olesen, J.M.; Braza, R.; Garcia, M. Plant performance in central and northern peripheral populations of the widespread *Plantago coronopus*. *Ecography* **2013**, *36*, 136–145. [CrossRef]

14. Mimura, M.; Aitken, S.N. Local adaptation at the range peripheries of *Sitka spruce*. *J. Evol. Biol.* **2010**, *23*, 249–258. [CrossRef] [PubMed]

15. Safriel, U.N.; Volis, S.; Kark, S. Core and peripheral populations and global climate change. *Isr. J. Plant Sci.* **1994**, *42*, 331–345. [CrossRef]

16. Gibson, S.Y.; Van Der Marel, R.C.; Starzomski, B.M. Climate change and conservation of leading-edge peripheral populations. *Conserv. Biol.* **2009**, *23*, 1369–1373. [CrossRef]

17. Milad, M.; Schaich, H.; Bürgi, M.; Konold, W. Climate change and nature conservation in Central European forests: A review of consequences, concepts and challenges. *Ecol. Manag.* **2011**, *261*, 829–843. [CrossRef]

18. Koskela, J.; Lefèvre, F.; Schueler, S.; Kraigher, H.; Olrik, D.C.; Hubert, J.; Yrjänä, L.; Alizoti, P.; Rotach, P. Translating conservation genetics into management: Pan-European minimum requirements for dynamic conservation units of forest tree genetic diversity. *Biol. Conserv.* **2013**, *157*, 39–49. [CrossRef]

19. Choi, I.S.; Lee, J.H.; Choi, B.H. Isolation and characterization of 12 microsatellite loci from *Maackia fauriei* (Fabaceae), a large tree endemic to Jeju Island. *Conserv. Genet. Resour.* **2014**, *6*, 1027–1029. [CrossRef]

20. Hong, Y.S.; Kim, E.J.; Lee, E.P.; Lee, S.Y.; Cho, K.; Lee, Y.K.; Chung, S.; Jeong, H.; You, Y.H. Characteristics of vegetation succession on the *Pinus thunbergii* forests in warm temperate regions, Jeju Island, South Korea. *J. Ecol. Environ.* **2019**, *43*, 1–16. [CrossRef]

21. Han, E.K.; Cho, W.B.; Choi, G.; Yang, S.; Choi, H.J.; Song, G.P.; Lee, J.H. New polymorphic microsatellite markers for *Sarcandra glabra* (Chloranthaceae), an evergreen broad-leaved shrub endangered in South Korea. *J. For. Res.* **2020**, *25*, 364–368. [CrossRef]

22. Noshiro, S.; Sasaki, Y. Identification of Japanese species of evergreen *Quercus* and *Lithocarpus* (Fagaceae). *IAWA J.* **2011**, *32*, 383–393. [CrossRef]

23. Tanouchi, H.; Yamamoto, S. Structure and regeneration of canopy species in an old-growth evergreen broad-leaved forest in Aya district, southwestern Japan. *Vegetatio* **1995**, *117*, 51–60. [CrossRef]

24. Zeng, Q.-M.; Liu, B.; Lin, R.-Q.; Jiang, Y.-T.; Liu, Z.-J.; Chen, S.-P. The complete chloroplast genome sequence of *Quercus gilva* (Fagaceae). *Mitochondrial DNA Part B* **2019**, *4*, 2493–2494. [CrossRef]

25. Kim, G.U.; Jang, K.S.; Lim, H.; Kim, E.H.; Lee, K.H. Genetic Diversity of *Quercus gilva* in Je-ju Island. *J. Korean Soc. For. Sci.* **2018**, *107*, 151–157.

26. Sugiura, N.; Kurokochi, H.; Tan, E.; Asakawa, S.; Sato, N.; Saito, Y.; Ide, Y. Development of 13 polymorphic chloroplast DNA markers in *Quercus gilva*, a regionally endemic species in Japan. *Conserv. Genet. Resour.* **2014**, *6*, 961–965. [CrossRef]

27. Hyun, H.J.; Song, K.M.; Choi, H.S.; Kim, C.S. Dynamics and Distribution of *Quercus gilva* Blume Population in Korea. *Korean J. Environ. Ecol.* **2014**, *28*, 385–392. [CrossRef]

28. Ministry of the Environment of Korea. *Korean Red List of Threatened Species*; National Institute of Biological Resources: Incheon, Korea, 2012.

29. Suh, M.H.; Koh, K.S.; Ku, Y.B.; Kil, J.H.; Choi, T.B.; Suh, S.U.; Qiu, Y.X.; Liu, Y.H.; Oh, J.G. *Research on the Conservation Strategy for the Endangered and Reserved Plants Based on the Ecological and Genetic Characteristics (I)*; National Institute of Environmental Research: Incheon, Korea, 2001.

30. Chung, M.Y.; López-Pujol, J.; Maki, M.; Kim, K.J.; Chung, J.M.; Sun, B.Y.; Chung, M.G. Genetic diversity in the common terrestrial orchid *Oreorchis patens* and its rare congener *Oreorchis coreana*: Inference of species evolutionary history and implications for conservation. *J. Hered.* **2012**, *103*, 692–702. [CrossRef]

31. Haig, S.M.; Miller, M.P.; Bellinger, R.; Draheim, H.M.; Mercer, D.M.; Mullins, T.D. The conservation genetics juggling act: Integrating genetics and ecology, science and policy. *Evol. Appl.* **2016**, *9*, 181–195. [CrossRef]

32. Stojnić, S.; Avramidou, E.; Fussi, B.; Westergren, M.; Orlović, S.; Matović, B.; Trudić, B.; Kraigher, H.; Aravanopoulos, F.A.; Konnert, M. Assessment of Genetic Diversity and Population Genetic Structure of Norway Spruce (*Picea abies* (L) Karsten) at Its Southern Lineage in Europe. Implications for Conservation of Forest Genetic Resources. *Forests* **2019**, *10*, 258. [CrossRef]

33. Chung, M.Y.; López-Pujol, J.; Chung, M.G. The role of the Baekdudaegan (Korean Peninsula) as a major glacial refugium for plant species: A priority for conservation. *Biol. Conserv.* **2017**, *206*, 236–248. [CrossRef]

34. Aoki, K.; Tamaki, I.; Nakao, K.; Ueno, S.; Kamijo, T.; Setoguchi, H.; Murakami, N.; Kato, M.; Tsumura, Y. Approximate Bayesian computation analysis of EST-associated microsatellites indicates that the broadleaved evergreen tree *Castanopsis sieboldii* survived the Last Glacial Maximum in multiple refugia in Japan. *Heredity* **2019**, *122*, 326–340. [PubMed]

35. Park, J.S.; Takayama, K.; Suyama, Y.; Choi, B.-H. Distinct phylogeographic structure of the halophyte *Suaeda malacosperma* (Chenopodiaceae/Amaranthaceae), endemic to Korea–Japan region, influenced by historical range shift dynamics. *Plant Syst. Evol.* **2019**, *305*, 193–203. [CrossRef]

36. Chen, D.; Zhang, X.; Kang, H.; Sun, X.; Yin, S.; Du, H.; Yamanaka, N.; Gapare, W.; Wu, H.X.; Liu, C. Phylogeography of *Quercus variabilis* based on chloroplast DNA sequence in East Asia: Multiple glacial refugia and mainland-migrated island populations. *PLoS ONE* **2012**, *7*, e47268. [CrossRef] [PubMed]

37. Tang, C.Q.; Matsui, T.; Ohashi, H.; Dong, Y.-F.; Momohara, A.; Herrando-Moraira, S.; Qian, S.; Yang, Y.; Ohsawa, M.; Luu, H.T.; et al. Identifying long-term stable refugia for relict plant species in East Asia. *Nat. Commun.* **2018**, *9*, 4488. [CrossRef]

38. Cho, W.B.; So, S.K.; Han, E.K.; Myeong, H.H.; Park, J.S.; Hwang, S.H. Rear-edge, low-diversity, and haplotypic uniformity in cold-adapted *Bupleurum euphorbioides* interglacial refugia populations. *Ecol. Evol.* **2020**, in press.

39. Chung, M.Y.; López-Pujol, J.; Chung, M.G. Genetic homogeneity between Korean and Japanese populations of the broad-leaved evergreen tree *Machilus thunbergii* (Lauraceae): A massive post-glacial immigration through the Korea Strait or something else? *Biochem. Syst. Ecol.* **2014**, *53*, 20–28. [CrossRef]

40. Cho, W.B.; Choi, I.S.; Choi, B.H. Development of microsatellite markers for the endangered *Pedicularis ishidoyana* (Orobanchaceae) using next-generation sequencing. *Appl. Plant Sci.* **2015**, *3*, 1500083. [CrossRef]

41. Miller, M.P.; Knaus, B.J.; Mullins, T.D.; Haig, S.M. SSR_pipeline: A Bioinformatic Infrastructure for Identifying Microsatellites from Paired-End Illumina High-Throughput DNA Sequencing Data. *J. Hered.* **2013**, *104*, 881–885. [CrossRef]

42. Kearse, M.; Moir, R.; Wilson, A.; Stones-Havas, S.; Cheung, M.; Sturrock, S.; Cooper, A.; Markowitz, S.; Drummond, A. Geneious Basic: An integrated and extendable desktop software platform for the organization and analysis of sequence data. *Bioinformatics* **2012**, *28*, 1647–1649.

43. Rozen, S.; Skaletsky, H. Primer3 on the WWW for general users and for biologist programmers. In *Bioinformatics Methods and Protocols*; Humana Press: Totowa, NJ, USA, 2000; pp. 365–386.

44. Chybicki, I.J.; Burczyk, J. Simultaneous estimation of null alleles and inbreeding coefficients. *J. Hered.* **2009**, *100*, 106–113. [CrossRef]

45. Peakall, R.; Smouse, P.E. genalex 6: Genetic analysis in Excel. Population genetic software for teaching and research. *Mol. Ecol. Notes* **2006**, *6*, 288–295. [CrossRef]

46. Weir, B.S.; Cockerham, C.C. Estimating F-statistics for the analysis of population structure. *Evolution* **1984**, *38*, 1358–1370. [CrossRef] [PubMed]

47. Goudet, J. FSTAT (version 1.2): A computer program to calculate F-statistics. *J. Hered.* **1995**, *86*, 485–486. [CrossRef]

48. Rousset, F. genepop'007: A complete re-implementation of the genepop software for Windows and Linux. *Mol. Ecol. Resour.* **2008**, *8*, 103–106. [CrossRef]

49. Cornuet, J.M.; Luikart, G. Description and power analysis of two tests for detecting recent population bottlenecks from allele frequency data. *Genetics* **1996**, *144*, 2001–2014. [PubMed]

50. Luikart, G. Distortion of allele frequency distributions provides a test for recent population bottlenecks. *J. Hered.* **1998**, *89*, 238–247. [CrossRef] [PubMed]

51. Pritchard, J.K.; Stephens, M.; Donnelly, P. Inference of population structure using multilocus genotype data. *Genetics* **2000**, *155*, 945–959.

52. Earl, D.A.; vonHoldt, B.M. STRUCTURE HARVESTER: A website and program for visualizing STRUCTURE output and implementing the Evanno method. *Conserv. Genet. Resour.* **2012**, *4*, 359–361. [CrossRef]

53. Jakobsson, M.; Rosenberg, N.A. CLUMPP: A cluster matching and permutation program for dealing with label switching and multimodality in analysis of population structure. *Bioinformatics* **2007**, *23*, 1801–1806. [CrossRef]

54. Manni, F.; Guérard, E.; Heyer, E. Geographic patterns of (genetic, morphologic, linguistic) variation: How barriers can be detected by using Monmonier's algorithm. *Hum. Biol.* **2004**, *76*, 173–190. [CrossRef] [PubMed]

55. Monmonier, M.S. Maximum-Difference Barriers: An Alternative Numerical Regionalization Method. *Geogr. Anal.* **1973**, *5*, 245–261. [CrossRef]

56. Nei, M.; Tajima, F. Maximum Likelihood Estimation of the Number of Nucleotide Substitutions from Restriction Sites Data. *Genetics* **1983**, *105*, 207–217.

57. Dieringer, D.; Schlötterer, C. Two distinct modes of microsatellite mutation processes: Evidence from the complete genomic sequences of nine species. *Genome Res.* **2003**, *13*, 2242–2251. [CrossRef] [PubMed]

58. Felsenstein, J. *PHYLIP (Phylogeny Inference Package) Version 3.6*; Department of Genome Sciences, University of Washington: Washington, DC, USA, 2004.

59. Merow, C.; Smith, M.J.; Silander, J.A. A practical guide to MaxEnt for modeling species' distributions: What it does, and why inputs and settings matter. *Ecography* **2013**, *36*, 1058–1069. [CrossRef]

60. GBIF Secretariat. GBIF Backbone Taxonomy Checklist dataset. *GBIF Backbone Taxon.* **2019**. [CrossRef]

61. Brown, J.L.; Bennett, J.R.; French, C.M. SDMtoolbox 2.0: The next generation Python-based GIS toolkit for landscape genetic, biogeographic and species distribution model analyses. *PeerJ* **2017**, *5*, e4095. [CrossRef]

62. Karger, D.N.; Conrad, O.; Böhner, J.; Kawohl, T.; Kreft, H.; Soria-Auza, R.W.; Zimmermann, N.E.; Linder, H.P.; Kessler, M. Climatologies at high resolution for the earth's land surface areas. *Sci. Data* **2017**, *4*, 1–20. [CrossRef]

63. Danielson, J.J.; Gesch, D.B. Global Multi-Resolution Terrain Elevation Data 2010 (GMTED2010). In *Geological Survey*; Springer: Garretson, SD, USA, 2011.

64. Gent, P.R.; Danabasoglu, G.; Donner, L.J.; Holland, M.M.; Hunke, E.C.; Jayne, S.R.; Lawrence, D.M.; Neale, R.B.; Rasch, P.J.; Vertenstein, M.; et al. The Community Climate System Model Version 4. *J. Clim.* **2011**, *24*, 4973–4991. [CrossRef]

65. Watanabe, S.; Hajima, T.; Sudo, K.; Nagashima, T.; Takemura, T.; Okajima, H.; Nozawa, T.; Kawase, H.; Abe, M.; Yokohata, T.; et al. MIROC-ESM 2010: Model description and basic results of CMIP5-20c3m experiments. *Geosci. Model Dev.* **2011**, *4*, 845–872. [CrossRef]

66. Ujiié, Y.; Ujiié, H.; Taira, A.; Nakamura, T.; Oguri, K. Spatial and temporal variability of surface water in the Kuroshio source region, Pacific Ocean, over the past 21,000 years: Evidence from planktonic foraminifera. *Mar. Micropaleontol.* **2003**, *49*, 335–364. [CrossRef]

67. Kao, S.J.; Wu, C.R.; Hsin, Y.C.; Dai, M. Effects of sea level change on the upstream Kuroshio Current through the Okinawa Trough. *Geophys. Res. Lett.* **2006**, *33*. [CrossRef]

68. Zheng, X.; Li, A.; Kao, S.; Gong, X.; Frank, M.; Kuhn, G.; Cai, W.; Yan, H.; Wan, S.; Zhang, H.-H.; et al. Synchronicity of Kuroshio Current and climate system variability since the Last Glacial Maximum. *Earth Planet. Sci. Lett.* **2016**, *452*, 247–257. [CrossRef]

69. Vogt-Vincent, N.S.; Mitarai, S. A persistent Kuroshio in the glacial East China Sea and implications for coral paleobiogeography. *Paleoceanogr. Paleoclimatol.* **2020**, e2020PA003902. [CrossRef]

70. Janzen, D.H. Seed predation by animals. *Annu. Rev. Ecol. Syst.* **1971**, *2*, 465–492.

71. Xiao, Z.; Zhibin, Z.; Wang, Y. Dispersal and germination of big and small nuts of *Quercus serrata* in a subtropical broad-leaved evergreen forest. *Ecol. Manag.* **2004**, *195*, 141–150. [CrossRef]

72. Tsukada, M. Vegetation and Climate during the Last Glacial Maximum in Japan. *Quat. Res.* **1983**, *19*, 212–235. [CrossRef]

73. Lee, J.H.; Choi, B.-H. Distribution of broad Leaved evergreen plants on islands of Incheon, middle part of Yellow Sea. *Korean J. Plant Taxon.* **2008**, *38*, 315–332. [CrossRef]

74. Aoki, K.; Suzuki, T.; Hsu, T.W.; Murakami, N. Phylogeography of the component species of broad-leaved evergreen forests in Japan, based on chloroplast DNA variation. *J. Plant Res.* **2004**, *117*, 77–94.

75. Aoki, K.; Matsumura, T.; Hattori, T.; Murakami, N. Chloroplast DNA phylogeography of *Photinia glabra* (Rosaceae) in Japan. *Am. J. Bot.* **2006**, *93*, 1852–1858. [CrossRef]

76. Liu, H.-Z.; Takeichi, Y.; Kamiya, K.; Harada, K. Phylogeography of *Quercus phillyraeoides* (Fagaceae) in Japan as revealed by chloroplast DNA variation. *J. Res.* **2013**, *18*, 361–370. [CrossRef]

77. Litkowiec, M.; Lewandowski, A.; Rączka, G. Spatial Pattern of the Mitochondrial and Chloroplast Genetic Variation in Poland as a Result of the Migration of *Abies alba* Mill. From Different Glacial Refugia. *Forests* **2016**, *7*, 284. [CrossRef]

78. Hou, Z.; Wang, Z.; Ye, Z.; Du, S.; Liu, S.; Zhang, J. Phylogeographic analyses of a widely distributed *Populus davidiana*: Further evidence for the existence of glacial refugia of cool-temperate deciduous trees in northern East Asia. *Ecol. Evol.* **2018**, *8*, 13014–13026. [CrossRef] [PubMed]

79. Di Pasquale, G.; Saracino, A.; Bosso, L.; Russo, D.; Moroni, A.; Bonanomi, G.; Allevato, E. Coastal Pine-Oak Glacial Refugia in the Mediterranean Basin: A Biogeographic Approach Based on Charcoal Analysis and Spatial Modelling. *Forests* **2020**, *11*, 673. [CrossRef]

80. Ito, S.; Ohtsuka, K.; Yamashita, T. Ecological distribution of seven evergreen *Quercus* species in southern and eastern Kyushu, Japan. *Veg. Sci.* **2007**, *24*, 53–63.

81. Nakashizuka, T.; Iida, S. Composition, dynamics and disturbance regime of temperate deciduous forests in Monsoon Asia. *Vegetatio* **1995**, *121*, 23–30. [CrossRef]

82. Nakao, K.; Matsui, T.; Horikawa, M.; Tsuyama, I.; Tanaka, N. Assessing the impact of land use and climate change on the evergreen broad-leaved species of *Quercus acuta* in Japan. *Plant Ecol.* **2011**, *212*, 229–243.

83. Moritz, C. Defining 'Evolutionarily Significant Units' for conservation. *Trends Ecol. Evol.* **1994**, *9*, 373–375. [CrossRef]

84. Crandall, K.A.; Bininda-Emonds, O.R.; Mace, G.M.; Wayne, R.K. Considering evolutionary processes in conservation biology. *Trends Ecol. Evol.* **2000**, *15*, 290–295. [CrossRef]

85. Fraser, D.J.; Bernatchez, L. Adaptive evolutionary conservation: Towards a unified concept for defining conservation units. *Mol. Ecol.* **2001**, *10*, 2741–2752. [CrossRef]

86. Palsbøll, P.J.; Bérubé, M.; Allendorf, F.W. Identification of management units using population genetic data. *Trends Ecol. Evol.* **2006**, *22*, 11–16. [CrossRef] [PubMed]

87. Waples, R.S.; Lindley, S.T. Genomics and conservation units: The genetic basis of adult migration timing in Pacific salmonids. *Evol. Appl.* **2018**, *11*, 1518–1526. [CrossRef] [PubMed]

Article

Adaptive Divergence under Gene Flow along an Environmental Gradient in Two Coexisting Stickleback Species

Thijs M. P. Bal [1],*, Alejandro Llanos-Garrido [2], Anurag Chaturvedi [3,4], Io Verdonck [4], Bart Hellemans [4] and Joost A. M. Raeymaekers [1]

[1] Faculty of Biosciences and Aquaculture, Nord University, N-8049 Bodø, Norway; joost.raeymaekers@nord.no
[2] Department of Organismic & Evolutionary Biology, Harvard University, Cambridge, MA 02138, USA; allanosgarrido@fas.harvard.edu
[3] Department of Ecology and Evolution, University of Lausanne, Biophore Building, 1015 Lausanne, Switzerland; anurag.chaturvedi@unil.ch
[4] Laboratory of Biodiversity and Evolutionary Genomics, KU Leuven, B-3000 Leuven, Belgium; Iomessanga@gmail.com (I.V.); bart.hellemans@kuleuven.be (B.H.)
* Correspondence: thijs.m.bal@nord.no

Abstract: There is a general and solid theoretical framework to explain how the interplay between natural selection and gene flow affects local adaptation. Yet, to what extent coexisting closely related species evolve collectively or show distinctive evolutionary responses remains a fundamental question. To address this, we studied the population genetic structure and morphological differentiation of sympatric three-spined and nine-spined stickleback. We conducted genotyping-by-sequencing and morphological trait characterisation using 24 individuals of each species from four lowland brackish water (LBW), four lowland freshwater (LFW) and three upland freshwater (UFW) sites in Belgium and the Netherlands. This combination of sites allowed us to contrast populations from isolated but environmentally similar locations (LFW vs. UFW), isolated but environmentally heterogeneous locations (LBW vs. UFW), and well-connected but environmentally heterogenous locations (LBW vs. LFW). Overall, both species showed comparable levels of genetic diversity and neutral genetic differentiation. However, for all three spatial scales, signatures of morphological and genomic adaptive divergence were substantially stronger among populations of the three-spined stickleback than among populations of the nine-spined stickleback. Furthermore, most outlier SNPs in the two species were associated with local freshwater sites. The few outlier SNPs that were associated with the split between brackish water and freshwater populations were located on one linkage group in three-spined stickleback and two linkage groups in nine-spined stickleback. We conclude that while both species show congruent evolutionary and genomic patterns of divergent selection, both species differ in the magnitude of their response to selection regardless of the geographical and environmental context.

Keywords: landscape genomics; local adaptation; population genetics; species-specific properties; three-spined stickleback; nine-spined stickleback

Citation: Bal, T.M.P.; Llanos-Garrido, A.; Chaturvedi, A.; Verdonck, I.; Hellemans, B.; Raeymaekers, J.A.M. Adaptive Divergence under Gene Flow along an Environmental Gradient in Two Coexisting Stickleback Species. *Genes* **2021**, *12*, 435. https://doi.org/10.3390/genes12030435

Academic Editor: Delphine Legrand

Received: 22 January 2021
Accepted: 15 March 2021
Published: 18 March 2021

Publisher's Note: MDPI stays neutral with regard to jurisdictional claims in published maps and institutional affiliations.

1. Introduction

The role of gene flow on the evolution of wild populations is a topic that has received substantial attention from biologists for over five decades [1,2]. Evidence from theoretical models as well as empirical studies show that gene flow has the potential to either enhance or disrupt local adaptation, specifically through the distribution of advantageous alleles or by homogenisation of the gene pool [3]. In addition to being essential for our comprehension of evolution in general, understanding the role of gene flow in local adaptation has become exceedingly important for biodiversity conservation in a time of increasing anthropogenic influence on the natural world [4,5]. Targeted gene flow for instance, the method of translocating individuals with predicted advantageous alleles to populations with low genetic diversity, has been getting more attention as a strategy in conservation efforts [6,7]. This conservation approach may become essential in situations of strong habitat fragmentation in which unassisted gene flow has become impossible. However, it remains challenging to reliably predict how wild populations respond to various levels of gene flow as this can be dependent on the initial levels of genetic diversity within the metapopulation, historical distribution patterns and species-specific properties [8]. It is therefore important to study adaptive divergence in systems in which population connectivity and environmental differences have been characterised, as this allows us to make more reliable predictions about the role of gene flow in local adaptation.

Studies on the relative contribution of gene flow on the emergence of adaptive divergence show disparate results and reveal that species' evolutionary responses can vary considerably. Populations may show evident signatures of local adaptation regardless of gene flow [9–12]. The maintenance of adaptive variation under gene flow is possible in situations where the selection for locally adapted alleles is stronger than the influx of non-locally adapted alleles. Here, the specific genomic architecture of a species likely also plays a substantial role, for instance, in the case of large effect loci that are clustered together [3]. However, the homogenizing effects of gene flow are often still observed at relatively small spatial scales, allowing for high migration rates even across strong environmental gradients [13–16]. In such cases, it is expected that the strength of adaptive divergence increases with distance.

Empirical studies and theoretical models addressing this question often focus on a single species along a single spatial scale or environmental contrast [9,10,17–19]. This approach can provide novel insights for a species in a specific geographical context, but also has its limitations. First, as the strength of adaptive divergence is affected by both divergent selection and the level of spatial isolation, it is challenging to discern the relative contribution of these two factors. A more direct approach would be to have a study design where environmental variation and distance among sites are not intrinsically confounded. Second, the distribution of genomic and phenotypic variation is affected by the current properties of the landscape, but is also a result of historical events and species-specific properties [20,21]. Here, the study of phenotypic and genetic variation among populations of multiple coexisting species helps us to infer both shared and unique features of population divergence, enabling broader conclusions regarding the dynamics between gene flow and selection across the landscape [22].

In this study, we investigate the spatial and environmental drivers of population divergence in two ecologically similar stickleback species along both a small-scale and large-scale brackish water–freshwater transition. The three-spined stickleback (*Gasterosteus aculeatus*) and the nine-spined stickleback (*Pungitius pungitius*) are phylogenetically related fishes and thus are excellent model species for a comparative analysis of population structure [22,23]. Both species are euryhaline and share a short similar life cycle with often only one breeding season, and their ecological [24,25], behavioural [26] and genomic [27–30] properties have been studied extensively. Phylogeographic studies show that the species have different ancestral environments. Three-spined stickleback ancestry can be traced back to mainly marine and coastal areas [31], while the nine-spined stickleback has mainly evolved in freshwater [32].

Both species of stickleback are sympatric in the coastal lowlands and upland rivers of Belgium and the Netherlands (Figure 1A). This allows us to study adaptive divergence under gene flow in each species under comparable conditions. Specifically, we selected sampling sites that vary in environmental conditions as well as in spatial connectivity, including both nearby lowland brackish water (LBW) and lowland freshwater sites (LFW), as well as more isolated upland freshwater (UFW) sites.

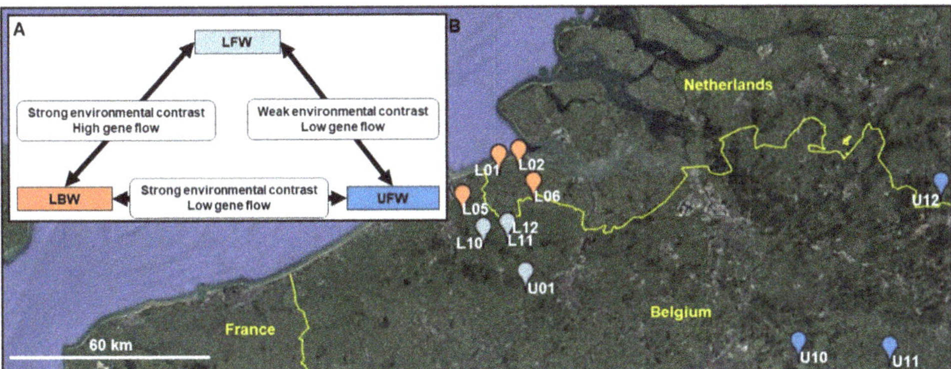

Figure 1. (**A**) Triangle representing the contrast among the three geographical scales. (**B**) Overview of the study area, lowland brackish water (LBW), lowland freshwater (LFW), and upland freshwater (UFW) are represented by orange-red, light-blue and dark-blue dots, respectively.

Importantly, the contrasts between these three sites form the sides of a triangle (Figure 1B) for which we anticipate varying degrees of the relative effects of selection and gene flow. We expect selection to be the predominant evolutionary force between the populations from the LBW and LFW sites, but also between the populations from the LBW and UFW sites. Yet, the response to selection might depend on levels of gene flow, which can either constrain or fuel adaptive divergence, and which we anticipate to be lower between LBW and UFW sites than between LBW and LFW sites. Furthermore, given their similar selective environments, we expect selection to be weak between the populations from the LFW and UFW sites. Finally, in order to locate selective processes within the evolutionary history of each species, we also investigate whether patterns of selection can be mostly attributed to the older or younger branches of the population genealogy of the two species.

2. Materials and Methods

2.1. Study Area and Sampling Design

Belgium and the Netherlands harbour diverse brackish and freshwater habitats, including estuaries, creeks, rivers, ditches and ponds. The connectivity with the open sea and between these sites is variable. This study includes data from eleven sampling locations, of which eight have been the focus of previous research [22,33]. These eight sites are all located in the Belgian-Dutch coastal lowlands and comprise four LBW (L01, L02, L05, L06) and four LFW sites (L10, L11, L12 and U01). These sites were sampled in the spring of 2009. The new sites for this study include three UFW sites (U10, U11 and U12) and are all located in the upland area of Belgium. These sites were sampled in the spring of 2012. Full sampling procedures are described in detail in [22,33]. In short, a minimum of 24 individuals per species were obtained using a dipnet. Sticklebacks were killed with a lethal dose of MS222 following directions of the KU Leuven Animal Ethics Committee (https://admin.kuleuven.be/raden/en/animal-ethics-committee), and flash-frozen in dry ice. The salinity of the water was determined using a Hach field-monitoring kit at different dates throughout the year. Sites with consistent conductivities < 1000 µS/cm were classified as freshwater sites. For each site, we determined the shortest Euclidean distance to the coast (DTC).

2.2. Morphological Characterisation

We scored fifteen morphological traits in both species including standard length, four armour traits, five body traits, and five gill traits. These fifteen morphological traits were recorded by carrying out linear measurements and trait counts. In addition, we performed geometric morphometric analysis to quantify differences in body shape.

The fish bodies were thawed on ice, measured for body size (standard length (SL); ± 0.1 cm), and weighed (0.01 g). Subsequently, the left side of each individual was photographed next to a linear scale using a standard camera position. After photographing, the caudal fin was collected and stored in 100% ethanol. All bodies were then stored on a 4% formalin solution. After 2 months, the formalin processed bodies were rinsed with water for 72 h and then bleached for 4 h using a 1% KOH bleach solution. After bleaching, the fish bodies were stained using an Alizarin Red solution to facilitate plate count and the characterisation of gill raker morphology. After staining, the number of lateral plates (Plates) on the left side was determined. The presence of a keel, a small modification of the caudal lateral plates, was noted, but not included in the plate count. Subsequently, the length of the pelvic plate (PP), the left pelvic spine (PS) and the first dorsal spine (DS) were measured using a digital caliper (±0.01 mm). Body depth (BD), the diameter of the eye (EYE), dorsal fin length (DF), anal fin length (AF) and tail length (Tail) were measured digitally using the TPS software v.2.18 [34]. Finally, the gill cover was removed to dissect the left part of the gills. With the aid of a stereomicroscope, the number of large gill rakers (NLGR) on the frontal and distal part of the first gill arch was determined. The length of the first branchial arch (GA), as well as the length of its second (LGR2), third (LGR3) and fourth (LGR4) gill raker, were measured under a stereomicroscope.

Variation in body shape was characterised based on geometric morphometric analysis following Sharpe et al., 2008 [35]. For both species, a total of fifteen homologous landmarks (including 12 landmarks and 3 semi-landmarks; Figure S1) were placed on the photograph using the TPS software v.2.18 [34]. We used (semi-)landmarks 1, 7 and 15 to perform digital unbending of the landmark coordinates to correct for potential bending of the caudal fin when the photographs were taken. The fifteen (semi-)landmarks were then transformed using a least-square Procrustes superimposition, resulting in 26 relative warp (RW1–RW26) scores per individual.

2.3. DNA Extraction, Genotyping-by-Sequencing and SNP Filtering

A total of 264 individuals per species (i.e., 24 individuals per site and species) were selected for sequencing. Fin clips were used for genomic DNA extraction using the Nucleospin 96 Tissue DNA Extraction kit (Macherey-Nagel, Düren, Germany) according to the manufacturer's protocol. The DNA extracts were then treated with the methylation-sensitive restriction enzyme *ApeKI* (GCWGC) and subsequently a unique and common bar code adapter was attached by ligation. All samples were pooled, purified and size-selected using a PCR reaction with Illlumina (San Diego, CA, USA) primers. Single-nucleotide polymorphisms (SNPs) were generated using genotyping-by-sequencing [36] on an Illumina HiSeq 2000 sequencing platform generating paired-end 100 base pair reads. SNP genotyping was performed using the TASSEL-GBS v5.2.52 [37] pipeline by setting the restriction enzyme-e *ApeKI*, requiring a minimum tag output of -c 5 and k-mer length between 20 and 64. In running the pipeline, converted sequencing read tags of the three-spined and nine-spined stickleback that were saved in the SQLite database were aligned to their respective reference genomes [29,38] using Bowtie v2.3.5.1 [39]. Alignment rates for the three-spined and nine-spined stickleback were 85.54% and 95.09%, respectively. Individuals with less than 500,000 reads were removed from the database. SNPs from the database were subsequently converted to a Variant Call Format (VCF).

Further SNP filtering was performed using VCFtools v0.1.13 [40] and we removed SNPs based on read depth (RD < 5), genotype quality (GQ < 20), non bi-allelic, variant coverage ($\geq$0.9), minimum allele frequency (<0.05), linkage disequilibrium (r2 $\geq$ 0.8; -genochisq) and heterozygosity (H_O > 0.5; removing potential paralogs). These filtering criteria were equal to the SNP filtering as applied in Raeymaekers et al. 2017 and were chosen to decrease the false positive rate while maintaining a number of SNPs that allowed for population genomics analyses at a high resolution of genomic data. Finally, we retained

10,836 SNPs in 239 individuals and 15,033 SNPs in 241 individuals for three-spined and nine-spined stickleback, respectively (Table 1). We used PGDSpider v2.1.1.5 [41] together with population definition files for the conversion to other data formats, as well as to assign population labels to each individual.

Table 1. Overview of population genomics statistics per site for populations of three-spined ([3s]) and nine-spined ([9s]) stickleback. DTC: distance to the coast in kilometers, n: number of individuals per population retained for downstream analyses, *Ho*: observed heterozygosity, *He*: expected heterozygosity, F_{IS}: inbreeding coefficient. N_e: effective population size.

Site	DTC	n[3s]	*Ho*[3s]	*He*[3s]	F_{IS}[3s]	N_e[3s]	n[9s]	*Ho*[9s]	*He*[9s]	F_{IS} [9s]	N_e[9s]
L01	3.94	21	0.266	0.292	0.0805	213.4	22	0.246	0.303	0.167	441.7
L02	4.30	24	0.266	0.290	0.0769	166.4	23	0.299	0.307	0.032	429.7
L05	10.90	23	0.282	0.299	0.0550	198.4	18	0.240	0.304	0.187	142.0
L06	11.14	23	0.294	0.298	0.0186	166.5	23	0.299	0.309	0.031	185.0
L10	21.75	23	0.229	0.234	0.0249	120.8	22	0.266	0.282	0.052	690.2
L11	22.84	24	0.229	0.273	0.141	72.2	23	0.291	0.288	0.001	127.4
L12	22.84	13	0.159	0.269	0.344	814.0	22	0.283	0.299	0.052	508.3
U01	36.20	24	0.232	0.259	0.097	219.5	22	0.283	0.289	0.022	575.5
U10	77.10	23	0.239	0.262	0.081	137.2	22	0.163	0.215	0.213	177.1
U11	99.20	20	0.261	0.271	0.034	138.8	21	0.219	0.250	0.114	344.5
U12	99.80	21	0.253	0.268	0.053	94.8	23	0.175	0.244	0.251	271.2

2.4. Population Structure and Genetic Diversity Statistics

We assessed population structure using the Bayesian approach implemented in fastStructure v1.0 [42]. For both species, we ran the algorithm for K = 1 to K = 11 under a simple prior using a default starting seed of 100. The generated output was subsequently used by StructureSelector v1.0 [43] to find the most likely population structure based on the maximal marginal likelihood, and to generate structure barplots.

Further downstream analyses were conducted in R 3.6.3 [44], specifically making use of the options and functions under the Hierfstat v0.4.22 [45] and Adegenet v2.1.2 [46] packages. First, we assessed the genetic diversity per species and site by calculating the expected heterozygosity (H_E). Simple linear regression was used to test for the significance of the decrease of H_E with distance to the coast. Overall and pairwise F_{ST} values (N = 55 pairwise combinations) were calculated using Adegenet v2.1.2 [46]. In order to test for isolation by distance, the correlation between pairwise F_{ST} and geographical distances along waterways among the sites was tested using a simple Mantel test [47] with 9999 permutations. Finally, we performed two-dimensional classical multidimensional scaling (MDS) on the pairwise F_{ST} values as an additional way of visualizing population structure.

2.5. Signatures of Adaptive Divergence

In order to compare the level of phenotypic differentiation directly with the level of genetic differentiation in each species, we calculated P_{ST}, an index that quantifies the proportion of population phenotypic variance in quantitative traits [48,49]. P_{ST} values along with 95% Bayesian confidence intervals were estimated following [48]. Specifically, traits were assumed to be normally distributed, and a linear model was fitted to each trait separately. Population was included in the model as a random effect, and body size as a covariate. The models were fitted to the data using a Gibbs sampler, implemented in WinBUGS v1.4.3 [50]. Prior distributions for each trait were uninformative, and posterior distributions were obtained by running five independent chains (50.000 iterations) after a burn-in of 1000 iterations.

We utilised two distinct methods for identifying genomic signatures of selection. First, we used BayeScan v2.1 [51] with the prior odds of neutrality set at 100 and starting from an initial 10 pilot runs of 5000 iterations followed by an additional 150,000 iterations and a burn-in of 50,000 iterations. For both species, these settings were applied globally, using all 11 sites, as well as for the three possible combinations of the three spatial scales, i.e., LBW–LFW (8 sites), LBW–UFW (7 sites), and LFW–UFW (7 sites). In each analysis, BayeScan v2.1

considered the included populations separately. The posterior probabilities calculated by BayeScan 2.1 were transformed into q-values corresponding to the FDR (False Discovery Rate) of the P-value, and the cut-off for determining statistical significance was set to 0.05.

Second, we used GRoSS v1.0 [52] to assign signatures of positive selection to the branches of the population genealogy, which for each species of stickleback corresponded to the three spatial groups (LBW, LFW, UFW). Specifically, GRoSS v1.0 aims to attribute genomic signatures of positive selection to the branches of an admixture graph. To do so, GRoSS v1.0 uses population phylogenies to determine allele frequencies across hierarchical groups, and tests which of those deviate from what is expected given the population genealogy. In addition, GRoSS v1.0 indicates along which specific evolutionary lineages these allelic variants were most likely selected. Using this method, we were able to identify the relative importance of local and regional selection pressures in the two species. The population phylogeny of each species was inferred using a neighbour-joining tree based on the first two dimensions of a multidimensional scaling analysis. Individuals within populations were bootstrapped 1000 times to generate a consensus tree based on the clusters with the most support. For both species, the consensus tree corresponded to the three spatial groups (LBW, LFW, UFW), and was in agreement with the Bayesian analysis of population structure using fastStructure v1.0 [42].

3. Results

3.1. Morphological Divergence

Across the full geographical scale of our study (i.e., all eleven locations), populations of the three-spined stickleback were morphologically more divergent than populations of the nine-spined stickleback. First, we found higher P_{ST} values for 13 out of 15 traits in three-spined stickleback, with only the length of the first branchial arch and the length of the first dorsal spine showing stronger divergence in nine-spined stickleback (Figure 2). Second, P_{ST} values had a range of 0.037–0.49 (average = 0.24 ± 0.14) and 0.013–0.27 (average = 0.10 ± 0.075) for three-spined stickleback and nine-spined stickleback, respectively.

For both species, the first two relative warps explained more than 50% of the variation in body shape, with successive RW scores only increasing the total variation explained by 12.67% or less. In line with the other morphological traits, the diversification of RW1 and RW2 was larger in three-spined stickleback than in nine-spined stickleback, although the differences in P_{ST} values were smaller than for the other morphological traits (Figure 2).

3.2. Population Structure and Genetic Diversity Statistics

Genetic diversity—The genetic diversity per population, calculated as H_E, showed similar patterns for both species, with generally lower genetic diversity for populations that are further away from the coast (Table 1; Figure 3). However, both LFW and UFW populations showed similar levels of H_E in three-spined stickleback, while UFW populations had substantially lower H_E levels than LFW populations in nine-spined stickleback. As a result, the relationship between genetic diversity and distance to the coast was only significant in nine-spined stickleback (3s: slope = −0.00020; P = 0.242; 9s: slope = −0.00076, P = 0.0002) (Figure 3).

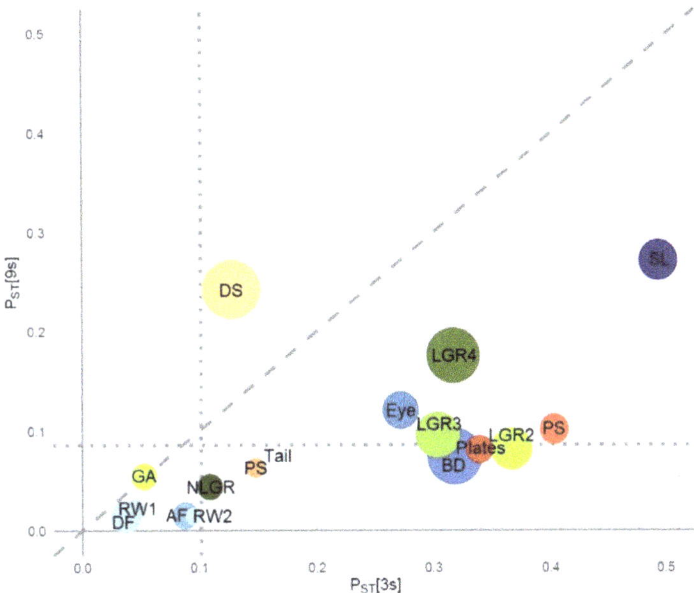

Figure 2. Levels of phenotypic differentiation (P_{ST}) among three-spined ([3s]) and nine-spined ([9s]) stickleback populations for 17 morphological traits (see Section 2.2 for description of trait codes), with the diagonal dashed line representing the 1:1-line. The dotted vertical and horizontal lines represent the level of neutral genetic divergence in [3s] (F_{ST} = 0.102) and [9s] (F_{ST} = 0.086), respectively. Circle sizes indicate the importance of parallel versus non-parallel effects. These effect sizes were determined using ANOVAs attributing the variation in each trait to the effect of site (degree of parallelism), the effect of species, and the effect of site by species interaction (degree of non-parallelism). Shades of *red-orange*, *blue* and *green* represent traits related to body armour, body shape and gill morphology, respectively.

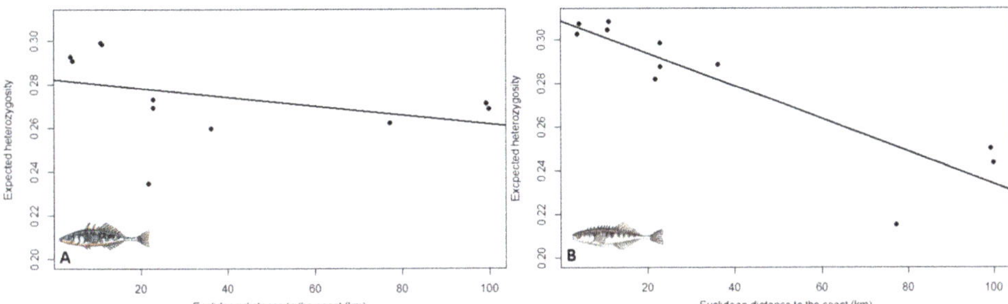

Figure 3. The relationship between Euclidean distance to the coast and expected heterozygosity in three-spined stickleback (A; slope = −0.00020, P = 0.242) and nine-spined stickleback (B; slope = −0.00075, P = 0.0002).

Genetic differentiation—Overall neutral F_{ST} was 0.102 and 0.086 in three-spined stickleback and nine-spined stickleback, respectively, and a significant isolation-by-distance pattern was observed in both species (3s: r = 0.42, P = 0.0487; 9s: r = 0.72, P = 0.0055) (Figure S2). Based on the maximal marginal likelihood scores, the most likely number of clusters to explain the population structure was K = 6 for both species (Figures 4 and 5). Yet, in three-spined stickleback there was strong support for population structuring between populations from LBW and LFW sites (Figure 4). Nine-spined stickleback showed less population divergence among populations from the LBW and LFW sites, while populations from UFW sites were genetically more isolated (Figure 5). MDS plots (Figure S3) based on pairwise F_{ST} (Table S1) confirmed the Bayesian structure analyses.

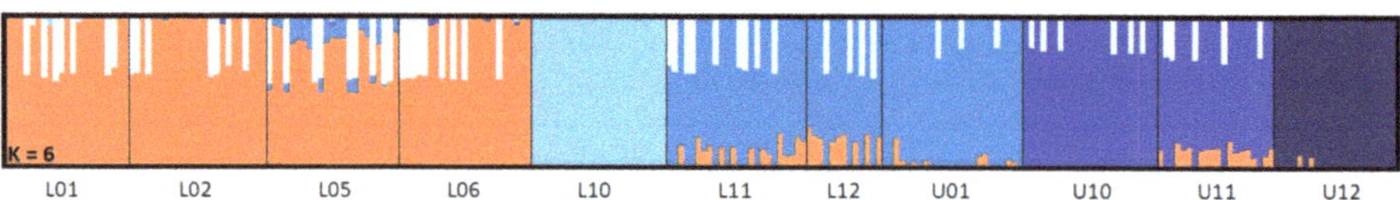

Figure 4. Barplots of population structure with K = 6 clusters in three-spined stickleback inferred by Bayesian analysis using fastStructure v1.0.

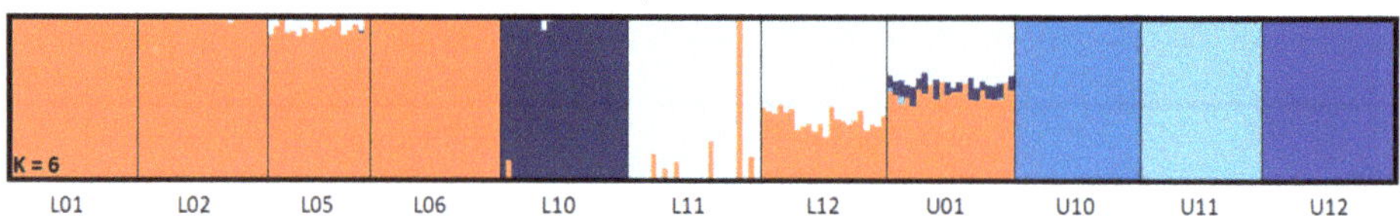

Figure 5. Barplots of population structure with K = 6 clusters in nine-spined stickleback inferred by Bayesian analysis using fastStructure v1.0.

3.3. Signatures of Adaptive Divergence

Morphological signatures of selection—On all four geographical scales, the proportion of morphological traits for which P_{ST} significantly exceeded F_{ST} was higher in three-spined stickleback (23–41%) than in nine-spined stickleback (6–12%; Table 2). The morphological traits that showed evidence for adaptive divergence largely overlap among the four geographical scales and the two species (Tables S2–S5). In three-spined stickleback, the lowest number of significant P_{ST} values were found on the LFW–UFW scale (4/17 traits). On the other geographical scales, three-spined stickleback showed equally strong (7/17) signals of adaptive divergence, primarily due to stronger divergence in the length of gill rakers (traits LGR2, LGR3 and LGR4). In nine-spined stickleback, signals of adaptive divergence varied from 1/17 to 2/17 traits across the four geographical scales considered (Table 2).

Table 2. Proportion of P_{ST} values that significantly exceed neutral genetic divergence F_{ST} at four geographical scales. Included are the 14 morphological traits (excluding standard length) and relative warp 1 and relative warp 2 defining variation in body shape. LFW: lowland freshwater, LBW: lowland brackish water, and UFW: upland freshwater.

Geographical Scale	F_{ST} [3s]	Mean P_{ST} [3s]	$P_{ST} > F_{ST}$ [3s]	F_{ST} [9s]	Mean P_{ST} [9s]	$P_{ST} > F_{ST}$ [9s]
LBW–LFW–UFW	0.102	0.22	7/17 (41%)	0.086	0.090	2/17 (12%)
LBW–LFW	0.081	0.19	7/17 (41%)	0.040	0.071	1/17 (6%)
LBW–UFW	0.071	0.20	7/17 (41%)	0.097	0.093	2/17 (12%)
LFW–UFW	0.140	0.19	4/17 (23%)	0.118	0.100	1/17(6%)

Genomic signatures of selection—Our first outlier detection approach using BayeScan identified 142 and 70 outliers across the eleven sites in three-spined and nine-spined stickleback, respectively. (Figure 6). Across all geographical scales, the proportion of outlier SNPs was higher in three-spined stickleback (0.77–1.31%) than in nine-spined stickleback (0.21–0.46%) (Table 3). In three-spined stickleback, the proportion of outliers was comparable across the sides of the LBW–LFW–UFW triangle (Table 3). In contrast, in nine-spined stickleback, the proportion of outlier SNPs was markedly higher across the LBW–UFW side of the triangle (Table 3). Our second outlier detection approach using GRoSS v1.0 was more stringent and identified an overall lower proportion of outlier SNPs. Yet, the strongest signal of adaptive divergence was again found in three-spined stickleback (3s: 0.0383%; 9s: 0.0333%). The overall genomic signatures of selection showed largely congruent patterns for the two species with respect to the distribution of outlier SNPs along the branches of the admixture graph. First, the majority of outlier SNPs were detected along the local site-specific branches (3s: 71/83; 9s: 92/100) of the admixture graph. Second, the majority (3s: 67/71; 9s: 80/92) of these outlier SNPs were detected along branches specifically associated with the freshwater sites, including both the LFW and UFW sites (Figure 7). Finally, only few outlier SNPs (N = 2 in both species) were detected along the branch splitting the freshwater from the brackish water populations (w-s). These outlier SNPs were associated with linkage group IV in three-spined stickleback (Figure S4, Table S6), and with linkage groups V and XII in nine-spined stickleback (Figure S5, Table S6). These genomic positions were also detected as outliers in the BayeScan v2.1 analyses carried out on the geographical scales that included a brackish water—freshwater transition. However, GRoSS v1.0 also revealed some dissimilarities in genomic signatures of selection between the two species. In the nine-spined stickleback, the LBW sites accounted for a larger proportion of outlier SNPs than in three-spined stickleback. Additionally, the relative contribution of outlier SNPs associated with lowland and upland freshwater sites was found to be reversed. In three-spined stickleback the largest proportion of outlier SNPs was associated with LFW sites, while in nine-spined stickleback the largest proportion of outlier SNPs was associated with UFW sites (Figure 7).

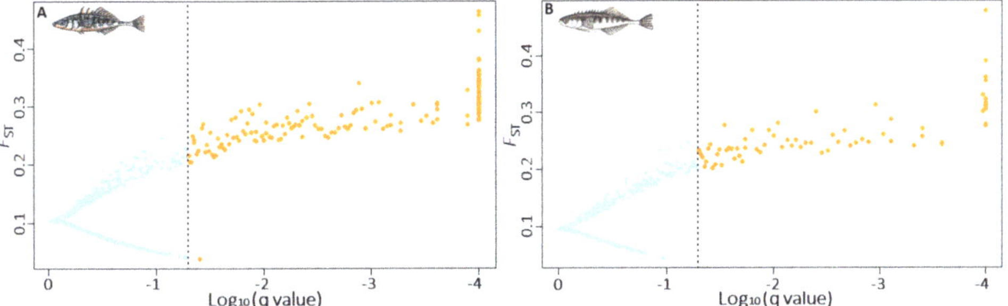

Figure 6. Proportion of outlier SNPs across the geographical scale LBW-LFW-UFW (eleven sites) detected by BayeScan v2.1 for (**A**) 10,836 SNPs in three-spined stickleback and (**B**) 15,033 SNPs in nine-spined stickleback.

Table 3. Proportion of outlier SNPs on four geographical scales as detected by BayeScan v2.1 for 10,836 SNPs in three-spined stickleback ([3s]) and for 15,033 SNPs in nine-spined stickleback ([9s]). LFW: lowland freshwater, LBW: lowland brackish water, and UFW: upland freshwater.

Geographical Scale	Total Outliers [3s]	Proportion [3s]	Total Outliers [9s]	Proportion [9s]
LBW–LFW–UFW	142	1.31%	70	0.46%
LBW–LFW	98	0.90%	44	0.29%
LBW–UFW	87	0.80%	61	0.41%
LFW–UFW	83	0.77%	31	0.21%

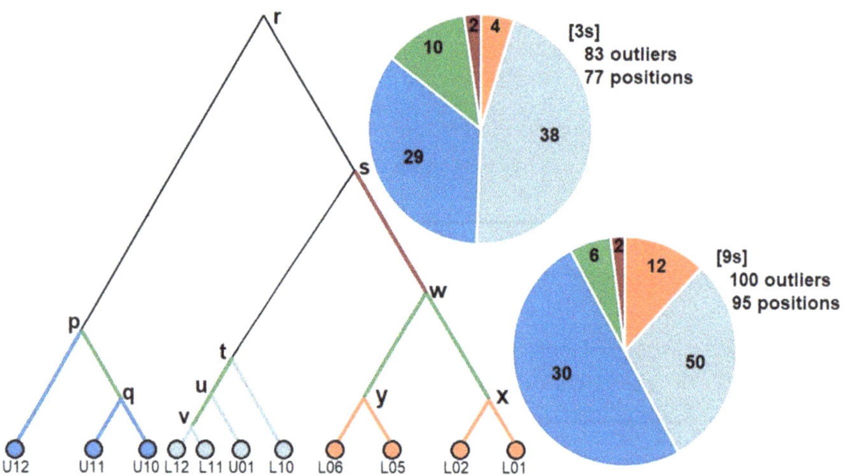

Figure 7. Consensus admixture graph used by GRoSS v1.0 for the detection of outlier SNPs in the two species, along with pie charts visualising the distribution of outlier SNPs across five categories of specific branches of this admixture graph. *Orange*: lowland brackish water terminal branches. *Light blue*: lowland freshwater terminal branches. *Dark blue*: upland freshwater terminal branches. *Green*: branches within the geographical subgroups. *Dark red*: branch splitting the freshwater from the brackish water sites. In total, the 10,836 (3s) and 15,033 SNPs (9s) were tested for signatures of selection along 20 branches, resulting in a proportion of 0.0383% (83/[20*10,836]) and 0.0333% (100/[20*15,033]) outliers in three-spined stickleback and nine-spined stickleback, respectively. The distribution of outlier SNPs over the five categories differs significantly between species (Chi-squared test: χ^2 = 10.03, P = 0.040).

4. Discussion

In this study, we investigated the population structure, levels of divergence and signatures of selection in coexisting three-spined and nine-spined stickleback populations from Belgium and the Netherlands. In each species, we compared three geographic contexts, i.e., a short-range brackish water-freshwater transition (LBW–LFW), a long-range brackish water-freshwater transition (LBW–UFW), and long-range spatial isolation (LFW–UFW). Since selection is expected to be strong across the brackish water-freshwater transitions (LBW–LFW and LBW–UFW), and gene flow is expected to be weak across the long-range comparisons (LBW–UFW and LFW–UFW), our sampling design allowed us to assess the single and joint effects of selection and gene flow on genetic and morphological variation in both species. While the two species share evident similarities in terms of population genetic diversity and structure, we found clear differences in the observed levels of adaptive divergence across the three geographical contexts and the two species. First, morphological and genomic adaptive divergence in three-spined stickleback was strong, in any geographical context. Second, morphological and genomic adaptive divergence in nine-spined stickleback was weaker, but the strongest genomic signatures of selection were observed across the long-range brackish water—freshwater transition (LBW–UFW). Third, for both species signals of positive genomic selection were low or missing for the genealogical branches associated with brackish water sites (LBW), while these signals were stronger for the genealogical branches associated with freshwater sites (LFW and UFW).

The locations in this study vary in salinity, which is known to impose important selection pressures in both the three-spined stickleback [21,33,38] and the nine-spined stickleback [53,54]. Thus, we expected salinity to be one of the important common drivers of adaptive divergence among the populations of both species. Accordingly, we generally observed stronger signatures of adaptive divergence in both species when contrasting populations from brackish water sites (LBW) with populations from freshwater sites (LFW and UFW), than among populations that were isolated by distance, but not by a different salinity environment (LFW vs. UFW). Most outlier SNPs in the two species were assigned to local terminal branches, and most of those involved freshwater sites. This may indicate that freshwater habitats are ecologically heterogeneous, imposing diverse selection pressures on local populations of both species. For both species, only two outlier SNPs were assigned to the root of the split between brackish water and freshwater populations (w-s). These outliers were located on different linkage groups (LG) for the two species (3s: LG IV; 9s: LG V and LG XII). In three-spined stickleback, one of these outlier SNPs is located within a gene (ID: ENSGACG00000018958) previously described as a candidate for adaptation to variation in salinity [38], while this does not apply to the gene associated with one of the two outlier SNPs in nine-spined stickleback. It has been demonstrated previously that freshwater adaptation in three-spined and nine-spined stickleback has different genomic origins [53,55]. However, the comparable distribution of outlier SNPs along the branches of the population admixture graph, may indicate that selection can leave similar genomic signatures on species that are subject to the same environmental contrast and to similar degrees of spatial isolation.

In line with our previous assessment of adaptive divergence in the two species [22], the most obvious difference between the two stickleback species was the overall stronger signals of adaptive divergence in the three-spined stickleback. We here confirm this observation at three different geographical scales. Thus, the stronger tendency of three-spined stickleback to adapt to local selective environments seems to be independent of the geographical and environmental context. However, our comparison of the three sides of the triangle revealed another important difference between the two species. In the three-spined stickleback, a comparable proportion of outlier SNPs on the LBW–UFW and the LBW–LFW scale indicates that enhanced gene flow at the LBW–LFW scale does not disrupt the effects of selection, and might even promote local adaptation [3]. In contrast, in the nine-spined stickleback, the proportion of outlier SNPs at the LBW–UFW scale was considerably higher than on the LBW–LFW scale, suggesting that gene flow among the LBW–LFW scale may constrain adaptive divergence [3]. In summary, the two species seem to differ in evolutionary potential in the face of high gene flow. They are therefore positioned differently at the migration-selection balance [56], with three-spined stickleback more tilted towards selection, and nine-spined stickleback towards migration [57].

Patterns of phenotypic and genetic divergence might be influenced by past demographic processes [58], thereby limiting our ability to compare the relative importance of recent selection, gene flow and genetic drift in both species. However, our analyses suggest that historical patterns are unlikely to be the dominant driver of the observed levels of phenotypic and genetic divergence in our study system. First, the brackish water and freshwater sites in this system are likely of a relatively young postglacial origin [59], and phylogeographic studies in Europe suggest that both species colonised these areas following the retreat of the Late Pleistocene ice sheets [60,61]. Such postglacial origins also have been confirmed for three-spined stickleback in our study area [62]. Second, the two species show clear similarities in population structure, with evolutionary relationships among populations that correspond well with the three defined spatial scales (LBW, LFW, UFW). More recently, aquatic habitats in Belgium and the Netherlands have been influenced by strong anthropogenic change in the form of the construction of canals, dykes and drainages. This has likely affected the distribution and genetic diversity of both species in comparable ways.

Importantly, a similar population structure in the two species implies that we may expect similar performance of outlier detection methods, both in detecting genomic signatures of selection based on departures from a baseline model under genetic drift, as well as in avoiding false positives. The various outlier detection methods used in this study (BayeScan, GROSS) as well as in our previous assessment of adaptive divergence in the two stickleback species (LOSITAN, Arlequin, BayeScan) [22] have consistently revealed a higher proportion of outliers in three-spined stickleback than in nine-spined stickleback, in line with the stronger morphological divergence in this species. Nevertheless, a recent view is that without estimates of local recombination rate, interpreting genome scan results is difficult [63]. Caution with the interpretation of the results is thus warranted until the incorporation of estimates of local recombination rate become feasible for non-model species.

The difference in evolutionary potential between both species is of particular significance because we compare them in exactly the same environmental matrix. Yet, it is important to remind that the absence of a strong signature of adaptive divergence in the nine-spined stickleback does not imply that the populations of this species are not adapted, since they might already be preadapted to the ecological gradients in the landscape [22]. Additionally, it is important to acknowledge that nine-spined stickleback may harbour stronger signals of adaptive divergence for other forms of variation that were not assessed in this study. These forms of variation might, for example, be expressed by differential epigenetic patterns or copy number variations in genomic regions under selection. However, just focusing on our current findings, there are various species-specific properties such as genomic architecture, dispersal capacities and life history that may underlie the different evolutionary responses in the two species that have been discussed previously [22].

However, in this study, we could also shed light on how these species-specific properties may lead to disparate evolutionary dynamics at different geographical scales. In particular, three-spined stickleback, genetic diversity was highest for the LBW populations, and comparably lower for the LFW and UFW populations. In contrast, genetic diversity in nine-spined stickleback was similar for LBW and LFW populations, but substantially lower for the UFW populations. This pattern suggests that the two species are unequally affected by genetic drift in different parts of the landscape. In LFW sites, nine-spined stickleback populations are possibly more resilient than the three-spined stickleback populations, as they may be able to cope better with hypoxic conditions during summer droughts [22,64]. In UFW sites, however, we speculate that three-spined stickleback populations are more stable than nine-spined stickleback populations, because their better swimming capacity might enable them to better cope with high flow rates following heavy rainfall [24,65]. In summary, species-specific properties may interact with regional environmental factors that differentially affect population persistence in the two species.

5. Conclusions

In conclusion, our landscape-level comparison between two closely related coexisting species revealed that the interaction between species-specific properties and regional landscape features might lead to distinct evolutionary responses and levels of adaptive divergence. This finding has profound implications for conservation biology and biodiversity management, as it emphasises that although adaptive potential is to a large extent species-specific, how much adaptive potential is realised may vary throughout the landscape.

Supplementary Materials: The following are available online at https://www.mdpi.com/2073-4425/12/3/435/s1. Figure S1: landmarks used for geometric morphometric analysis, Figure S2: The relationship between Euclidean distance between sites and pairwise F_{ST}, Figure S3: Two-dimensional classical multidimensional scaling of pairwise F_{ST}, Figure S4: Manhattan plot of GRoSS v1.0 outlier detection in three-spined stickleback, Figure S5: Manhattan plot of GRoSS v1.0 outlier detection in nine-spined stickleback, Table S1: pairwise F_{ST} values among the eleven sites, Table S2: P_{ST} values for the morphological traits on the LBW–LFW–UFW geographical scale, Table S3: P_{ST} values for the morphological traits on the LBW–LFW geographical scale, Table S4: P_{ST} values for the morphological traits on the LBW–UFW geographical scale, Table S5: P_{ST} values for the morphological traits on the LFW–UFW geographical scale, Table S6: Genomic positions of outlier SNPs associated with the w-s branch of the population genealogy.

Author Contributions: Conceptualization, T.M.P.B. and J.A.M.R.; Formal analysis, T.M.P.B., A.L.-G. and I.V.; Funding acquisition, J.A.M.R.; Investigation, T.M.P.B.; Methodology, T.M.P.B., A.L.-G., A.C., I.V., B.H. and J.A.M.R.; Supervision, J.A.M.R.; Validation, T.M.P.B.; Writing—original draft, T.M.P.B.; Writing—review & editing, T.M.P.B., A.L.-G., A.C. and J.A.M.R. All authors have read and agreed to the published version of the manuscript.

Funding: Research was sponsored by the Research Foundation—Flanders (Grant 1526712N to J.A.M.R.) and by the University of Leuven (KU Leuven Centre of Excellence PF/10/07).

Institutional Review Board Statement: Sampling was done in accordance to the European directive 2010/63/EU and explicit permission of the Agency for Nature and Forests. The individuals were euthanised following directions of the KU Leuven Animal Ethics Committee (https://admin.kuleuven.be/raden/en/animal-ethics-committee). Ethical review and approval were waived for this study, because sampling and handling fish in the field was not considered an experimental manipulation at the time of sampling (2008–2012).

Informed Consent Statement: Not applicable.

Data Availability Statement: Raw data, VCF files and morphological measurements are available upon request. Data files containing all outlier SNP positions are available online at: https://doi.org/10.6084/m9.figshare.14210651.v1 (uploaded on 12/03/2021).

Acknowledgments: We thank the reviewers for their thoughtful comments. We thank Camille de Raedemaekers, Jonas Merckx, and Filip Volckaert for data collection and logistic support, and Ingrid Oline Berg Sivertsen for the stickleback artwork. Finally, we would like to thank RAVON for field assistance during sampling.

Conflicts of Interest: The authors declare no conflict of interests.

References

1. Levins, R. The theory of fitness in a heterogeneous environment. IV. The Adaptive Significance of Gene Flow. *Evolution* **1964**, *18*, 635. [CrossRef]
2. Endler, J.A. Gene flow and population differentiation: Studies of clines suggest that differentiation along environmental gradients may be independent of gene flow. *Science* **1973**, *179*, 243–250. [CrossRef]
3. Tigano, A.; Friesen, V.L. Genomics of local adaptation with gene flow. *Mol. Ecol.* **2016**, *25*, 2144–2164. [CrossRef] [PubMed]
4. Flanagan, S.P.; Forester, B.R.; Latch, E.K.; Aitken, S.N.; Hoban, S. Guidelines for planning genomic assessment and monitoring of locally adaptive variation to inform species conservation. *Evol. Appl.* **2018**, *11*, 1035–1052. [CrossRef] [PubMed]
5. Hällfors, M.H.; Liao, J.; Dzurisin, J.; Grundel, R.; Hyvärinen, M.; Towle, K.; Wu, G.C.; Hellmann, J.J. Addressing potential local adaptation in species distribution models: Implications for conservation under climate change. *Ecol. Appl.* **2016**, *26*, 1154–1169. [CrossRef]
6. Kelly, E.; Phillips, B.L. Targeted gene flow and rapid adaptation in an endangered marsupial. *Conserv. Biol.* **2019**, *33*, 112–121. [CrossRef] [PubMed]
7. Kelly, E.; Phillips, B.L. Targeted gene flow for conservation. *Conserv. Biol.* **2016**, *30*, 259–267. [CrossRef] [PubMed]
8. Bernatchez, L. On the maintenance of genetic variation and adaptation to environmental change: Considerations from population genomics in fishes. *J. Fish Biol.* **2016**, *89*, 2519–2556. [CrossRef]
9. Pfeifer, S.P.; Laurent, S.; Sousa, V.C.; Linnen, C.R.; Foll, M.; Excoffier, L.; Hoekstra, H.E.; Jensen, J.D. The evolutionary history of Nebraska deer mice: Local adaptation in the face of strong gene flow. *Mol. Biol. Evol.* **2018**, *35*, 792–806. [CrossRef]
10. Diopere, E.; Vandamme, S.G.; Hablützel, P.I.; Cariani, A.; Van Houdt, J.; Rijnsdorp, A.; Tinti, F.; Volckaert, F.A.M.; Maes, G.E. FishPopTrace Consortium Seascape genetics of a flatfish reveals local selection under high levels of gene flow. *ICES J. Mar. Sci.* **2018**, *75*, 675–689. [CrossRef]
11. Moody, K.N.; Hunter, S.N.; Childress, M.J.; Blob, R.W.; Schoenfuss, H.L.; Blum, M.J.; Ptacek, M.B. Local adaptation despite high gene flow in the waterfall-climbing Hawaiian goby. *Sicyopterus Stimpsoni. Mol. Ecol.* **2015**, *24*, 545–563. [CrossRef] [PubMed]
12. Hablützel, P.I.; Grégoir, A.F.; Vanhove, M.P.M.; Volckaert, F.A.M.; Raeymaekers, J.A.M. Weak link between dispersal and parasite community differentiation or immunogenetic divergence in two sympatric cichlid fishes. *Mol. Ecol.* **2016**, *25*, 5451–5466. [CrossRef] [PubMed]
13. Moore, J.-S.; Hendry, A.P. Both selection and gene flow are necessary to explain adaptive divergence: Evidence from clinal variation in stream stickleback. *Evol. Ecol. Res.* **2005**, *7*, 871–886.
14. Bachmann, J.C.; Jansen van Rensburg, A.; Cortazar-Chinarro, M.; Laurila, A.; Van Buskirk, J. Gene flow limits adaptation along steep environmental gradients. *Am. Nat.* **2020**, *195*, E67–E86. [CrossRef]
15. Cornwell, B.H. Gene flow in the anemone *Anthopleura elegantissima* limits signatures of local adaptation across an extensive geographic range. *Mol. Ecol.* **2020**, *29*, 2550–2566. [CrossRef] [PubMed]

16. Kalske, A.; Leimu, R.; Scheepens, J.F.; Mutikainen, P. Spatiotemporal variation in local adaptation of a specialist insect herbivore to its long-lived host plant. *Evolution* **2016**, *70*, 2110–2122. [CrossRef] [PubMed]
17. Dennenmoser, S.; Vamosi, S.M.; Nolte, A.W.; Rogers, S.M. Adaptive genomic divergence under high gene flow between freshwater and brackish-water ecotypes of prickly sculpin (Cottus asper) revealed by Pool-Seq. *Mol. Ecol.* **2017**, *26*, 25–42. [CrossRef] [PubMed]
18. Pinho, C.; Hey, J. Divergence with gene Flow: Nodels and data. *Annu. Rev. Ecol. Evol. Syst.* **2010**, *41*, 215–230. [CrossRef]
19. Feder, J.L.; Egan, S.P.; Nosil, P. The genomics of speciation-with-gene-flow. *Trends Genet.* **2012**, *28*, 342–350. [CrossRef]
20. Oke, K.B.; Rolshausen, G.; LeBlond, C.; Hendry, A.P. How parallel is parallel evolution? A comparative analysis in fishes. *Am. Nat.* **2017**, *190*, 1–16. [CrossRef]
21. Ferchaud, A.-L.; Hansen, M.M. The impact of selection, gene flow and demographic history on heterogeneous genomic divergence: Three-spine sticklebacks in divergent environments. *Mol. Ecol.* **2016**, *25*, 238–259. [CrossRef]
22. Raeymaekers, J.A.M.; Chaturvedi, A.; Hablützel, P.I.; Verdonck, I.; Hellemans, B.; Maes, G.E.; De Meester, L.; Volckaert, F.A.M. Adaptive and non-adaptive divergence in a common landscape. *Nat. Commun.* **2017**, *8*, 267. [CrossRef] [PubMed]
23. DeFaveri, J.; Shikano, T.; Ghani, N.I.A.; Merilä, J. Contrasting population structures in two sympatric fishes in the Baltic Sea basin. *Mar. Biol.* **2012**, *159*, 1659–1672. [CrossRef]
24. Copp, G.H.; Kováč, V. Sympatry between threespine *Gasterosteus aculeatus* and ninespine *Pungitius pungitius* sticklebacks in English lowland streams. In *Annales Zoologici Fennici*; Finnish Zoological and Botanical Publishing Board: Helsinki, Finland, 2003; Volume 40, pp. 341–355.
25. Raeymaekers, J.A.M.; Huyse, T.; Maelfait, H.; Hellemans, B.; Volckaert, F.A.M. Community structure, population structure and topographical specialisation of *Gyrodactylus* (monogenea) ectoparasites living on sympatric stickleback species. *Folia Parasitol.* **2008**, *55*, 187–196. [CrossRef] [PubMed]
26. Copp, G.H.; Edmonds-Brown, V.R.; Cottey, R. Behavioural interactions and microhabitat use of stream-dwelling sticklebacks *Gasterosteus aculateus* and *Pungitius pungitius* in the laboratory and field. *Folia Zool. Praha* **1998**, *47*, 275–286.
27. Lenz, T.L.; Eizaguirre, C.; Kalbe, M.; Milinski, M. Evaluating patterns of convergent evolution and trans-species polymorphism at MHC immunogenes in two sympatric stickleback species: Mhc evolution in two sympatric stickleback species. *Evolution* **2013**, *67*, 2400–2412. [CrossRef]
28. Guo, B.; Chain, F.J.J.; Bornberg-Bauer, E.; Leder, E.H.; Merilä, J. Genomic divergence between nine- and three-spined sticklebacks. *BMC Genom.* **2013**, *14*, 756. [CrossRef]
29. Varadharajan, S.; Rastas, P.; Löytynoja, A.; Matschiner, M.; Calboli, F.C.F.; Guo, B.; Nederbragt, A.J.; Jakobsen, K.S.; Merilä, J. A high-quality assembly of the nine-Spined stickleback (*Pungitius pungitius*) Genome. *Genome Biol. Evol.* **2019**, *11*, 3291–3308. [CrossRef]
30. Fang, B.; Merilä, J.; Matschiner, M.; Momigliano, P. Estimating uncertainty in divergence times among three-spined stickleback clades using the multispecies coalescent. *Mol. Phylogenet. Evol.* **2020**, *142*, 106646. [CrossRef] [PubMed]
31. Mäkinen, H.S.; Merilä, J. Mitochondrial DNA phylogeography of the three-spined stickleback (*Gasterosteus aculeatus*) in Europe—Evidence for multiple glacial refugia. *Mol. Phylogenet. Evol.* **2008**, *46*, 167–182. [CrossRef]
32. Wang, C.; Shikano, T.; Persat, H.; Merilä, J. Mitochondrial phylogeography and cryptic divergence in the stickleback genus *Pungitius*. *J. Biogeogr.* **2015**, *42*, 2334–2348. [CrossRef]
33. Raeymaekers, J.A.M.; Konijnendijk, N.; Larmuseau, M.H.D.; Hellemans, B.; De Meester, L.; Volckaert, F.A.M. A gene with major phenotypic effects as a target for selection vs. homogenizing gene flow. *Mol. Ecol.* **2014**, *23*, 162–181. [CrossRef]
34. Rohlf, F.J. The tps series of software. In *Hystrix*; Academia: San Francisco, CA, USA, 2015; Volume 26.
35. Sharpe, D.M.T.; Räsänen, K.; Berner, D.; Hendry, A.P. Genetic and environmental contributions to the morphology of lake and stream stickleback: Implications for gene flow and reproductive isolation. *Evol. Ecol. Res.* **2008**, *10*, 849–866.
36. Elshire, R.J.; Glaubitz, J.C.; Sun, Q.; Poland, J.A.; Kawamoto, K.; Buckler, E.S.; Mitchell, S.E. A robust, simple genotyping-by-sequencing (GBS) approach for high diversity species. *PLoS ONE* **2011**, *6*, e19379. [CrossRef]
37. Glaubitz, J.C.; Casstevens, T.M.; Lu, F.; Harriman, J.; Elshire, R.J.; Sun, Q.; Buckler, E.S. TASSEL-GBS: A high capacity genotyping by sequencing analysis pipeline. *PLoS ONE* **2014**, *9*, e90346. [CrossRef] [PubMed]
38. Jones, F.C.; Grabherr, M.G.; Chan, Y.F.; Russell, P.; Mauceli, E.; Johnson, J.; Swofford, R.; Pirun, M.; Zody, M.C.; White, S.; et al. The genomic basis of adaptive evolution in threespine sticklebacks. *Nature* **2012**, *484*, 55–61. [CrossRef] [PubMed]
39. Langmead, B.; Salzberg, S.L. Fast gapped-read alignment with Bowtie 2. *Nat. Methods* **2012**, *9*, 357–359. [CrossRef] [PubMed]
40. Danecek, P.; Auton, A.; Abecasis, G.; Albers, C.A.; Banks, E.; DePristo, M.A.; Handsaker, R.E.; Lunter, G.; Marth, G.T.; Sherry, S.T.; et al. The variant call format and VCFtools. *Bioinformatics* **2011**, *27*, 2156–2158. [CrossRef] [PubMed]
41. Lischer, H.E.L.; Excoffier, L. PGDSpider: An automated data conversion tool for connecting population genetics and genomics programs. *Bioinformatics* **2012**, *28*, 298–299. [CrossRef]
42. Raj, A.; Stephens, M.; Pritchard, J.K. fastSTRUCTURE: Variational inference of population structure in large SNP data sets. *Genetics* **2014**, *197*, 573–589. [CrossRef]
43. Li, Y.L.; Liu, J.X. StructureSelector: A web-based software to select and visualize the optimal number of clusters using multiple methods. *Mol. Ecol. Resour.* **2018**, *18*, 176–177. [CrossRef] [PubMed]
44. R Core Team. *R: A Language and Environment for Statistical Computing*; R foundation for statistical computing: Vienna, Austria, 2013.
45. Goudet, J. Hierfstat, a package for r to compute and test hierarchical F-statistics. *Mol. Ecol. Notes* **2005**, *5*, 184–186. [CrossRef]

46. Jombart, T. Adegenet: A R package for the multivariate analysis of genetic markers. *Bioinformatics* **2008**, *24*, 1403–1405. [CrossRef]
47. Mantel, N. The detection of disease clustering and a generalized regression approach. *Cancer Res.* **1967**, *27*, 209–220. [PubMed]
48. Leinonen, T.; Cano, J.M.; Mäkinen, H.; Merilä, J. Contrasting patterns of body shape and neutral genetic divergence in marine and lake populations of threespine sticklebacks. *J. Evol. Biol.* **2006**, *19*, 1803–1812. [CrossRef]
49. Raeymaekers, J.A.M.; Van Houdt, J.K.J.; Larmuseau, M.H.D.; Geldof, S.; Volckaert, F.A.M. Divergent selection as revealed by P_{ST} and QTL-based P_{ST} in three-spined stickleback (*Gasterosteus aculeatus*) populations along a coastal-inland gradient. *Mol. Ecol.* **2007**, *16*, 891–905. [CrossRef]
50. Lunn, D.J.; Thomas, A.; Best, N.; Spiegelhalter, D. WinBUGS – a Bayesian modelling framework: Concepts, structure and extensibility. *Statistics Comput.* **2000**, *10*, 325–337. [CrossRef]
51. Foll, M.; Gaggiotti, O. A genome-scan method to identify selected loci appropriate for both dominant and codominant markers: A Bayesian perspective. *Genetics* **2008**, *180*, 977–993. [CrossRef]
52. Refoyo-Martínez, A.; da Fonseca, R.R.; Halldórsdóttir, K.; Árnason, E.; Mailund, T.; Racimo, F. Identifying loci under positive selection in complex population histories. *Genome Res.* **2019**, *29*, 1506–1520. [CrossRef]
53. Wang, Y.; Zhao, Y.; Wang, Y.; Li, Z.; Guo, B.; Merilä, J. Population transcriptomics reveals weak parallel genetic basis in repeated marine and freshwater divergence in nine-spined sticklebacks. *Mol. Ecol.* **2020**, *29*, 1642–1656. [CrossRef]
54. Kemppainen, P.; Li, Z.; Rastas, P.; Löytynoja, A.; Fang, B.; Guo, B.; Shikano, T.; Yang, J.; Merilä, J. Genetic population structure constrains local adaptation and probability of parallel evolution in sticklebacks. *bioRxiv* **2020**. [CrossRef]
55. Shapiro, M.D.; Summers, B.R.; Balabhadra, S.; Aldenhoven, J.T.; Miller, A.L.; Cunningham, C.B.; Bell, M.A.; Kingsley, D.M. The genetic architecture of skeletal convergence and sex determination in ninespine sticklebacks. *Curr. Biol.* **2009**, *19*, 1140–1145. [CrossRef]
56. Yeaman, S.; Whitlock, M.C. The genetic architecture of adaptation under migration–selection balance. *Evolution* **2011**, *656*, 1897–1911. [CrossRef]
57. Fang, B.; Kemppainen, P.; Momigliano, P.; Merilä, J. Population structure limits parallel evolution. *bioRxiv* **2021**. [CrossRef]
58. Lowe, W.H.; Kovach, R.P.; Allendorf, F.W. Population Genetics and Demography Unite Ecology and Evolution. *Trends Ecol. Evol.* **2017**, *32*, 141–152. [CrossRef]
59. Van der Molen, J.; De Swart, H.E. Holocene tidal conditions and tide-induced sand transport in the southern North Sea. *J. Geophys. Res. C Oceans* **2001**, *106*, 9339–9362. [CrossRef]
60. Teacher, A.G.F.; Shikano, T.; Karjalainen, M.E.; Merilä, J. Phylogeography and genetic structuring of european nine-spined sticklebacks (*Pungitius pungitius*)—mitochondrial DNA evidence. *PLoS ONE* **2011**, *6*, e19476. [CrossRef]
61. Bell, M.A.; Andrews, C.A. Evolutionary consequences of postglacial colonization of fresh water by primitively anadromous fishes. In *Evolutionary Ecology of Freshwater Animals: Concepts and Case Studies*; Streit, B., Städler, T., Lively, C.M., Eds.; Birkhäuser Basel: Basel, Switzerland, 1997; pp. 323–363. ISBN 9783034888806.
62. Raeymaekers, J.A.M.; Maes, G.E.; Audenaert, E.; Volckaert, F.A.M. Detecting Holocene divergence in the anadromous–freshwater three-spined stickleback (*Gasterosteus aculeatus*) system. *Mol. Ecol.* **2005**, *14*, 1001–1014. [CrossRef] [PubMed]
63. Booker, T.R.; Yeaman, S.; Whitlock, M.C. Variation in recombination rate affects detection of outliers in genome scans under neutrality. *Mol. Ecol.* **2020**, *29*, 4274–4279. [CrossRef] [PubMed]
64. Lewis, D.B.; Walkey, M.; Dartnall, H.J.G. Some effects of low oxygen tensions on the distribution of the three-spined stickleback *Gasterosteus aculeatus* L. and the nine-spined stickleback *Pungitius pungitius* (L.). *J. Fish Biol.* **1972**, *4*, 103–108. [CrossRef]
65. Kovac, V.; Copp, G.H.; Dimart, Y.; Uzikova, M. Comparative morphology of threespine *Gasterosteus aculeatus* and ninespine *Pungitius pungitius* sticklebacks in lowland streams of southeastern England. *Folia Zool. Praha* **2002**, *51*, 319–336.

Article

Congruent Genetic and Demographic Dispersal Rates in a Natural Metapopulation at Equilibrium

Delphine Legrand [1,*], Michel Baguette [1,2], Jérôme G. Prunier [1], Quentin Dubois [3], Camille Turlure [3] and Nicolas Schtickzelle [3]

[1] Theoretical and Experimental Ecology Station (UMR 5371), National Centre for Scientific Research (CNRS), Paul Sabatier University (UPS), 09200 Moulis, France; michel.baguette@mnhn.fr (M.B.); jerome.prunier@gmail.com (J.G.P.)
[2] Institut Systématique, Evolution, Biodiversité (ISYEB), UMR 7205 Museum d'Histoire Naturelle, CNRS, Sorbonne Université, EPHE, Université des Antilles, F-75005 Paris, France
[3] Biodiversity Research Centre, Earth and Life Institute, Université Catholique de Louvain, 1348 Louvain-la-Neuve, Belgium; quent.dubois@gmail.com (Q.D.); turlure_camille@hotmail.com (C.T.); nicolas.schtickzelle@uclouvain.be (N.S.)
* Correspondence: delphine.legrand@sete.cnrs.fr

Abstract: Understanding the functioning of natural metapopulations at relevant spatial and temporal scales is necessary to accurately feed both theoretical eco-evolutionary models and conservation plans. One key metric to describe the dynamics of metapopulations is dispersal rate. It can be estimated with either direct field estimates of individual movements or with indirect molecular methods, but the two approaches do not necessarily match. We present a field study in a large natural metapopulation of the butterfly *Boloria eunomia* in Belgium surveyed over three generations using synchronized demographic and genetic datasets with the aim to characterize its genetic structure, its dispersal dynamics, and its demographic stability. By comparing the census and effective population sizes, and the estimates of dispersal rates, we found evidence of stability at several levels: constant inter-generational ranking of population sizes without drastic historical changes, stable genetic structure and geographically-influenced dispersal movements. Interestingly, contemporary dispersal estimates matched between direct field and indirect genetic assessments. We discuss the eco-evolutionary mechanisms that could explain the described stability of the metapopulation, and suggest that destabilizing agents like inter-generational fluctuations in population sizes could be controlled by a long adaptive history of the species to its dynamic local environment. We finally propose methodological avenues to further improve the match between demographic and genetic estimates of dispersal.

Keywords: butterfly metapopulation; dispersal; genetic structure; demography; spatio-temporal stability; environmental fluctuations; *Boloria eunomia*

Citation: Legrand, D.; Baguette, M.; Prunier, J.G.; Dubois, Q.; Turlure, C.; Schtickzelle, N. Congruent Genetic and Demographic Dispersal Rates in a Natural Metapopulation at Equilibrium. *Genes* **2021**, *12*, 362. https://doi.org/10.3390/genes12030362

Academic Editors: Didier Jollivet and Simon Blanchet

Received: 5 January 2021
Accepted: 24 February 2021
Published: 3 March 2021

Publisher's Note: MDPI stays neutral with regard to jurisdictional claims in published maps and institutional affiliations.

1. Introduction

The metapopulation concept provides an operational framework for both (evolutionary) ecologists and conservation managers [1]. Classically defined as a set of interacting populations for which frequent local extinctions are balanced by recolonization [2,3], a metapopulation can also broadly refer to patchy populations [4], that is, to any set of local populations potentially related by movements of individuals in a landscape [5]. While retaining the fundamental aspect of the classical metapopulation concept, i.e., a biological structure linking local and regional-scale processes, the latter definition reflects a wider variety of biological situations (e.g., [6,7]). Classical metapopulations indeed correspond to a narrow range of ecological parameters (i.e., patch occupancy, turnover, etc., [8]) and many metapopulations do not experience local population extinctions at each generation [7]. In any case however, describing the long-term functioning of natural metapopulations, a necessary step to accurately predict how they would respond to contemporary environmental changes, relies on an accurate knowledge of the dispersal process.

Dispersal, individual movements potentially leading to gene flow [9,10], is the eco-evolutionary process creating biological links within metapopulations [11], and thereby within ecological networks [12]. It affects local adaptation (either positively or negatively, [13,14]) and spatial synchrony among populations [15], buffers the risk of local extinctions through (genetic) rescue effects [16] and allows recolonization of vacant patches [4]. As such, dispersal is central for metapopulation stability [17]. Any biotic or abiotic change occurring at the local or regional scale and affecting dispersal (e.g., landscape modification, decrease in individuals' movement ability) may result in the collapse of the whole metapopulation. Both the frequent short-distance and the rarest long-distance dispersal movements may have a significant impact on metapopulations' equilibrium [18]. Dispersal rate is thus a fundamental metric of metapopulation functioning, which needs to be measured at accurate spatial and temporal scales in nature [19].

Dispersal rate is the proportion of individuals that move from a population to another. It can be estimated on the field through two approaches relying on distinct theoretical frameworks: demography and genetics. Demographic estimates of dispersal rates are obtained either through the direct monitoring of individual movements using visual observation or telemetric systems depending on landscapes and taxa [20–22], or through Capture-Mark-Recapture (CMR) approaches [23]. From such movement data a dispersal kernel can be computed, i.e., the probability density function that dispersing individuals move a certain distance. The quality of dispersal kernels, a key element to describe and predict the dynamics of a metapopulation, is strongly dependent on the quality of the raw movement data, but there are a number of limitations with these direct measurements of dispersal rates. First, they often require laborious and costly field sessions [24,25]. Second, there is a recurrent bias toward the recording of short-distance movements to the detriment of long-distance movements, meaning that dispersal kernels might only model processes occurring at small spatial scales (see [26] for an example of a mathematical correction of biased dispersal kernels). Third, dispersal is context-dependent, which means that "the existence of a species-specific dispersal function is probably a myth" [27]. This poses limits to the transfer of dispersal kernels, even between metapopulations of the same species (e.g., [28]). Finally, whatever the precision of the recording of dispersal movements and hence of dispersal kernels, direct measures of dispersal do not provide information on how much the movements contribute to the gene pool at the next generations. However, reliable estimates of effective dispersal are key to understand the evolutionary dynamics and long-term stability of metapopulations.

The genetic approach indirectly estimates effective dispersal rates through its impact on the genetic structure at the metapopulation level, determined from allelic composition and frequencies in the local populations (from microsatellites, Amplified Fragment Length Polymorphisms AFLPs, Single-Nucleotid Polymorphisms SNPs, etc.). Most often, the genetic differentiation index F_{ST} is used as a proxy for historical effective number of dispersers (often called "migrants" in the genetic literature) between pairs of populations [29]. A series of unrealistic assumptions, including symmetrical fluxes and equal population sizes, are however required to apply Wright's island model [30]. Refinements have been proposed (e.g., [31]), and coalescent-based methods have notably been developed to directly infer asymmetrical effective dispersal rates (see reviews in, e.g., [12,32,33]). The choice between these different methods often results from a balance between accepting some violation of model assumptions and analysis complexity (determining the time required to run the analysis). Since genetic estimates of dispersal rates might not reflect contemporary gene flow, especially because of the time lag between the contemporary processes affecting dispersal and the actual setting-up of genetic differentiation among populations [34], methods such as Bayesian assignment (e.g., [35,36]) or parentage analyses [37] have been developed to identify dispersers among individual genotypes. As for direct observation of movements, these latter methods inform on the origin and the target populations of each detected disperser. Nonetheless, as for demography, estimates of effective dispersal present some limitations. For instance, and as discussed above, natural situations rarely fulfill all genetic models' assumptions. Furthermore, indirect measures of dispersal do not inform on the phenotypic traits of dispersers. It thus has recurrently been proposed to combine, when possible, demographic and genetic approaches to understand metapopulation dynamics and estimate their stability over time (e.g., [12,25,32,33,38]). Such integrative studies remain however rare, and there is a general call for empiricists to compare dispersal estimates with different methodologies at appropriate spatio-temporal scales ([12,32], see [25,39] for examples).

Here, we present a field study in which we analyzed the population structure, the dynamics, and the stability of a natural butterfly metapopulation in the Belgian Ardenne combining demographic and genetic datasets obtained at a large spatial scale (~200 km^2) over three generations (three years). The bog fritillary *Boloria eunomia* has long been used as a model species in the metapopulation literature because of its patchy distribution and sensitivity to habitat fragmentation. Over the last three decades, important knowledge has accumulated regarding *B. eunomia* habitat use [40], dispersal behaviour and demography [41–44], metapopulation functioning [45–48]), and genetic structure at local, regional and continental scales [49–52]. However, none of these studies have synthesized both demographic and genetic data within the same large network of populations over several generations. Our objectives were thus to:

(i)   determine the genetic structure within a *B. eunomia* metapopulation based on genetic material collected on more than 1000 individuals over three generations across nine local populations;
(ii) describe the dynamics of the whole-metapopulation through a thorough comparison between genetic and demographic estimates of dispersal rates;
(iii) evaluate the long-term demographic stability of the metapopulation.

Despite important inter-annual fluctuations in census population sizes, local population extinctions were very rarely observed over two decades of field sampling in this metapopulation [45]. We thus predicted that the metapopulation should harbour high degree of genetic stability, both in terms of genetic structure and long-term effective population sizes. Predictions are more uncertain regarding the congruence between direct and indirect dispersal estimates. We predicted that the correlation between genetic and demographic estimates of dispersal should be strong only provided that: (i) the assumptions of demographic and genetic models are not largely violated within the studied *B. eunomia* metapopulation, (ii) long-distance dispersal movements, particularly difficult to record in the field, are not too frequent; (iii) a relatively high proportion of the dispersal movements are effective in terms of gene flow among populations.

2. Materials and Methods

2.1. Model Species and Study Area

B. eunomia (formerly known as *Proclossiana eunomia*) is a Holarctic butterfly species occurring in wet meadows and some peat bogs in middle Europe. In this area, the species is a specialist, strictly associated with the bistort *Polygonum bistorta* (Figure 1), its host plant at larval stage and the only source of nectar for adults in the study area. The species is univoltine: it spends the winter as a caterpillar and adults fly around for about one month from late-May to early-July. In Belgium, *B. eunomia* is protected because it suffers, as in other European countries, from the transformation of its habitat (bistort meadows) into improved pastures or spruce *Picea abies* plantations (e.g., [45]). We collected samples in nine local populations in a ~200 km^2 landscape in the Belgian Ardenne (Figure 1) in 2009, 2010 and 2011 (three successive generations) using both a Capture-Mark-Recapture approach and a genetic sampling approach, described in more details below. The Euclidian distances between the centroid of each site are available in Table S1.

Figure 1. Model species and study area. The studied metapopulation is situated in the Ardenne region, southeastern Belgium (up-left map); the location of the nine local populations are represented in red on an aerial view of the region. On the left of the map, we provide a picture of one marked *Boloria eunomia* butterfly on an inflorescence of *P. bistorta*, its host plant.

2.2. Demographic Modelling: Population Size, Survival and Dispersal from CMR Campaigns

For the three successive years of sampling, each site was visited daily during the flight period (excluding rainy days), and Capture-Mark-Recapture (CMR) sessions performed with a sampling effort normalized according to habitat patch area. We captured butterflies with a net, sexed and marked each of them with a unique identifier using a thin pen on the left hindwing at the first capture (Figure 1). We marked 5481 individuals in total. These CMR data were first used to compute the number of dispersers between each pair of populations over the three years. CMR data were then analysed using Jolly–Seber models, as implemented in the POPAN analysis in MARK software [53]. Based on capture histories of the different individuals recorded in a population, the probability of an individual to be (re)captured, a measure of detectability, is estimated (data per population are given in Table S2), and subsequently used to correct estimates of survival, birth rates, daily and total (seasonal) census population sizes (N) [54]. For details of the analytical method, see [45]. To test for the congruence between the ranking of the nine population sizes across the three years (a proxy of global demographic stability), we performed a Kruskal-Wallis rank test between 2009–2010, 2010–2011 and 2009–2011. We finally compiled all recapture data to count the number of dispersers between each pair of populations over the three years.

2.3. Molecular Markers, Genetic Structure and Effective Dispersal

2.3.1. Laboratory Work

Among the 5481 individuals marked during CMR, one leg was preserved from 1217 of them in absolute ethanol and total genomic DNA later extracted using the column version of the DNeasy Blood and Tissue kit (Qiagen, Venlo, The Netherlands). From these 1217 DNA extractions, we multiplexed and amplified by PCR 12 microsatellite loci as described before [55]. Tests for linkage disequilibrium and the presence of null alleles were performed in [55] and showed the presence of a few null alleles that did not significantly impacted population genetic indices. The genotyping was performed on an ABI3730 sequencer (Applied Biosystems, GeT GenoToul platform, Toulouse, France). Fragment sizes for each locus were determined with the GeneMapper software.

2.3.2. Population Structure

Genetic differentiation between all pairs of populations was estimated with Wright's pairwise fixation index F_{ST} [56] using Arlequin v.3.5.2.2 [57] for each generation separately. Significance of F_{ST} values was determined based on 10,000 random permutations, and *p*-values were adjusted with the Benjamini-Yekutieli correction [58]. Genetic clustering of individuals was assessed using the Bayesian clustering method Structure v.2.3.4 [35] including all generations in a full analysis. We used the admixture model, which allows mixed ancestries of individuals, and the correlated allele frequency model (*F* model), which assumes that allele frequencies in different populations are likely to be similar. Five independent runs for each value of *K* (the number of clusters) ranging from one to nine were performed, using 500,000 iterations and a burn-in period of 50,000 steps. To detect the number of clusters that best fit the data, we estimated the rate of change in the log probability of the data between successive *K* values and the corresponding variance of log probabilities [59]. We completed this analysis with a global Analysis of Molecular Variance (AMOVA) performed per generation using Arlequin to determine if the genetic variance was effectively significantly structured by the inter-cluster differences detected with Structure. The *p*-values were calculated using 10,000 permutations. Despite the existence of two significant major genetic clusters, a large part of the genetic variance remained unexplained (see Results). We therefore performed supplementary clustering runs within each first-order cluster to search for significant sub-structuring, and continued the process across hierarchical levels until no more structure was detected [60]. Significant sub-genetic structure was detected with Bayesian assignation, and F_{ST} methods revealed that genetic structuring was significant at the level of our nine predefined populations (see Results). We thus continued with the genetic analyses for the nine populations separately, thus matching the partition used in the demographic estimates described above.

2.3.3. Isolation by Distance

To test for Isolation By Distance (IBD), that is the effect of Euclidian distance between populations on gene flow, a Mantel test regressing $F_{ST}/(1-F_{ST})$ against the natural logarithm of geographic distance between all pairs of samples is generally performed [61]. However, since measures of F_{ST} stem from the balance between gene flow on the one hand and genetic drift on the other hand, F_{ST} estimates cannot be considered a proper proxy for gene flow, especially when population sizes are unequal [30]. We thus corrected F_{ST} values using the method developed in [31]. For each pair of populations, we first computed the *di* metric of Spatial Heterogeneity in Effective population sizes (*SHNe*) as the sum of the inverse of population census sizes *N* inferred from demographic data. For each generation separately, we then computed the residuals of the linear regression between $F_{ST}/(1-F_{ST})$ and *di* across all pairs of populations, and used a simple Mantel test with 10,000 permutations to assess significance of the relationship between these residuals and the natural logarithm of geographic distances.

2.3.4. Allelic Diversity and Effective Population Sizes

For the three generations and each of the nine populations, the number of alleles (*Na*), the expected (*He*) and observed (*Ho*) heterozygosities, and the F_{IS} were calculated using Arlequin and averaged over all loci. The number of private alleles (*Np*) was estimated using Convert v.1.31 [62]. The Allelic Richness (*A*) and Private Allelic Richness (*Ap*) based on the minimum sample size were estimated by the rarefaction method implemented in HP-Rare v.1.1. We estimated contemporary effective population sizes (*Ne*) for each generation using the bias-corrected version of the linkage disequilibrium method described in [63] implemented in NeEstimator v.2 [64]. With a sufficient number of microsatellites, sample sizes generally higher than 25 individuals as well as non-overlapping generations, we fulfilled most of the prerequisites of the method. We set the critical threshold value of rare alleles to 0.02 and used the jackknife method to estimate confidence intervals. The significance of correlations between *Ne* (effective population size) and *N* (census population size) estimates for the nine local populations was assessed using a Spearman correlation test for each year separately.

2.3.5. Effective Dispersal and Statistical Comparison with Demographic Dispersal

We determined contemporary dispersal by estimating the number of first-generation dispersers between populations using Geneclass2 [36]. Using Monte Carlo resampling algorithms, the program computes the probability for each individual to be a resident in the population where it was sampled or to descend from a disperser at the preceding generation. We used the Bayesian criterion described in [65] and detection was done using the ratio of the likelihood computed from the population where the individual was sampled over the highest likelihood value among all sampled populations [66]. The probability for an individual to be a resident was computed using the resampling algorithm of [66] with 10,000 simulated individuals and a level of Type I error set to 0.01 as advised by these authors. To assess the relationship between the demographic and the genetic estimates of contemporary dispersal rates, we used a simple Mantel test with 10,000 permutations based on the Spearman's correlation coefficient between the pairwise numbers of first-generation dispersers inferred from Geneclass2 (genetic matrix) and the pairwise numbers of dispersers inferred from CMR data over the three years (demographic matrix). Significance was assessed by permuting 10,000 times the names of the nine populations in one of the two matrices.

2.3.6. Past Demographic Events

We finally searched for the existence of past demographic events, i.e., bottlenecks and exponential changes in population sizes. We used Bottleneck v.1.2.02 [67] to calculate, for each population and each locus, the distribution of the expected gene diversity from the observed number of alleles given the sample size and assuming mutation-drift equilibrium. In neutral conditions, the number of loci showing *He* excess should be equal to the number of loci showing *He* deficit. Conversely, *He* excess across loci is expected after a bottleneck whereas *He* deficiency is expected after exponential change in population size. Expected *He* under mutation-drift equilibrium was determined by coalescent simulations under the stepwise mutation model (SMM), and the two-phased Mutation model (TPM), with more than one-repeat mutations occurring at frequencies of either 10 and 20%. Wilcoxon sign-rank tests were used to determine significant departures from null distributions, with 10,000 iterations, and they were adjusted with the Benjamini-Yekutieli correction [58]. As a qualitative complement to this analysis, we also used the Bottleneck software to examine the shape of the allele-frequency distribution. L-shaped distributions are expected under mutation-drift equilibrium while mode-shift distributions are expected in cases of bottlenecks [68].

3. Results

3.1. Demography from CMR Data

From the 5481 individuals marked during the three years of CMR campaigns, we estimated the total local census population sizes (N) of each of the nine populations separately (Table 1). We provide details on male and female population sizes, catchability and survival in Table S2. The Pisserotte and Prés de la Lienne populations were the largest populations; Bérisménil, Bièvres, Bihain and Grand Fange populations had intermediate population sizes; and Chapons, Mormont and Langlire had lower abundances. The ranking of population sizes across years was congruent (no rejection of the null hypothesis, Kruskal-Wallis test $\chi^2 = 8$, df = 8, p-value = 0.433 for each pair of years comparison). Regarding individual dispersal movements, we recorded a total of 49 inter-population movements over the three years, of which only one was considered as a long-distance movement, i.e., more than 5 km (dashed arrow in Figure 2A).

Table 1. Estimates of population sizes for each year of sampling. *N* = census population size recorded from Capture-Mark-Recapture data, *Ne* = effective population size inferred from microsatellite data analyzed with NeEstimator with their confidence intervals (*CI Ne*).

	2009			2010			2011		
Population	*N*	*Ne*	*CI Ne*	*N*	*Ne*	*CI Ne*	*N*	*Ne*	*CI Ne*
Bérismenil	196	58.6	27.4–410.1	191	35	23.6–54.5	235	102	39.8–∞
Chapons	14	NA	NA	13	1.9	1.1–6.1	8	14.6	1.7–∞
Mormont	25	4.7	1.7–32.2	30	9.9	3.9–24.6	56	12.8	4.4–59.1
Bièvres	195	50.5	19.2–∞	259	44.8	26.4–92.5	304	46	26.9–109.9
Bihain	136	22.4	12.3–54.3	419	64.4	40.6–118.5	289	111	52.4–990.9
Grande Fange	70	17	3–∞	166	55.8	26.8–240.2	336	113	34–∞
Langlire	57	91.2	12.5–∞	101	56.5	23–∞	99	95.3	23.8–∞
Pisserotte	978	34.3	14.2–522.8	1428	142.6	72.7–657.9	1139	NA	84.9–∞
Prés de la Lienne	291	46.4	26.8–105.1	940	57.7	24.6–∞	1131	122	75.4–241.1

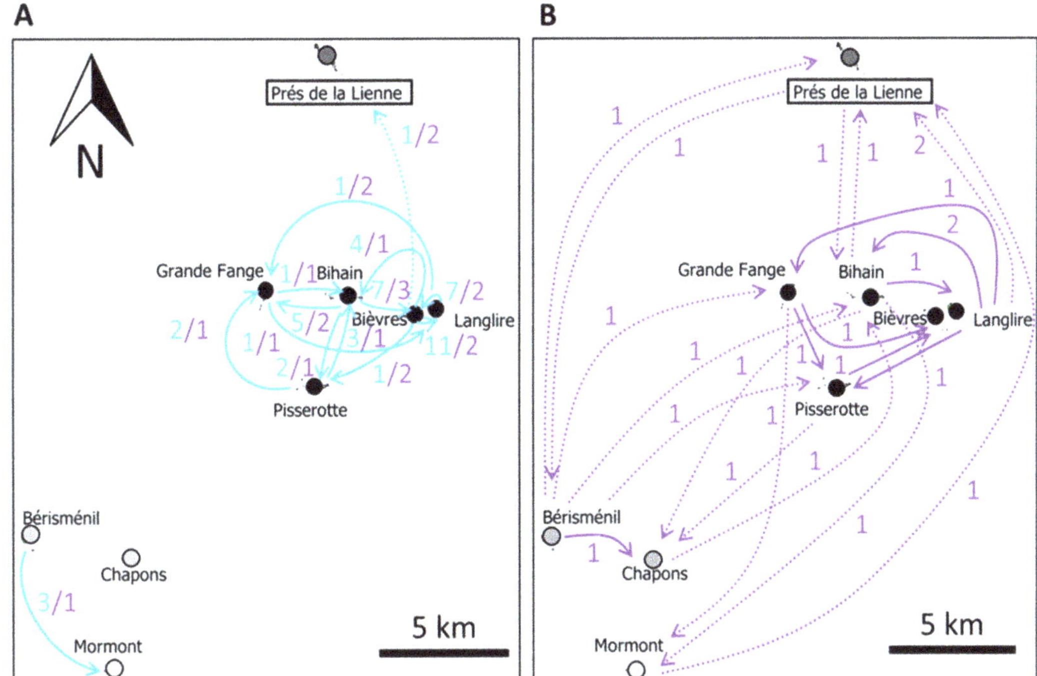

Figure 2. Contemporary dispersal movements from demographic (blue arrows) and genetic (purple arrows) data cumulated over the three sampling years. (**A**) movements directly recorded in the field via Capture-Mark-Recapture (blue), complemented by the number of first-generation effective movements inferred from genetic data (purple) for the same pairs of populations. (**B**) All the other movements inferred from genetic data for other pairs of patches, separated to gain in map clarity. Long-distance movements are symbolized by dashed arrows.

3.2. Genetic Structure, Isolation by Distance and Diversity

Using the DNA extracted from the legs of 1217 butterflies, the first step of Bayesian clustering revealed the existence of two major genetic clusters repeatedly found across years of sampling (Figure 3, Figure S1A). Pisserotte and Grande Fange were the two populations with the highest mixed ancestry origin, but they were nonetheless well assigned to cluster 2. This was confirmed by the AMOVA analysis, which showed that a significant amount of genetic variance was explained by the among-clusters structure (Table S3). However, this represented only 3.1%, 1.7% and 3.6% of the total genetic variance for 2009, 2010 and 2011 respectively, most of it being explained by suborder structuring. We thus ran again Structure within cluster 1 and 2 separately, and detected sub-structuring distinguishing the Mormont and Prés de la Lienne populations (Figure S1B,C). By continuing the procedure, we detected sub-structuring that was congruent with the delineation of local populations (data not shown). We confirmed this result with F_{ST} analyses. We detected significant genetic differentiation for each pair of the nine populations sampled in the metapopulation in at least one out of the three years (Table S4). We detected a significant IBD pattern for the three years, with correlation coefficients between genetic distances and Euclidian distances of 0.502 ($p < 0.0001$), 0.536 ($p < 0.0001$) and 0.467 ($p = 0.049$) for 2009, 2010 and 2011 respectively after *SHNe* correction (Figure S2). Usual genetic diversity indices averaged across the 12 loci are presented in Table 2. Populations had rather similar levels of genetic diversity, except the Mormont and Chapons populations, which were less diversified. Overall, populations had a recurrent deficit in heterozygotes across years as indicated by significant positive F_{IS} values.

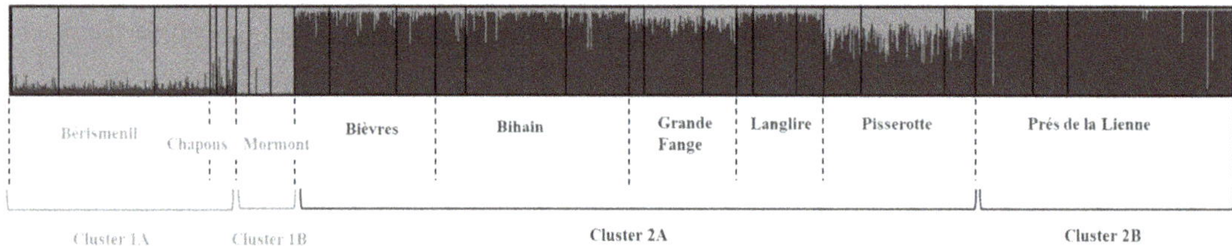

Figure 3. Genetic clustering across the whole metapopulation of *Boloria eunomia* using the three years of sampling. Bayesian clustering shows two major clusters separating Bérismenil, Chapons and Mormont (light grey) from the other populations (dark grey). Each bar represents the membership assignment to the two clusters of each individual. Within each population, vertical black traits separate the three years of sampling in the same order (2009, 2010 and 2011). A second run of analysis in the two clusters reveals the sub-structuring as indicated at the bottom of the figure.

Table 2. Genetic diversity. n = genetic sample size, Na = number of alleles, Np = number of private alleles, A = Allelic richness, Ap = Private allelic richness, Ho = observed heterozygosity, He = expected heterozygosity, significant (< 0.05) p-values for F_{IS} values are in bold in the last column.

Population	Year	n	Na	Np	A	Ap	Ho	He	F_{IS}	p-Value
	2009	47	5.3	2	3.54	0.18	0.44	0.64	0.17	**0.0003**
Bersimenil	2010	97	5.1	0	3.69	0.16	0.47	0.61	0.08	**0.002**
	2011	56	5.2	0	3.64	0.11	0.52	0.67	0.09	**0.004**
	2009	6	2.9	0	2.5	0.09	0.6	0.61	−0.15	0.6
Chapons	2010	11	3	0	2.69	0.05	0.41	0.45	−0.01	0.59
	2011	9	3.5	1	3.13	0.14	0.43	0.61	0.24	**0.01**
	2009	13	3	2	2.74	0.27	0.46	0.57	0.008	0.51
Mormont	2010	22	3.6	0	2.94	0.18	0.34	0.58	0.26	**<0.0001**
	2011	24	3.6	0	2.84	0.17	0.39	0.55	0.16	**0.006**
	2009	36	5.6	1	3.74	0.12	0.43	0.68	0.18	**<0.0001**
Bièvres	2010	67	6.4	2	3.91	0.15	0.48	0.68	0.15	**<0.0001**
	2011	38	6.1	2	3.8	0.15	0.48	0.69	0.13	**0.0003**
	2009	31	5.6	2	3.79	0.21	0.46	0.69	0.18	**<0.0001**
Bihain	2010	99	6.7	2	3.87	0.18	0.47	0.68	0.13	**<0.0001**
	2011	63	6.4	3	3.81	0.18	0.48	0.7	0.15	**<0.0001**

Table 2. *Cont.*

Population	Year	*n*	*Na*	*Np*	*A*	*Ap*	*Ho*	*He*	F_{IS}	*p*-Value
Grande Fange	2009	15	4.4	4	3.52	0.4	0.49	0.64	0.13	0.47
	2010	57	6.5	5	4.06	0.21	0.48	0.69	0.17	**<0.0001**
	2011	34	5.8	4	3.76	0.21	0.47	0.68	0.12	**0.002**
Langlire	2009	17	5.1	2	3.69	0.17	0.45	0.68	0.19	**0.004**
	2010	42	5.4	2	3.99	0.09	0.46	0.68	0.19	**<0.0001**
	2011	28	5.6	0	3.75	0.09	0.41	0.68	0.18	**<0.0001**
Pisserotte	2009	36	5.3	2	3.56	0.06	0.47	0.67	0.09	**0.03**
	2010	83	6	0	3.94	0.15	0.49	0.69	0.13	**<0.0001**
	2011	31	5.3	0	3.74	0.04	0.49	0.69	0.15	**0.0008**
Prés Lienne	2009	57	5.1	2	3.31	0.12	0.42	0.6	0.14	**<0.0001**
	2010	35	4.6	1	3.27	0.11	0.41	0.62	0.15	**0.0003**
	2011	163	5.4	3	3.3	0.08	0.41	0.6	0.23	**<0.0001**

3.3. Comparison between Estimates of Population Sizes and Dispersal

We estimated contemporary effective population sizes (Table 1) using NeEstimator and found that they were significantly correlated with estimates of census population sizes for the three years (Spearman's *rho* = 0.28 in 2009, *p* = 0.050; 0.83 in 2010, *p* = 0.008; and 0.83 in 2011, *p* = 0.015). Over all years and populations, the *Ne/N* ratio was 0.38. At this contemporary timescale, we also detected significant correlation between the number of effective dispersal movements at the preceding generation (we detected 46 movements with GenClass2) and the number of movements observed by CMR, when pooling the three years (Spearman's *rho* = 0.547, *p* < 0.0001, Figure 2A,B). Despite this significant correlation, it is noteworthy that genetic data revealed a number of long-distance dispersal movements that were not captured by CMR data (Figure 2B).

3.4. Past Demographic Events

In a last analysis, we searched for genetic footprints of past demographic events. By comparing the number of microsatellite loci presenting deficit or excess in heterozygotes to the expected number under mutation-drift equilibrium, we were unable to detect any demographic event after correction for multiple testing (Table S5). The Chapons population was excluded from the analysis because of a too low number of individuals to obtain reliable outputs. This result was confirmed by the qualitative exploration of the shape of the allele-frequency distribution, which was normal for every year in every population, except for Grande Fange in 2009 (Table S5). This sample presented a shifted distribution toward too many loci with heterozygote deficiency, revealing a potential population expansion.

4. Discussion

Understanding the dynamics of natural metapopulations at relevant spatial and temporal scales is of prime importance to accurately feed both theoretical eco-evolutionary models and conservation plans [69]. Adding new demographic and genetic data to a well-studied European metapopulation of *B. eunomia*, we were able to highlight the general stability of the genetic structure and effective sizes of its local populations. Building on a trans-generational sampling at both demographic and genetic levels, we highlighted congruence in the estimates of dispersal rates and local population sizes across years, suggesting that these two approaches can fruitfully be used to infer metapopulation functioning. This also suggests that, at least in some cases, ecological parameters can be used as reliable proxies of evolutionary parameters, and the other way round. Based on the comparison between demographic and genetic parameters, we discuss below the ecological modalities that underlie this metapopulation equilibrium (two first sections). We also emphasize crucial methodological limits to our study (third section), which could affect the congruence between demographic and genetic parameters, notably due to the lack of integration of the effect of landscape structure on dispersal.

4.1. Metapopulation Functioning: An Integrative Story in B. eunomia over Space and Time

We estimated several important metrics summarizing the structure and dynamics of a natural butterfly metapopulation aiming at evaluating its stability. First, the ranking of local population sizes as estimated from intensive CMR campaigns (i.e., census adult population sizes) was consistent across generations. This means that, despite large observed fluctuations in population size between years in many of the local populations [47,48], they all fluctuated with some synchrony overtime. Genetic estimates of population sizes (i.e., effective population sizes) confirmed this finding as they were correlated with census adult population sizes. Such congruent ranking of population sizes in a metapopulation network can be interpreted as a global control of demographic fluctuations, with putatively low temporal variation in the mean $Npop_i/Npop_j$ ratio. Comparing census and effective populations sizes, we found the mean Ne/N ratio to be 0.38 over years and populations. This is higher than the value usually observed in many systems (~0.1–0.2, [70]). We suggest that this could be due to features of the life history of *B. eunomia* in this region allowing a large proportion of the individuals to effectively reproduce, as opposed to cases where reproductive success is very high for a few individuals, or very low for the majority. In *B. eunomia*, males mate multiple times along their life (females only once), increasing their chance to reproduce at least once. Besides, the large availability of the host plant, combined with continuous egg laying in small batches over the whole female lifetime, increases reproductive success by creating a situation where mortality risks for eggs and larvae are spread over space and time. Although relatively high given the census population sizes, Ne were systematically far below the threshold of 1000 reproductive individuals ensuring high evolutionary potential to local populations [70]. Accordingly, we detected a recurrent heterozygosity deficit at the local population scale, most probably resulting from inbreeding depression. Such indicators would suggest the metapopulation to be at risk of future extinction. Searching for past demographic events from genetic data, we were yet unable to detect any drastic reduction in local population sizes in neither of the three generations, which could suggest a progressive erosion rather than an abrupt collapse in population sizes. Such scenario could explain our limited ability to detect recent changes in population sizes through molecular methods. Putting all these elements together, we suggest that the studied metapopulation remained stable over the last decades despite small local population sizes, with a low probability of undetected recent population size changes. Accordingly, stochastic local population extinctions have very rarely been observed in populations monitored for decades now, even when their local population sizes were recurrently small like Mormont.

Second, we observed a stable genetic structure of the metapopulation over the three years of sampling. Bayesian clustering (Structure analysis) revealed a first level of genetic partition separating the southern populations of Bérismenil, Mormont and Chapons from the six northern other populations. The Pisserotte and Grande Fange populations had the highest mixed ancestry among all populations, which agrees with their central positions in the network. However, Grande Fange had higher level of private allelic richness and F_{ST} values than Pisserotte. While these two populations probably exchange a lot with the other populations of the network, Grande Fange could be more isolated than Pisserotte, and could have particular local dynamics allowing private genetic diversity to be maintained. Grande Fange could have experienced a recent founding effect followed by an expansion (as suggested by its deficit in heterozygous loci), but could also have acted as a past local refugia [71] or function nowadays as a sink [72], the two hypotheses being non-exclusive. Such genetic particularities could reflect the existence of (dis)assortative mating or local selection. On the contrary, Pisserotte had the highest census and effective population sizes over years, which suggests it is a hub for immigrants coming almost equally from the two clusters. A second Structure run revealed significant sub-structuring, with on the one hand the distinction of Mormont, the southernmost population among the three composing cluster 1, and on the other hand the distinction of the Prés de la Lienne, the northernmost population among the six composing cluster 2. By continuing the process, we detected genetic partition at the level of the nine local populations, which was confirmed by the general significance of F_{ST} values. The IBD analysis confirmed the suspected impact of geographic distance on genetic differentiation: about 25% of the genetic variance between populations could be attributed to Euclidian distances. It is noteworthy that both genetic structure and IBD were highly consistent over the three generations, meaning that the above-described temporal stability in the ranking of population sizes goes along with equilibrium in genetic structure. Nevertheless, the time lag between demographic processes and the genetic response could have hindered our ability to detect recent changes in population structure, even with fast evolving markers such as microsatellites [34]. Future sampling could help bring this possible scenario to light.

Third, we measured contemporary (inter-annual pool of effective first-generation dispersers versus inter-annual pool of direct movements) estimates of dispersal rates. There was a significant correlation between the demographic and genetic approaches, showing that dispersal movements were more frequent between close populations. This pattern agrees with the significant IBD pattern. As expected, CMR campaigns detected less long-distance dispersal movements (more than 5 km) than the genetic approach [12,73]. We yet detected similar numbers of dispersal movements (49 through CMR versus 46 through microsatellite analysis), which suggests that the individual monitoring of *B. eunomia* dispersal movements overestimated the real number of effective short-distance movements (Figure 2), as expected from [12]. Overall, dispersal estimates showed that populations are all connected by dispersal movements but with frequencies contingent upon their geographic distance. However, the significance of F_{ST} values indicates that effective dispersal is not high enough to genetically homogenize the metapopulation.

Metapopulation functioning might rely on source/sink dynamics, the permanent dispersal of individuals from a source population of good quality to a receiving population with demographic deficit living in habitats of poorer quality [74]. On the long-term, source/sink dynamics may lead to the demographic stabilization of the overall system and could explain some aspects of the described metapopulation equilibrium. However, the qualitative analysis of dispersal movements did not show any obvious source/sink pattern. This suggests that strong dispersal asymmetry might not be predominant in the functioning of this *B. eunomia* metapopulation, although we cannot exclude a role of weak source/sink patterns, like the putative case of Grande Fange (a population harboring a high level of private alleles, see above). We hypothesize that the observed stability probably results from other eco-evolutionary processes and discuss this possibility hereafter.

4.2. Eco-Evolutionary Perspectives: Controlled Ecological Fluctuations Lead to Long-Term Equilibrium at the Metapopulation Scale

The focal metapopulation has been surveyed for decades, and we accumulated knowledge about various eco-evolutionary processes that might further explain its stability. Population parameters are often subject to oscillations as a result of the confrontation of intrinsic (phenotypes) to extrinsic (environments) factors (see review in [75]), and *B. eunomia* is no exception. For instance, its larvae are attacked by a tiny hymenopteran parasitoid, which abundance regulates the butterfly local population size [45,76,77]. Besides, local population sizes are correlated with climatic factors (temperature and humidity) acting differently over the year according to the life history stages of the butterfly [46]. Together with the host- and food-plant abundance and other factors (see [40,78] for detailed reviews), the habitat quality varies over time on a yearly basis and across space on a few hundred meters [46]. The habitat of the butterfly may thus be seen as a moving mosaic of patches of low and high quality that support different adult densities [48]. Adult males and females move differently between high and low quality patches. Due to male harassment, females of *B. eunomia* emigrate seeking for patches with low male density, while males emigrate seeking for patches with high female density [41]. Besides, male harassment in combination with life-history characteristics (differential reproductive success, detectability by predators) probably explains the long-term maintenance of female color polymorphism, i.e., existence of an andromorph wing coloration in some females [79]. Such selective patterns may control local demographic fluctuations and trait variability and suggest a long-adaptive history of *B. eunomia* to its local environment, including its predictable oscillations. Habitat quality and butterfly density within local populations are fine-grained at the scale of a few hundred meters [46], which should favor local efficient response to microhabitat fluctuations as experienced by the species in this area over thousands of generations. Adult movements among patches of different quality prevent in turn strong spatial synchrony. Altogether these two processes (fine grained adaptation and adult movements within and among local populations) mitigate the risk of metapopulation collapse caused by synchronous environmental variations. Thus, we argue that, in concert with the control of ecological fluctuations at the local scale, the current rate of dispersal and gene flow (strongly dependent upon geographic distance between populations) is a key mechanism conferring stability to the system. Indeed, when dispersal is sufficiently high to prevent very high consanguinity and local extinctions, and sufficiently low to avoid region-wide synchrony and genetic pool homogenization, such equilibrium metapopulations should maintain on the long-term [80]. However, the persistence of such metapopulations at equilibrium in cases of catastrophic events is not guaranteed if the amplitude of the catastrophe exceeds the regulatory eco-evolutionary feedbacks between local adaptation and dispersal. This hypothesis could unfortunately be formally tested now in our studied metapopulation. The large Prés de la Lienne local population has indeed recently gone extinct, very likely as a direct consequence of poorly prepared reintroduction of beavers (*Castor spp.*) in Belgium. The construction of beaver dams along the Lienne river created frequent and long-lasting floods during the winter in the wet meadows inhabited by *Boloria eunomia*. Although this is still an unproven but likely hypothesis, these floods should lead to very high mortality rates of the diapausing caterpillars, susceptible to cause this large local population to vanish over a couple of years: from over 1000 individuals in 2010 and 2011, only 6 males were captured in 2016, and none in the next years. We hope to be able to estimate the impact of the Prés de la Lienne extinction on the stability of the whole metapopulation in the forthcoming years.

4.3. Methodological Perspectives: Congruence between Demographic and Genetic Estimates of Dispersal

The ability of demographic approaches such as CMR to estimate effective dispersal is still a matter of debate. It supposedly suffers from several biases: difficulties in the acquisition of field data, adequation of demographic models to natural situations, definition of dispersal movements, scale effects, etc. [12,32,33,81]. However, we here found a noticeable congruence between demographic and genetic approaches ($r = 0.5$). We highlight below a few critical points that could improve the match between demographic and genetic approaches in *B. eunomia* (and beyond) in future works.

Despite intensive field work covering the whole flight period in a well-studied butterfly system, we were unable to capture the vast majority of long-distance dispersal movements using CMR data. This recurrent difficulty in the monitoring of individual movements has strong consequences on the ability to construct appropriate dispersal kernels, where long-distance movements play a key role, but also to define the exact contours of natural metapopulations. In *B. eunomia*, unsampled distant populations nonetheless connected by rare dispersal movements to our metapopulation could serve as reservoirs feeding the general stability of the whole-system, i.e., undetected source/sink dynamics. The eco-evolutionary mechanisms we proposed above to explain equilibrium (fine-grained local adaptation and dispersal) could in this case be less predominant. Underestimation of long-distance dispersal and overestimation of short-distance movements with the demographic approach raise one important question about the origin of the uncommon variance between the two approaches (~75% in our case). Can we attribute most of this unexplained variance to this CMR campaign effect, or should we look for other explanations? A number of other methodological biases can be mentioned. They include for instance the ability of genetic methods to effectively detect first-generation dispersers, incomplete genetic sampling, or asymmetric reproductive success. It will be difficult to tackle these issues in our current dataset. An easy improvement would nonetheless be the use of more numerous molecular markers, such as SNPs, to get more accurate genetic estimates.

Finally, we observed a strong IBD pattern, which makes sense given that long-distance dispersal movements are less numerous than short-distance movements in the studied metapopulation. Nonetheless, ~75% of the genetic variance remains unexplained by the Euclidian distance between populations. *Boloria eunomia* presents distinct behaviors when encountering different matrix types [82]. Although beyond the scope of this study, we will probably need to incorporate functional connectivity indices in a landscape genetic approach to better explain our observed pattern of genetic differentiation.

5. Conclusions

There is no doubt that classical metapopulations do exist, but they might not be as widespread as generally supposed, because extinction/recolonization cycles are not necessary characteristics of metapopulation functioning [6,7]. This study on *B. eunomia* provides an example of a metapopulation where destabilizing agents like inter-generational fluctuations in population sizes seem to be controlled by a long adaptive history of the species to its dynamic local environment, including the evolution of appropriate rates of dispersal. In such a case, population extinction should be the result of rare catastrophic events, whose consequences on short-term dynamics and long-term stability are of prime interest to study in this butterfly of conservation concern. In our study system, genetic and demographic approaches provided congruent estimates of dispersal rates. This comfortable situation may not hide the necessity to integrate, e.g., functional connectivity to fully capture the functioning of our metapopulation. Indeed, a non-negligible part of the variance in genetic differentiation remained unexplained. We hope that other case studies of such stable 'unclassical natural metapopulations' will be available to test for the generality of the mechanisms we have proposed to explain metapopulation equilibrium in *B. eunomia*.

Supplementary Materials: The following are available online at https://www.mdpi.com/2073-4425/12/3/362/s1, Figure S1: Determination of the number of clusters in the Structure analysis, Figure S2: Isolation by distance patterns per year, Table S1: Euclidean distance between pairs of population in meters, Table S2: Demographic results from CMR modelling, Table S3: AMOVA analyses considering clusters 1 and 2 as the unit of genetic partitioning, Table S4: Fst values and significance, Table S5: Results of the detection of past demographic events using Bottleneck. Raw microsatellite data are provided as a word file entitled '*microsat_data_structure_format*'.

Author Contributions: M.B., C.T. and N.S. conceived the study. D.L., M.B., N.S. and C.T. acquired field data. Q.D. and D.L. performed the molecular laboratory work. D.L., N.S. and J.G.P. analyzed the data. D.L. wrote the first draft of the manuscript. All authors approved the final version. All authors have read and agreed to the published version of the manuscript.

Funding: This research was funded by the EU FP6 Biodiversa ERANET TenLamas and by the EU FP7 SCALES project (project no. 226852) (D.L., M.B. and C.T.) and by the Fonds de la Recherche Scientifique-FNRS (N.S.).

Institutional Review Board Statement: Not applicable.

Informed Consent Statement: Not applicable.

Data Availability Statement: Data is contained within the article and Supplementary Materials.

Acknowledgments: D.L., M.B. and J.G.P. are part of the Laboratoire d'Excellence TULIP (ANR–10–LABX–41). N.S. is Senior Research Associate of the F.R.S.–FNRS. This paper is contribution BRC368 of the Biodiversity Research Center (UCLouvain).

Conflicts of Interest: The authors declare no conflict of interest.

References

1. Akçakaya, H.R.; Mills, G.; Doncaster, C.P. The role of metapopulations in conservation. In *Key Topics in Conservation Biology*; Macdonald, D., Service, K., Eds.; Blackwell Publishing: Oxford, UK, 2007; pp. 64–84.
2. Levins, R. Extinction. In *Some Mathematical Problems in Biology*; Gesternhaber, M., Ed.; American Mathematical Soc.: Providence, RI, USA, 1970; pp. 77–107.
3. Levins, R. Some Demographic and Genetic Consequences of Environmental Heterogeneity for Biological Control. *Bull. Entomol. Soc. Am.* **1969**, *15*, 237–240. [CrossRef]
4. Harrison, S. Local Extinction in a Metapopulation Context: An Empirical Evaluation. *Biol. J. Linn. Soc.* **1991**, *42*, 73–88. [CrossRef]
5. Hanski, I.; Simberloff, D. The Metapopulation Approach, Its History, Conceptual Domain, and Application to Conservation. In *Metapopulation Biology, Ecology, Genetics, and Evolution*; Hanski, I., Simberloff, D., Eds.; Academic Press: Cambridge, MA, 1997; pp. 5–26.
6. Kritzer, J.P.; Sale, P.F. Metapopulation Ecology in the Sea: From Levins' Model to Marine Ecology and Fisheries Science. *Fish Fish.* **2004**, *5*, 131–140. [CrossRef]
7. Baguette, M. The Classical Metapopulation Theory and the Real, Natural World: A Critical Appraisal. *Basic Appl. Ecol.* **2004**, *5*, 213–224. [CrossRef]
8. Fronhofer, E.A.; Kubisch, A.; Hilker, F.M.; Hovestadt, T.; Poethke, H.J. Why Are Metapopulations so Rare? *Ecology* **2012**, *93*, 1967–1978. [CrossRef] [PubMed]
9. Ronce, O. How Does It Feel to Be Like a Rolling Stone? Ten Questions About Dispersal Evolution. *Annu. Rev. Ecol. Evol. Syst.* **2007**, *38*, 231–253. [CrossRef]
10. Clobert, J.; Baguette, M.; Benton, T.G.; Bullock, J.M. *Dispersal Ecology and Evolution*; Oxford University Press: Oxford, UK, 2012.
11. Hanski, I. Habitat Connectivity, Habitat Continuity, and Metapopulations in Dynamic Landscapes. *Oikos* **1999**, *87*, 209. [CrossRef]
12. Baguette, M.; Blanchet, S.; Legrand, D.; Stevens, V.M.; Turlure, C. Individual Dispersal, Landscape Connectivity and Ecological Networks: Dispersal, Connectivity and Networks. *Biol. Rev.* **2013**, *88*, 310–326. [CrossRef] [PubMed]
13. Lenormand, T. Gene Flow and the Limits to Natural Selection. *Trends Ecol. Evol.* **2002**, *17*, 183–189. [CrossRef]
14. Jacob, S.; Legrand, D.; Chaine, A.S.; Bonte, D.; Schtickzelle, N.; Huet, M.; Clobert, J. Gene Flow Favours Local Adaptation under Habitat Choice in Ciliate Microcosms. *Nat. Ecol. Evol.* **2017**, *1*, 1407–1410. [CrossRef] [PubMed]
15. Kendall, B.E.; Bjørnstad, O.N.; Bascompte, J.; Keitt, T.H.; Fagan, W.F. Dispersal, Environmental Correlation, and Spatial Synchrony in Population Dynamics. *Am. Nat.* **2000**, *155*, 628–636. [CrossRef] [PubMed]
16. Hufbauer, R.A.; Szűcs, M.; Kasyon, E.; Youngberg, C.; Koontz, M.J.; Richards, C.; Tuff, T.; Melbourne, B.A. Three Types of Rescue Can Avert Extinction in a Changing Environment. *Proc. Natl. Acad. Sci. USA* **2015**, *112*, 10557–10562. [CrossRef] [PubMed]
17. Wang, S.; Haegeman, B.; Loreau, M. Dispersal and Metapopulation Stability. *PeerJ* **2015**, *3*, e1295. [CrossRef] [PubMed]
18. Bohrer, G.; Nathan, R.; Volis, S. Effects of Long-Distance Dispersal for Metapopulation Survival and Genetic Structure at Ecological Time and Spatial Scales. *J. Ecol.* **2005**, *93*, 1029–1040. [CrossRef]
19. Perry, G.L.W.; Lee, F. How Does Temporal Variation in Habitat Connectivity Influence Metapopulation Dynamics? *Oikos* **2019**, *128*, 1277–1286. [CrossRef]
20. Cain, M.L.; Milligan, B.G.; Strand, A.E. Long-Distance Seed Dispersal in Plant Populations. *Am. J. Bot.* **2000**, *87*, 1217–1227. [CrossRef]
21. Osborne, J.L.; Loxdale, H.D.; Woiwod, I.P. Monitoring Insect Dispersal: Methods and Approaches. In *Dispersal Ecology*, Blackwell Science: Oxford, UK, 2002; pp. 24–49.
22. Katzner, T.E.; Arlettaz, R. Evaluating Contributions of Recent Tracking-Based Animal Movement Ecology to Conservation Management. *Front. Ecol. Evol.* **2020**, *7*, 519. [CrossRef]
23. Lebreton, J.-D.; Burnham, K.P.; Clobert, J.; Anderson, D.R. Modeling Survival and Testing Biological Hypotheses Using Marked Animals: A Unified Approach with Case Studies. *Ecol. Monogr.* **1992**, *62*, 67–118. [CrossRef]
24. Berry, O.; Tocher, M.D.; Sarre, S.D. Can Assignment Tests Measure Dispersal? *Mol. Ecol.* **2004**, *13*, 551–561. [CrossRef] [PubMed]
25. Griesser, M.; Halvarsson, P.; Sahlman, T.; Ekman, J. What Are the Strengths and Limitations of Direct and Indirect Assessment of Dispersal? Insights from a Long-Term Field Study in a Group-Living Bird Species. *Behav. Ecol. Sociobiol.* **2014**, *68*, 485–497. [CrossRef]
26. Chadœuf, J.; Millon, A.; Bourrioux, J.; Printemps, T.; Hecke, B.; Lecoustre, V.; Bretagnolle, V. Modelling Unbiased Dispersal Kernels over Continuous Space by Accounting for Spatial Heterogeneity in Marking and Observation Efforts. *Methods Ecol. Evol.* **2018**, *9*, 331–339. [CrossRef]
27. Clobert, J.; Ims, R.; Rousset, F. Causes, mechanisms and consequences of dispersal. In *Ecology, Genetics and Evolution of Metapopulation*; Hanski, I., Gaggiotti, O.E., Eds.; Elsevier Academic Press: Amsterdam, The Netherlands, 2004; pp. 307–335.
28. Schtickzelle, N.; WallisDeVries, M.F.; Baguette, M. Using Surrogate Data in Population Viability Analysis: The Case of the Critically Endangered Cranberry Fritillary Butterfly. *Oikos* **2005**, *109*, 89–100. [CrossRef]
29. Wright, S. Evolution in Mendelian Populations. *Genetics* **1931**, *16*, 97–159. [CrossRef]
30. Whitlock, M.C.; Mccauley, D.E. Indirect Measures of Gene Flow and Migration: FST≠1/(4Nm+1). *Heredity* **1999**, *82*, 117–125. [CrossRef]
31. Prunier, J.G.; Dubut, V.; Chikhi, L.; Blanchet, S. Contribution of Spatial Heterogeneity in Effective Population Sizes to the Variance in Pairwise Measures of Genetic Differentiation. *Methods Ecol. Evol.* **2017**, *8*, 1866–1877. [CrossRef]

32. Broquet, T.; Petit, E.J. Molecular Estimation of Dispersal for Ecology and Population Genetics. *Annu. Rev. Ecol. Evol. Syst.* **2009**, *40*, 193–216. [CrossRef]
33. Cayuela, H.; Rougemont, Q.; Prunier, J.G.; Moore, J.-S.; Clobert, J.; Besnard, A.; Bernatchez, L. Demographic and Genetic Approaches to Study Dispersal in Wild Animal Populations: A Methodological Review. *Mol. Ecol.* **2018**, *27*, 3976–4010. [CrossRef] [PubMed]
34. Anderson, C.D.; Epperson, B.K.; Fortin, M.-J.; Holderegger, R.; James, P.M.A.; Rosenberg, M.S.; Scribner, K.T.; Spear, S. Considering Spatial and Temporal Scale in Landscape-Genetic Studies of Gene Flow: Scale in landscape genetics. *Mol. Ecol.* **2010**, *19*, 3565–3575. [CrossRef] [PubMed]
35. Pritchard, J.K.; Stephens, M.; Donnelly, P. Inference of Population Structure Using Multilocus Genotype Data. *Genetics* **2000**, *155*, 945–959.
36. Piry, S.; Alapetite, A.; Cornuet, J.-M.; Paetkau, D.; Baudouin, L.; Estoup, A. GENECLASS2: A Software for Genetic Assignment and First-Generation Migrant Detection. *J. Hered.* **2004**, *95*, 536–539. [CrossRef] [PubMed]
37. Jones, A.G.; Small, C.M.; Paczolt, K.A.; Ratterman, N.L. A Practical Guide to Methods of Parentage Analysis: TECHNICAL REVIEW. *Mol. Ecol. Resour.* **2010**, *10*, 6–30. [CrossRef] [PubMed]
38. Lowe, W.H.; Allendorf, F.W. What Can Genetics Tell Us about Population Connectivity?: GENETIC AND DEMOGRAPHIC CONNECTIVITY. *Mol. Ecol.* **2010**, *19*, 3038–3051. [CrossRef]
39. Vandewoestijne, S.; Baguette, M. Demographic versus Genetic Dispersal Measures. *Popul. Ecol.* **2004**, *46*, 281–285. [CrossRef]
40. Turlure, C.; Van Dyck, H.; Schtickzelle, N.; Baguette, M. Resource-Based Habitat Definition, Niche Overlap and Conservation of Two Sympatric Glacial Relict Butterflies. *Oikos* **2009**, *118*, 950–960. [CrossRef]
41. Baguette, M.; Vansteenwegen, C.; Convi, I.; Nève, G. Sex-Biased Density-Dependent Migration in a Metapopulation of the Butterfly *Proclossiana eunomia*. *Acta Oecol.* **1998**, *19*, 17–24. [CrossRef]
42. Baguette, M.; Schtickzelle, N. Negative Relationship between Dispersal Distance and Demography in Butterfly Metapopulations. *Ecology* **2006**, *87*, 648–654. [CrossRef]
43. Turlure, C.; Schtickzelle, N.; Van Dyck, H.; Seymoure, B.; Rutowski, R. Flight Morphology, Compound Eye Structure and Dispersal in the Bog and the Cranberry Fritillary Butterflies: An Inter- and Intraspecific Comparison. *PLoS ONE* **2016**, *11*, e0158073. [CrossRef]
44. Schtickzelle, N.; Baguette, M. Behavioural Responses to Habitat Patch Boundaries Restrict Dispersal and Generate Emigration-Patch Area Relationships in Fragmented Landscapes: *Restricted Dispersal and Fragmentation*. *J. Anim. Ecol.* **2003**, *72*, 533–545. [CrossRef] [PubMed]
45. Schtickzelle, N.; Le Boulengé, E.; Baguette, M. Metapopulation Dynamics of the Bog Fritillary Butterfly: Demographic Processes in a Patchy Population. *Oikos* **2002**, *97*, 349–360. [CrossRef]
46. Schtickzelle, N.; Baguette, M. Metapopulation Viability Analysis of the Bog Fritillary Butterfly Using RAMAS/GIS. *Oikos* **2004**, *104*, 277–290. [CrossRef]
47. Schtickzelle, N.; Mennechez, G.; Baguette, M. Dispersal Depression with Habitat Fragmentation in the Bog Fritillary Butterfly. *Ecology* **2006**, *87*, 1057–1065. [CrossRef]
48. Baguette, M.; Clobert, J.; Schtickzelle, N. Metapopulation Dynamics of the Bog Fritillary Butterfly: Experimental Changes in Habitat Quality Induced Negative Density-Dependent Dispersal. *Ecography* **2011**, *34*, 170–176. [CrossRef]
49. Nève, G.; Barascud, B.; Descimon, H.; Baguette, M. Genetic Structure of *Proclossiana eunomia* Populations at the Regional Scale (Lepidoptera, Nymphalidae). *Heredity* **2000**, *84*, 657–666. [CrossRef]
50. Nève, G.; Mousson, L.; Baguette, M. Adult Dispersal and Genetic Structure of Butterfly Populations in a Fragmented Landscape. *Acta Oecol.* **1996**, *17*, 621–626.
51. Maresova, J.; Habel, J.C.; Neve, G.; Sielezniew, M.; Bartonova, A.; Kostro-Ambroziak, A.; Fric, Z.F. Cross-Continental Phylogeography of Two Holarctic Nymphalid Butterflies, *Boloria eunomia* and Boloria Selene. *PLoS ONE* **2019**, *14*, e0214483. [CrossRef] [PubMed]
52. Vandewoestijne, S.; Baguette, M. Genetic Population Structure of the Vulnerable Bog Fritillary Butterfly: Genetic Population Structure of the Bog Fritillary Butterfly. *Hereditas* **2005**, *141*, 199–206. [CrossRef] [PubMed]
53. White, G.C.; Burnham, K.P. Program MARK: Survival Estimation from Populations of Marked Animals. *Bird Study* **1999**, *46*, S120–S139. [CrossRef]
54. Cooch, E.; White, G.C. *Program Mark: A Gentle Introduction*; Colorado State University: Fort Collins, CO, USA, 2002; Available online: http://www.phidot.org/software/mark/docs/book/.
55. Legrand, D.; Chaput-Bardy, A.; Turlure, C.; Dubois, Q.; Huet, M.; Schtickzelle, N.; Stevens, V.M.; Baguette, M. Isolation and Characterization of 15 Microsatellite Loci in the Specialist Butterfly *Boloria eunomia*. *Conserv. Genet. Resour.* **2014**, *6*, 223–227. [CrossRef]
56. Weir, B.S.; Cockerham, C.C. Estimating F-Statistics for the Analysis of Population Structure. *Evolution* **1984**, *38*, 1358. [CrossRef] [PubMed]
57. Excoffier, L.; Lischer, H.E.L. Arlequin Suite Ver 3.5: A New Series of Programs to Perform Population Genetics Analyses under Linux and Windows. *Mol. Ecol. Resour.* **2010**, *10*, 564–567. [CrossRef]
58. Narum, S.R. Beyond Bonferroni: Less Conservative Analyses for Conservation Genetics. *Conserv. Genet.* **2006**, *7*, 783–787. [CrossRef]

59. Evanno, G.; Regnaut, S.; Goudet, J. Detecting the Number of Clusters of Individuals Using the Software Structure: A Simulation Study. *Mol. Ecol.* **2005**, *14*, 2611–2620. [CrossRef]

60. Coulon, A.; Morellet, N.; Goulard, M.; Cargnelutti, B.; Angibault, J.-M.; Hewison, A.J.M. Inferring the Effects of Landscape Structure on Roe Deer (Capreolus Capreolus) Movements Using a Step Selection Function. *Landsc. Ecol.* **2008**, *23*, 603–614. [CrossRef]

61. Rousset, F. Genetic Differentiation and Estimation of Gene Flow from FStatistics Under Isolation by Distance. *Genet* **1997**, *145*, 1219–1228. [CrossRef]

62. Glaubitz, J.C. Convert: A User-Friendly Program to Reformat Diploid Genotypic Data for Commonly Used Population Genetic Software Packages. *Mol. Ecol. Notes* **2004**, *4*, 309–310. [CrossRef]

63. Waples, R.S.; Do, C. Linkage Disequilibrium Estimates of Contemporary N_e Using Highly Variable Genetic Markers: A Largely Untapped Resource for Applied Conservation and Evolution. *Evol. Appl.* **2010**, *3*, 244–262. [CrossRef]

64. Do, C.; Waples, R.S.; Peel, D.; Macbeth, G.M.; Tillett, B.J.; Ovenden, J.R. NEESTIMATOR v2: Re-Implementation of Software for the Estimation of Contemporary Effective Population Size (N_e) from Genetic Data. *Mol. Ecol. Resour.* **2014**, *14*, 209–214. [CrossRef]

65. Rannala, B.; Mountain, J.L. Detecting Immigration by Using Multilocus Genotypes. *Proc. Natl. Acad. Sci. USA* **1997**, *94*, 9197–9201. [CrossRef]

66. Paetkau, D.; Slade, R.; Burden, M.; Estoup, A. Genetic Assignment Methods for the Direct, Real-Time Estimation of Migration Rate: A Simulation-Based Exploration of Accuracy and Power. *Mol. Ecol.* **2004**, *13*, 55–65. [CrossRef] [PubMed]

67. Cornuet, J.-M.; Luikart, G. Description and Power Analysis of Two Tests for Detecting Recent Population Bottlenecks From Allele Frequency Data. *Genetics* **1996**, *144*, 2001–2014. [CrossRef]

68. Luikart, G. Distortion of Allele Frequency Distributions Provides a Test for Recent Population Bottlenecks. *J. Hered.* **1998**, *89*, 238–247. [CrossRef] [PubMed]

69. Thrall, P.H.; Burdon, J.J.; Murray, B.R. The metapopulation paradigm: A fragmented view of conservation biology. In *Genetics, Demography and Viability of Fragmented Populations*; Young, A.G., Clarke, G.M., Eds.; Cambridge University Press: Cambridge, UK, 2000; pp. 75–96.

70. Frankham, R.; Bradshaw, C.J.A.; Brook, B.W. Genetics in Conservation Management: Revised Recommendations for the 50/500 Rules, Red List Criteria and Population Viability Analyses. *Biol. Conserv.* **2014**, *170*, 56–63. [CrossRef]

71. Maggs, C.A.; Castilho, R.; Foltz, D.; Henzler, C.; Jolly, M.T.; Kelly, J.; Olsen, J.; Perez, K.E.; Stam, W.; La, R.V.I. Evaluating signatures of glacial refugia for north atlantic benthic marine taxa. *Ecology* **2008**, *89*, S108–S122. [CrossRef] [PubMed]

72. Kennington, W.J.; Gockel, J.; Partridge, L. Testing for Asymmetrical Gene Flow in a *Drosophila melanogaster* Body-Size Cline. *Genetics* **2003**, *165*, 667–673. [PubMed]

73. Watts, P.C.; Rousset, F.; Saccheri, I.J.; Leblois, R.; Kemp, S.J.; Thompson, D.J. Compatible Genetic and Ecological Estimates of Dispersal Rates in Insect (Coenagrion Mercuriale: Odonata: Zygoptera) Populations: Analysis of 'Neighbourhood Size' Using a More Precise Estimator: Neighbourhood size in coenagrion mercuriale. *Mol. Ecol.* **2006**, *16*, 737–751. [CrossRef] [PubMed]

74. Dias, P.C. Sources and Sinks in Population Biology. *Trends Ecol. Evol.* **1996**, *11*, 326–330. [CrossRef]

75. Barraquand, F.; Louca, S.; Abbott, K.C.; Cobbold, C.A.; Cordoleani, F.; DeAngelis, D.L.; Elderd, B.D.; Fox, J.W.; Greenwood, P.; Hilker, F.M.; et al. Moving Forward in Circles: Challenges and Opportunities in Modelling Population Cycles. *Ecol. Lett.* **2017**, *20*, 1074–1092. [CrossRef]

76. Choutt, J.; Turlure, C.; Baguette, M.; Schtickzelle, N. Parasitism Cost of Living in a High Quality Habitat in the Bog Fritillary Butterfly. *Biodivers Conserv.* **2011**, *20*, 3117–3131. [CrossRef]

77. Choutt, J. *The Butterfly and the Wasp: Host Parasitoid Relationship between Boloria Eunomia and Cotesia Eunomiae*; UCLouvain: Louvain-la-Neuve, Belgium, 2011.

78. Turlure, C. *Habitat from a Butterfly's Point of View: How Specialist Butterflies Map onto Ecological Resources*; UCLouvain: Louvain-la-Neuve, Belgium, 2009.

79. Turlure, C.; Legrand, D.; Schtickzelle, N.; Baguette, M. Male Disguised Females: Costs and Benefits of Female-Limited Dimorphism in a Butterfly: Evolutionary Ecology of Female-Limited Dimorphism. *Ecol. Entomol.* **2016**, *41*, 572–581. [CrossRef]

80. Yaari, G.; Ben-Zion, Y.; Shnerb, N.M.; Vasseur, D.A. Consistent Scaling of Persistence Time in Metapopulations. *Ecology* **2012**, *93*, 1214–1227. [CrossRef]

81. Burgess, S.C.; Baskett, M.L.; Grosberg, R.K.; Morgan, S.G.; Strathmann, R.R. When Is Dispersal for Dispersal? Unifying Marine and Terrestrial Perspectives: When Is Dispersal for Dispersal? *Biol. Rev.* **2016**, *91*, 867–882. [CrossRef] [PubMed]

82. Turlure, C.; Baguette, M.; Stevens, V.M.; Maes, D. Species- and Sex-Specific Adjustments of Movement Behavior to Landscape Heterogeneity in Butterflies. *Behav. Ecol.* **2011**, *22*, 967–975. [CrossRef]

Article

Using Reciprocal Transplants to Assess Local Adaptation, Genetic Rescue, and Sexual Selection in Newly Established Populations

Jacques Labonne [1,*], Aurélie Manicki [1], Louise Chevalier [1], Marin Tétillon [1], François Guéraud [1] and Andrew P. Hendry [2,*]

1. Université de Pau et des Pays de l'Adour, UMR INRAE-UPPA, Ecobiop, FR-64310 Saint-Pée sur Nivelle, France; aurelie.manicki@inrae.fr (A.M.); louise.chevalier@inrae.fr (L.C.); marin.tetillon@inra.fr (M.T.); francois.gueraud@inrae.fr (F.G.)
2. Redpath Museum and Department of Biology, McGill University, Montreal, QC H3A 0C4, Canada
* Correspondence: jacques.labonne@inrae.fr (J.L.); andrew.hendry@mcgill.ca (A.P.H.)

Abstract: Small populations establishing on colonization fronts have to adapt to novel environments with limited genetic variation. The pace at which they can adapt, and the influence of genetic variation on their success, are key questions for understanding intraspecific diversity. To investigate these topics, we performed a reciprocal transplant experiment between two recently founded populations of brown trout in the sub-Antarctic Kerguelen Islands. Using individual tagging and genetic assignment methods, we tracked the fitness of local and foreign individuals, as well as the fitness of their offspring over two generations. In both populations, although not to the same extent, gene flow occurred between local and foreign gene pools. In both cases, however, we failed to detect obvious footprints of local adaptation (which should limit gene flow) and only weak support for genetic rescue (which should enhance gene flow). In the population where gene flow from foreign individuals was low, no clear differences were observed between the fitness of local, foreign, and F1 hybrid individuals. In the population where gene flow was high, foreign individuals were successful due to high mating success rather than high survival, and F1 hybrids had the same fitness as pure local offspring. These results suggest the importance of considering sexual selection, rather than just local adaptation and genetic rescue, when evaluating the determinants of success in small and recently founded populations.

Keywords: genetic rescue; local adaptation; mating success; gene flow; small population

Citation: Labonne, J.; Manicki, A.; Chevalier, L.; Tétillon, M.; Guéraud, F.; Hendry, A.P. Using Reciprocal Transplants to Assess Local Adaptation, Genetic Rescue, and Sexual Selection in Newly Established Populations. *Genes* **2021**, *12*, 5. https://dx.doi.org/10.3390/genes12010005

Received: 17 November 2020
Accepted: 18 December 2020
Published: 23 December 2020

Publisher's Note: MDPI stays neutral with regard to jurisdictional claims in published maps and institutional affiliations.

1. Introduction

Local adaptation (LA) happens when individuals have higher fitness in their local environment than do immigrant individuals [1]. LA is built via selective processes, wherein some individuals achieve higher survival and reproductive success than others. The genetic contribution of these individuals is therefore more likely to be passed on to the next generation, further shaping the new identity of the local gene pool and their phenotypic traits [2–4]. In some cases, however, LA can be compromised by limited genetic variation, especially in small populations and/or on colonization fronts (i.e., the margins of a distribution area where individuals are colonizing new habitats). In such cases, the effects of genetic drift can counteract the efficacy of LA [5]. But moderate gene flow sometimes can improve population fitness [6–8] and even rescue it from extinction ("genetic rescue" or GR, ref. [9–12]), which then enhance subsequent LA. On the other hand, gene flow can introduce non-adapted alleles into the population, increasing the risk of severe maladaptation that can lead to extinction [13,14].

In the context of the sixth biodiversity crisis [15], documenting the interaction between LA and GR is of major importance for understanding how organisms might cope with rapid environmental change in fluctuating demographic contexts [16]. That is, many species and populations that exist in a tenuous demographic state, such as low population size, must now also face a rapidly changing environment. Moreover, many species are shifting their ranges and colonizing new environments, either naturally as a response to changing environments or unnaturally through human-mediated introductions. In this complicated intersection of colonization, demography, and environmental change, rapid LA to new environmental conditions becomes critical [17]. It thus appears paramount to understand the speed at which LA arises, and whether or not GR is important, when organisms colonize (or experience) new environments.

The pace of LA can be investigated through the study of fitness across generations following the migration of a pool of individuals into a new environment. Provided founding genetic variation is not limiting, efficient LA by selection should rapidly improve the fitness of residents. After that period of rapid adaptation by residents, any new immigrants should incur a fitness disadvantage in terms of survival, reducing their odds of transmitting their alleles to the next generation, compared to locally adapted resident individuals. If the immigrants and residents interbred, LA should then translate in a lower overall fitness for hybrids compared to local individuals. By contrast, GR would be expected to be evident as an increase in hybrid fitness, though the benefits of increased genetic variation. This latter signature of GR is especially expected in the context of small populations on colonization fronts, wherein standing genetic variation can be reduced through potent genetic drift and where inbreeding can drastically impact the fitness of local individuals. Further, in each generation, fitness differences can be decomposed into mating success (sexual selection) and survival (natural selection), with the latter being more directly indicative of adaptive effects. These components of fitness can be related to the degree of mixture between local and foreign genes (Hybrid index, [12]).

Salmonid fishes have contributed actively to our understanding of LA [18–22], with strong evidence of LA being found in many populations of various species [23,24]. The mechanisms underlying these adaptations have been investigated through correlational approaches [25–28], genomic analyses [29,30], common garden experiments [31], and some reciprocal transplant experiments [32]. For most of these studies, however, the sampled populations were already established, and at a (presumably) stable equilibrium. As a result, we have very little understanding of the rate and determinants of LA in its earliest stage. One exception to this research gap is the study of LA in invasive brown trout in Newfoundland, wherein local—but recent—populations fared better in terms of survival compared to foreign introduced populations [33]. However, LA is not just about survival differences—but also about mating and reproductive success. Further, LA can also influence the success of hybrids and backcrosses. Hence, we also need studies of LA in new populations conducted on a multigenerational scale [22]. GR has been less well studied in salmonids, although some studies suggest that small isolated populations have very low genetic variation, and might therefore benefit from gene flow [34,35].

We studied how these processes played out following an introduction of brown trout (*Salmo trutta* L.) to the remote sub-Antarctic islands of Kerguelen [36]. Of particular interest were populations on the western side the colonization front [37], because these populations face increasingly challenging environmental conditions due to their close proximity to the melting ice cap [38,39]. Rapid LA might be especially important in such cases [40]. In an earlier study of this system, we investigated the fate of two populations introduced in 1993 from just a few founders, finding very high inbreeding levels and selection against homozygosity up to 2010 [41]. Here, then, we have an interesting intersection of LA (potentially favored in the novel environments) and GR (potentially favored owing to low genetic variation and inbreeding). Thus, in 2010, approximately 3 to 4 generations after their initial introduction, we conducted a reciprocal transplant between the two populations [42,43]. If important LA occurred on such a short time scale, introduced foreign individuals should show low survival and reproductive success—resulting in low gene flow (sensu [44]). If GR is important, however, hybrid offspring should be more successful than pure resident offspring, thus enhancing gene flow.

We monitored the fate of this experiment by sampling the two populations again in 2011, 2012, and 2018 to assesses the level and structure of gene flow, and to estimate the fitness of local, foreign, and potential F1, F2 and backcrosses hybrid individuals. In addition, we investigated whether the degree of hybridization (hybrid index) was related to possible components of selection, such as recapture proportions between local and foreign individuals following the transplantation protocol (indicative of adult survival), sired family size (indicative of offspring survival), or homozygosity level (as indicative of inbreeding load).

2. Materials and Methods

2.1. Sites and Populations Description

Two populations inhabiting different environments (henceforth "systems") were selected based on our knowledge of their recent past and introduction conditions [36,41]. The Val Travers system, located in the northern area of the main island of Kerguelen sub-Antartic archipelago (491,832″ S, 692,579″ E) is 9 km long, with a gradient of habitats and slopes from mountain to lowland landscapes. It empties into Lake Bontemps (700 ha), which is connected to the sea by a steep outlet. The Clarée system is 3 km long, and is located to the south of the main island (492,935″ S, 693,744″ E) on a plain featuring several interconnected arms originating from Lake Hermance (350 ha) and also from a tributary flowing from a nearby glacier (River Galets). The Clarée empties directly into a shallow marine bay.

The trout populations in both systems were artificially introduced in 1993 from two other Kerguelen systems [36]. The Val Travers population was founded with 2000 six months old juveniles from a single cross between one male and one female from the River Chateau, which was first colonized in 1962. The Clarée population was founded with 1700 six months old juveniles from a cross between one female and two males that were captured while migrating upstream in the River Armor, which at that time was not yet colonized by brown trout (no natural reproduction observed [36], see Figure 1). The genetics of these populations have been investigated [41], revealing a high initial level of inbreeding in both populations. However, subsequent selection against homozygotes was also detected in the first generations, especially in Val Travers.

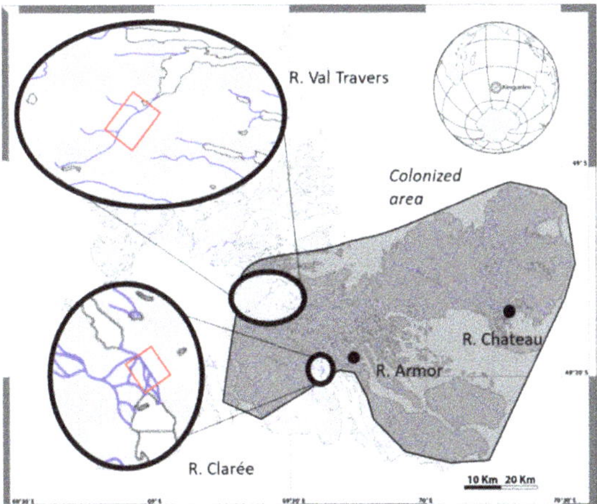

Figure 1. Location of the two studied systems in the Kerguelen islands. In each system, the red rectangle indicates the area where sampling took place. The grey polygon indicates the extent of the colonized area by brown trout in 2018.

2.2. Transplantation Experiment and Populations Sampling

In March 2010, in each system, we used electrofishing to sample 261 non-mature individuals aged from 1 two 3 years old (mean body size 13.55 cm for Val Travers, 11.46 cm for Clarée). Each fish was anaesthetized using phenoxy-ethanol, measured for body size, weighed, and individually tagged with PIT-tags. For each fish, we clipped a piece of caudal fin, and placed the clip in 96% ethanol for further genetic analysis. Fish were then placed in submerged cages for 14 days (3 cages per population) to ensure recovery from handling: one fish died in Val Travers, and three died in Clarée over this period. In each population, the surviving individuals were separated in two lots. The first lots (107 fish for Clarée, 109 fish for Val Travers) were released on site, so to have a proxy of resident (or local) fish survival through recapture in each population for the next years (2011 and 2012). The second lots (151 fish for both populations) were immediately transported by helicopter (15 min travelling time) to the other system (Val Travers origin to Clarée, and vice versa), where they were released as foreign individuals. We will refer to all these fish as cohort 0 (C0) hereafter, encompassing resident fish ("local") and transplanted fish ("foreign").

In three designated areas for each system, a sampling protocol using electrofishing (2-pass depletion method on a fixed area) was applied to estimate local densities (in 2010, 2011, and 2012, see Appendix A), and to potentially recapture C0 tagged fish (in 2011 and 2012). In 2012 and in 2018, we also sampled both systems specifically for young-of-the-year offspring (approximately 6 months old fish) at the same sites that were previously used for density estimation. We also extended these latter samplings to stretches of river between these sites, to minimize the risk of over-representing foreign or local parental contribution. These offspring were anaesthetized, then killed with an overdose of anesthetic, and kept in 96% ethanol for further genotyping. The 2012 sampling was conducted to detect the first potential F1 hybrid offspring between local and foreign parents from C0 (since transplanted individual ages ranged from 1+ to 3+ two years before, and most individual start reproducing at 5 years old in Kerguelen Is., [45]). These offspring will be referred as to C1 hereafter. The 2018 sampling was conducted to potentially detect not only F1 individuals, but also F2 of either local or foreign origin, and backcrosses with either local or foreign individuals. These offspring will be referred as to C2 hereafter.

2.3. Ethical Statement

At the time of the transplantation experiment (2010), no ethical committee was constituted and recognized in France. All procedures however were previously submitted to the scrutiny of the French Polar Institute as well as the Natural Reserve of the French sub Antarctic islands for evaluation, and were approved. For the 2018 sampling, authorization APAFIS#16249-201807241223324 was delivered by the French committee for ethics in animal experimentation $n°073$.

2.4. Genetic Analyses

To estimate the potential gene flow in each population after transplantation, we genotyped C0, C1, and C2 individuals using 15 microsatellite markers (see Appendix B for details regarding DNA extraction, markers amplification and genotyping methods, Appendix B Table A1 for markers error rates). These markers are located on different linkage groups [46], and provide satisfactory discriminant power to contrast the two populations (FST = 0.1426). The number of fish genotyped per population and per year is shown in Appendix B.

2.5. Genotypic Categories Assignation and Reconstruction of Families

We used the NewHybrids 1.1 software [47] to reconstruct the structure of gene flow in our transplantation experiment for each population. In essence, NewHybrids attempts to assign individuals to a mixture of genotypic categories representing the possible structure of the gene flow. Because the C1 samples could be only either from pure origin (local or foreign) or F1 hybrids, we ran a first analysis using only C0 genotypes and C1 genotypes.

C0 genotypes contributed to improve allelic frequencies estimation but were not used to estimate the π mixture, which was only assessed using C1 offspring genotypes (see [47]). Using these first assignments for the C1 offspring, we then ran a second analysis, integrating this time the C2 offspring samples, so as to determine their own specific π mixture, and to assign them to all possible genetic classes (pure local, pure foreign, F1, F2, backcross with local, backcross with foreign). This two-step analysis approach allowed us to better reflect the transplantation protocol and to benefit from our precise knowledge of the possible genetic categories that could potentially be found in 2012 and 2018, respectively. For each population, the analysis was realized using all samples genotyped on a minimum of 10 microsatellites. After 10,000 iterations for burning, 100,000 iterations were run to estimate the model's parameters on three different MCMC chains. We checked the stability of the estimates by running 3 different chains: the average difference in individual assignation probabilities to the various genotypic categories was 0.0003189 and 0.0001413 for Val Travers and Clarée, respectively.

To delineate families among C1 and C2 gene pools, an analysis was run separately in each river using COLONY 2.0.6.5 software [48]. In particular, COLONY uses multi-locus genotypes to infer sibship among samples. Potential parental genotypes were included in the analyses (84 females and 116 males for Val Travers, 51 males and 67 females for Clarée). We performed long runs, using weak priors and full likelihood method, assuming polygamy for males and females, with inbreeding, for diploid dioecious species. These tests were repeated three times to validate results manually. We then tagged families as either local, foreign, F1, F2 or backcrosses by simply matching the individual assignations obtained from NewHybrids with the family structure obtained from COLONY. In some cases, some families could not safely be assigned because they were composed of more than one type of offspring (for instance, both local individual and hybrid individual were found in the same family, for a same run of the analysis, or between runs). These families were not used in the following analyses (they represented 5.4% of families for Val Travers and 5.65% for Clarée).

All data files and additional settings for NewHybrids and Colony softwares are accessible online (https://doi.org/10.15454/NDFQJD).

2.6. Estimating Fitness

Our general approach to estimate relative fitness was to calculate the genetic contribution of the different genotypic categories of C0 individuals (local, foreign) to the C1 gene pool, and then the contribution the different genotypic categories of C1 individuals (local, F1, foreign) to the C2 gene pool. The approach is straightforward for C1 individuals: the data describing genotyping categories contain both C1 and C2 genotypic frequencies.

For C0 individuals, however, the initial proportions of transplanted (foreign) and resident (local) C0 individuals, in relationship with our field sampling protocol of C1 individuals, are not perfectly known. For instance, if foreign individuals move far from their release site and reproduce out of our sampling area, we might underestimate their total contribution to the next generation. To account for this, we here envisioned two different scenarios. In a first scenario, we assumed restricted dispersal, wherein foreign individuals would not move too far from their release site (and therefore from our sampling sites). To do so, we accounted for the usual home range known for brown trout, wherein most individuals remain within a 300 to 500 m linear of the river [49,50]. We multiplied this length by the average width of each river to obtain the surface area (which yielded about one hectare in both systems). We then calculated the likely proportion of foreign individuals by dividing the number of transplanted individuals by the total number of individuals expected on the surface area. The second scenario, however, assumed that individuals could move in the whole system, making the likelihood for them to sire offspring in our sampling area much smaller. The surface area in this scenario was therefore much bigger, and accounted for all habitable area for each population. Using these two scenarios enabled us to account for uncertainty in the proportion of transplanted individuals among potential parents in our sample, each scenario representing an extreme situation for dispersal or sampling bias regarding transplanted individuals. This uncertainty is thus accounted for in the calculation of C0 individuals' fitness.

To test whether fitness was different between the genotypic categories (for C0 and C1 individuals), we compared the observed genetic contribution to the expected genetic contribution assuming full random association between gametes, and equal survival and capture probabilities among offspring up to sampling date. We used Goodness-of-Fit X^2 tests to assess the statistical significance of differences between observed and expected genetic contributions.

2.7. Components of Selection

We looked at different components of selection during the transplantation experiments. First, we compared recaptured proportions between local and foreign C0 individuals, between 2010 and 2012, using the Fisher exact-test, as a proxy of their respective survival until the first potential reproduction.

We then ranked all individuals and families according to their hybrid index ([12]: 0 for local, 0.25 for F1xlocal, 0.5 for F1 and F2, 0.75 for F1xforeign, and 1 for foreign). For C1 and C2 gene pools, we tested whether family size (a proxy of survival between birth and sampling date, at 6 months old) was related to the hybrid index, using a polynomial model with Gaussian distributed error. The polynomial approach allows to detect linear and non-linear relationships between the hybrid index and the variable of interest. The statistical significance of linear and non-linear components of the model was assessed using F-tests on variance ratios.

Finally, because the two studied populations have been founded by very small numbers of parents and because selection against homozygotes was shown to be active [41], we assessed the Homozygosity Level of each individual (HL, [51]) using the Rhh package in R [52]. For all individuals (C0, C1 and C gene pools), we tested whether HL was related to the hybrid index, using again a polynomial model with Gaussian distributed error, and F-tests on variance ratios.

3. Results

3.1. Gene Flow

Through NewHybrids assignment, we determined the most probable genotypic categories in both populations for C1 individuals and then C2 individuals (Figure 2). In Val Travers, among the 432 C1 individuals sampled, 72.6% were pure local, 2.3% were pure foreign, and 25% were F1 hybrids. In Clarée, among 528 C1 individuals, 99.6% were pure local, none were foreign, and only 0.4% individuals were assigned as F1 hybrids. These results indicate that, in both populations, foreign transplanted individuals achieved some mating success, although with contrasting efficiencies. For the C2 individuals in Val Travers ($N = 236$), 42.8% were assigned as pure locals, 15.7% as F1 Hybrids, 5.5% as F2 Hybrids, and 33.5% and 2.5% as backcrosses with local and foreign categories, respectively. In Clarée, among 183 individuals sampled, 92.3% were pure local, and 7.7% were assigned as backcrosses with the local category. No pure foreign individuals were detected in Clarée nor Val Travers in the C2 gene pool. To sum up, gene flow occurred in both populations, although not to the same extent: whereas non pure local individuals represented 2.64% of the population in Clarée overall, they amounted to 37.87% in Val Travers.

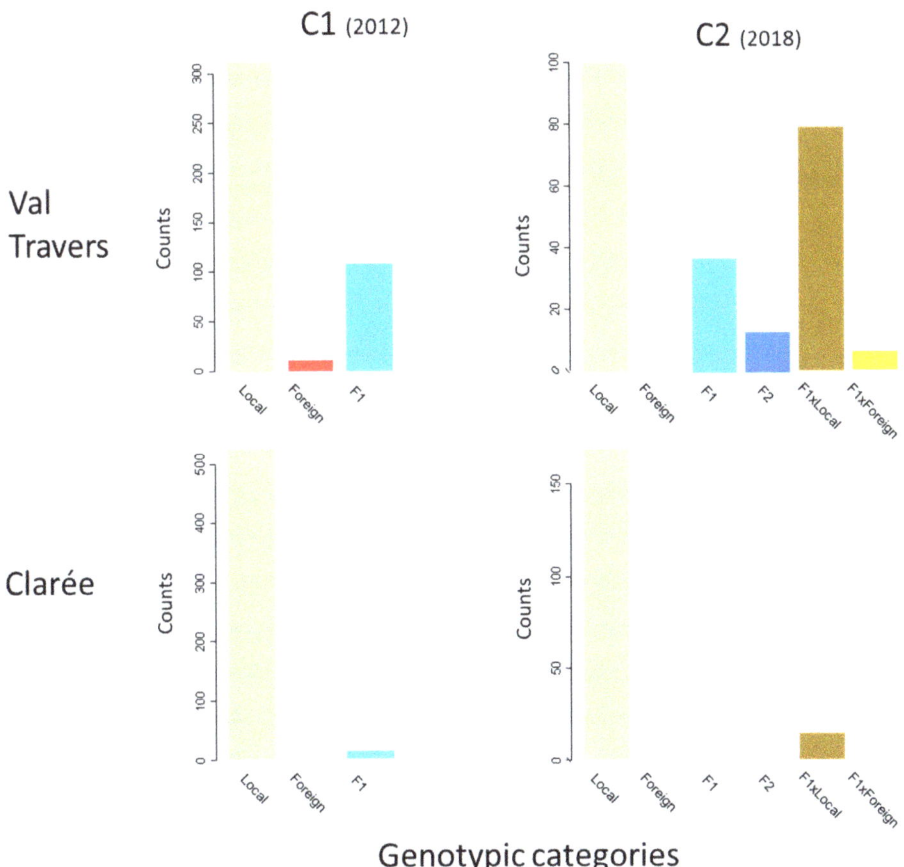

Figure 2. Counts of the different genotypic categories detected for C1 (**left**) and C2 (**right**) gene pools, in Val Travers (**top**) and Clarée (**bottom**) populations.

3.2. Fitness of the Different Genotypic Categories

To calculate the fitness of the C0 individuals from the different genetic groups, we first estimated the proportion of the foreign (transplanted) individuals in the populations relative to local individuals. To do so, we first estimated local densities to 2161 and 1475 individuals per hectare in Val Travers and Clarée, respectively (Appendices A and C). Under a restricted dispersal scenario, wherein the 151 foreign transplanted individuals would remain close from their release location, we estimated that they represented 6.98% and 10.24% of the sampled populations for Val Travers and Clarée, respectively (Appendix C, Tables A2–A5). Under this scenario, the genetic contribution of foreign C0 individuals was two-fold higher than expected in Val Travers (X^2 $_{1\,df}$, $p < 0.0001$, Table 1). On the contrary, the genetic contribution of foreign C0 individuals in Clarée was 36 times less than expected under random association of gametes ($p < 0.00001$).

When we relaxed the restricted dispersal assumption, assuming the foreign transplanted individuals could disperse in the whole system, the proportions of the foreign transplanted individuals were, respectively, 0.77% and 0.57% for Val Travers and Clarée (Appendix C, Tables A2–A5). Under this scenario, the genetic contribution of foreign C0 individuals was 20 times higher than expected in Val Travers (X^2 $_{1\,df}$, $p < 0.0001$, Table 1), whereas it was 2 times lower than expected in Clarée, although this latter difference was not statistically significant ($p = 0.08$).

Table 1. Goodness-of-fit Tests on the observed fitness (estimate of the number of offspring sired) for the C0 genotypic categories (local, foreign) in Val Travers and Clarée populations, assuming two contrasted dispersal scenarios (unrestricted and restricted) conditioning the initial percentage of transplanted individuals in each population.

Dispersal Scenario.(and Initial Percentage of Transplanted Individuals)	Population	Variables	Genotypic Category		Statistics	
			Local	Foreign	Sum	*p*-Value
(0.77%)	Val Travers	Expected fitness	428.6461	3.3539	432	
		Observed fitness	368	64	432	
		X^2 value	8.580	1096.612	**1105.192**	$p < 0.0001$
Unrestricted						
(0.57%)	Clarée	Expected fitness	524.9973	3.0027	528	
		Observed fitness	526.5	1.5	528	
		X^2 value	0.0043	0.7520	**0.7563**	$p = 0.08$
(6.98%)	Val Travers	Expected fitness	410.8147	30.1853	432	
		Observed fitness	368	64	432	
		X^2 value	2.8456	37.8806	**40.7262**	$p < 0.0001$
Restricted						
(10.24%)	Clarée	Expected fitness	473.9511	54.0489	528	
		Observed fitness	526.5	1.5	528	
		X^2 value	5.826	51.09	**56.916**	$p < 0.0001$

Therefore, foreign individuals clearly had a better fitness than local ones in Val Travers whatever the dispersal scenario considered. In Clarée, foreign individuals had a lower fitness under a restricted movement scenario, or were on par with local individuals under an unrestricted dispersal scenario.

C1 individuals, the observed genetic contribution of the two origins (foreign and local), differed from random expectation in Val Travers, with most of the deviation due to a greater contribution of foreign individuals, who produced 4 times more offspring than expected, whereas F1 individuals appeared to perform similarly to local individuals (X^2 $_{2\,df}$, $p < 0.0001$, Table 2). In Clarée, F1 hybrids appeared to outperform local individuals, producing 7 times more offspring than expected (X^2 $_{1\,df}$, $p < 0.0001$, Table 2).

Table 2. Goodness-of-fit Tests on the observed fitness (estimate of the number of offspring sired) for the C1 genotypic categories (local, F1, foreign) in Val Travers and Clarée populations.

Population	Variables	Genotypic Category			Statistics	
		Local	F1	Foreign	Sum	*p*-Value
Val Travers	Expected fitness	171.54	59	5.46	236	
	Observed fitness	159	55.5	21.5	236	
	Chi square value	0.916	0.207	47.078	**48.202**	$p < 0.0001$
Clarée	Expected fitness	181.96	1.04	0	183	
	Observed fitness	176	7	0	183	
	Chi square value	0.195	34.165	0	**34.360**	$p < 0.0001$

3.3. Components of Selection

We first considered potential difference in survival among C0 individuals after release. In Val Travers, 32 out of 109 local fish (29.36%) were recaptured, whereas only 1 out of 151 foreign fish (0.66%) was recaptured, suggesting a considerable apparent disadvantage for foreign individuals (Fisher's exact test, $p < 0.00001$). In Clarée, 5 out of 107 local fish (4.67%) were recaptured, and 7 out of 151 foreign fish (4.63%) were recaptured (Fisher's exact test, $p = 1$), indicating no apparent disadvantage for foreign individuals.

Investigating whether early family size (a proxy for offspring survival) could be related to Hybrid Index, we found no significant relationship between both variables in Val Travers ($p = 0.5424$ for the linear term, $p = 0.2898$ for the non-linear term) nor in Clarée ($p = 0.7906$ for the linear term, $p = 0.3385$ for the non-linear term, Table 3). We found a significant relationship between Homozygosity Level (HL) and Hybrid Index (Figure 3, Table 4) in Val Travers ($p < 0.0001$ for the linear term, $p < 0.0001$ for the non-linear term) and in Clarée ($p = 0.0231$ for the linear term, $p = 0.0337$ for the non-linear term): intermediate Hybrid Index individuals appeared to have lower HL values than extreme Hybrid Index individuals.

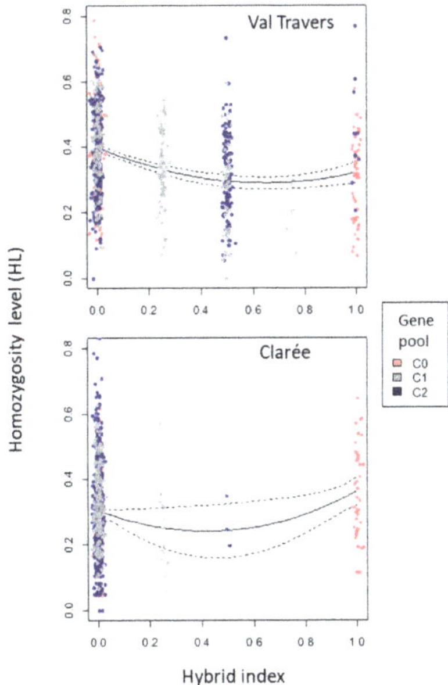

Figure 3. Relationship between Hybrid Index and Homozygosity Level for Val travers (**top** panel) and Clarée (**bottom** panel). The Hybrid index is 0 for local individuals, 0.25 for F1xlocal, 0.5 for F1 and F2, 0.75 for F1xforeign, and 1 for foreign individuals. Red, blue and green dots indicate C0, C1 and C2 individuals respectively. Jittering was added to the X coordinates to improve visibility. The full line represents the polynomial model prediction, the interrupted lines indicating the standard error around this fit.

Table 3. ANOVA analysis of the linear models testing for the linear and non-linear effects of Hybrid Index on family sizes in Val Travers and Clarée populations.

Population	Hybrid Index Effect	Degrees of Freedom	Mean Square	*F* Value	*p* Value
Val Travers	Linear term	1	0.064184	0.3717	0.5424
	Non-Linear term	1	0.194016	0.2898	0.2898
	Residuals	382	0.1726664		
Clarée	Linear term	1	0.004937	0.0706	0.7906
	Non-Linear term	1	0.064213	0.9179	0.3385
	Residuals	835	0.069953	1.2004	

Table 4. ANOVA analysis of the linear models testing for the linear and non-linear effects of Hybrid Index on individual Homozygosity Level in Val Travers and Clarée populations.

Population	Hybrid Index Effect	Degrees of Freedom	Mean Square	*F* Value	*p* Value
Val Travers	Linear term	1	1.19384	70.456	<0.0001
	Non-Linear term	1	0.45615	26.921	<0.0001
	Residuals	859	0.01694		
Clarée	Linear term	1	0.098087	5.1809	0.02309
	Non-Linear term	1	0.085691	4.5261	0.03367
	Residuals	835	0.018933	1.2004	

Additionally, when investigating the relationship between individual homozygosity HL and family size, we found that individuals originating from larger families had lower values of HL in Val Travers ($p < 0.0001$) and in Clarée, ($p < 0.0001$).

4. Discussion

Our study incorporated several key aspects that are usually hard to assemble in a single experiment: (1) both populations were founded at known dates by known numbers of individuals of known ages, (2) the transplantation experiment and our marker set were designed to efficiently distinguish the various genotypic categories over two generations, and, to some extent, (3) we could verify whether differences in fitness were related to particular life cycle stages or to genetic variation (heterozygosity). Of particular interest was the context of the recent foundations of our populations, which allowed us to simultaneously assess the speed at which LA arose and the potential role of GR. Beyond the usual complexities of reciprocal transplantation experiments [53], we also had to account for uncertainty in proportion of transplanted individuals relative to resident individuals. By envisioning two contrasting extreme scenarios for these proportions, we were able to explore the probable range of fitness differences for first generation foreign and local individuals. We also assessed the fitness of the second generation foreign, hybrid and local individuals. Our results are generally nuanced, with little support for LA, some hints that GR might be operating, and intriguing evidence that variation in mating success (i.e., sexual selection) was a key factor moderating gene flow.

4.1. Local Adaptation

LA is a widespread phenomenon in wild populations [44] although it is not always evident [54]. The speed at which such LA evolves is an active research topic, with some examples of substantial LA evolving in fewer than 10 generations [3,18,22]. In our study system, up to 4 generations had passed since the foundation of both populations, a length of time during which at least some selection has been at work. For instance, our previous work showed that selection against homozygosity was clearly active, especially in Val

Travers—which had been founded by the progeny of only two individuals. This form of selection was further confirmed in the present study, wherein we detected selection against homozygotes during early life, in both populations. Given ecological differences between their habitats and phenotypic divergence between the populations [55], LA was a logical candidate contributing to such selection. Indeed, evidence for LA evolving on such time scales has been reported for other salmonid systems [56–58].

Surprisingly, then, our experiment failed to observe clear footprints of LA that would have reduced gene flow after the transplantation. In the Clarée population, despite observing the same recapture rates between local and introduced foreign individuals, the genetic contribution of foreign individuals was potentially lower than that of local individuals. However, that result hinged on assumptions regarding dispersal of individuals in the system relative to our sampling efficiency. Additionally, although F1 hybrids were few in Claree, they had higher than expected fitness, indicating—at the least—they suffered no selective disadvantage overall. In Val Travers, we recaptured significantly fewer transplanted foreign individuals that expected. This difference could reflect LA, adaptive plasticity in the early stages of life (before the transplant took place), or a lack of local experience. The low recaptures rate of these individuals also could be related to behavioral response to transplantation, if transplanted individuals were more likely to disperse [35,59]. However, beyond recaptures of the transplanted individuals, most of our data suggested a lack of LA. In particular, transplanted foreign individuals seemingly had very high mating success, and the fitness of F1 hybrids was similar to that of local individuals. For instance, we did not find footprints of differential survival between local or hybrid individuals (a finding also present in [32]).

How can we explain this weak (if any) LA between our study populations? One possible explanation is that the spatial and temporal scale of LA is larger than the contrast examined in our experiment [22]. Indeed, the founding individuals all originated from the Kerguelen islands where the species was introduced and first reproduced naturally in 1962 (although they probably do not stem from the main strain in Europa though, possibly boosting available genetic variation through admixture, see [36]). Thus, prior to colonizing Val Travers and Claree, brown trout had been subject to about 10 generations of selection in these sub-Antarctic environments. This strong selection for adaptation to the overall Kerguelen conditions might have been the over-riding determinant of selection—as opposed to the finer-scale adaptation to local (stream-specific) conditions [60]. Another possible explanation is that the very low number of founders of the two study populations did not contain sufficient genetic variation to enable rapid LA—at least not within the 4 generations that we studied following establishment. Indeed, the general literature often points at reduced genetic variation as an obstacle for LA in small populations [10,61], which naturally leads us now to the "genetic rescue" hypothesis.

4.2. Genetic Rescue

GR had the potential to contribute substantially to the results of our experiment because both populations were strongly inbred, a situation where outbred hybrid offspring could be expected to have higher fitness than inbred resident offspring [10,61–64]. Although both populations provided tests of GR, the Val Travers population was especially informative due to substantial interbreeding that occurred between local and foreign individuals following the transplantation. Our assessment of GR combined three levels of insight: fitness of hybrids, footprints of selection favoring hybrids, and increased genetic variation in hybrids. Most importantly, the fitness of F1 hybrids was equal to the fitness of pure residents in Val Travers, and only slightly higher than the fitness of pure residents in Clarée, thus indicating weak (if any) footprints of GR. Additionally, we did not find any evidence of increased offspring survival (as assessed via the size of hybrid families relative to that of pure families). This absence of GR is not unprecedented in the literature on salmonids. For instance, Robinson et al. [35] also failed to find significant benefits of outbreeding for family size in brook trout (*Salvelinus fontinalis*). Interestingly, and like

Robinson et al. [35], we found some evidence of increased genetic variation as would be expected under GR, wherein individuals with intermediate hybrid index had lower homozygosities. This pattern was especially obvious in Val Travers, possible due to the higher sample size or perhaps because this population had initially very low effective size (Ne = 12, [41]). These patterns of selection could signal ongoing purging of inbreeding load in the population.

We emphasize that our study does not provide unequivocal evidence against the action of GR. First, although we do have demographic data on these two populations, they remain imprecise and it is too early to correctly assess a demographic change at the population scale related to the transplantation. Second, it is also too early to fully assess the fitness of the second generation of hybrid individuals (F2 and backcrosses) beyond finding similar family sizes compared to other genotypic categories. Instead, the benefits of increased genetic variation might become apparent later in the life cycle and might set the stage for further selection and adaptation processes that will perhaps be visible in future generations [10].

4.3. Other Drivers of Gene Flow

Having found only weak—if any—support for LA and GR as important drivers shaping patterns of fitness variation in our populations, we are left to suggest additional forces. We first note the different results obtained in the two populations. Clarée seems to conform adequately to a neutral scenario with no evident LA and only weak GR. In Val Travers, however, the fitness of foreign transplanted individuals was 2 to 20 times greater than expected, with the magnitude of the effect depending on the dispersal scenario. This pattern remained in the next generation where foreign individuals showed fitness five times greater than expected. This fitness is the product of two components in our experimental design: the mating success of a genotypic category, multiplied by the survival of the progeny until sampling (6 months-old here). But we demonstrated that family sizes were not different between the different genotypic categories, which implies equal offspring survival between these categories. Variation in fitness can thus be largely attributed to variation in mating success—that is, sexual selection.

These facts indicate that Clarée individuals introduced into the Val Travers may have either superior competitive ability to access potential sexual partners, and/or may be more attractive to local individuals. In general, it is often expected that mating preferences are related to LA, with a preference for locally adapted phenotypes thereby reinforcing the effect of LA on reproductive isolation [65–67]. However, this implies that local preference evolved quickly when dealing with newly founded populations on colonization front, a process possibly achieved by runaway selection, but often difficult to observe at the micro-evolutionary scale [68–71]. Alternatively, and more often, preference for dissimilar phenotype (i.e., inbreeding avoidance, MHC diversity, [72]) may offer a mating advantage to migrants [73] and potentially counterbalance the expected effect of LA. The average body size of introduced Claree individuals was also higher than the body size of Val Travers local individuals among C0 individuals, which could confer a competitive advantage to access sexual partners. However, such advantage should be limited, since they were introduced in an already established population, facing local competitors of higher body size and older ages that were not part of our C0 sampling.

Heterogamous sexual preference is certainly known in salmonids either related to phenotype and origin [74] or to difference in MHC genotypic variation [75,76], but see [77] for negative results in small Atlantic salmon populations. Paralleling that possibility, the effective number of breeders prior to transplantation in Val Travers (Ne = 12) was much lower than that in Clarée (Ne= 46, [41]). Although that difference might partly originate from the founding conditions, the former value is closer to monogamous mating systems expectation than the latter. Such potential difference in mating habits may possibly result in a higher mating success for introduced individuals originating from Clarée. Rare phenotypes can also be sexually favored over more common phenotypes (negative frequency dependent selection, [78,79]). The present nuance here is that it was observed only in one population. Alternatively, the Clarée individuals may bear a trait that would be attractive, notably for Val Travers individuals, but such a trait has to be non-plastic—since the transplanted individuals spent two years in the Val Travers environment before their first possible reproduction.

Finally, assortative mating is known to be context dependent in brown trout: Gauthey et al. [80] have shown that assortative mating was strong when river discharge was not predictable, whereas it could disappear (random mating) when river discharge became very predictable. Whereas the Val Travers watershed features a classic landscape from mountains brooks to a lowland plain, the Clarée system is under the strong influence of the upstream Hermance lake: wind variation on the lake can change the discharge in the river by a factor two or three in a matter of minutes. We correlatively observed that assortative mating with respect to genetic origins was strong in Clarée, a possibly very unpredictable system, whereas it did not occur in Val Travers, the most stable system.

In any case, mating success was probably the key component balancing the gene flow in this experiment: such gene flow could potentially erase any—undetected here—adaptation or founder effects in the next generations. Our finding adds to the growing evidence that sexual selection may have a tremendous effect on evolution [66,81–83]. It can possibly promote gene flow towards non-adaptive pathways [84–87], an outcome that we will endeavor to monitor in the next generations. In particular, changes in conditions could change the relative intensity of the two selection pressures, and upset the equilibrium between the strength of sexual and viability selection [86], shaping patterns of diversity along the colonization range.

Author Contributions: Conceptualization, J.L. and A.P.H.; methodology, J.L., A.P.H., A.M.; software, J.L and A.M.; lab analyses, F.G., A.M., M.T.; writing—original draft preparation, J.L., A.M., L.C.; writing—review and editing, J.L., L.C., A.P.H. All authors have read and agreed to the published version of the manuscript.

Funding: The present study is part of the SALMEVOL-1041 research project, member of the Zone Atelier Antarctique et Terres Australes LTSER, funded by the French Polar Institute and our own institute INRAE.

Acknowledgments: We also thank the French Polar Institute logistic staff (Y. LeMeur, R. Belleck, N. Marchand) for accompanying us and making this experiment possible and successful. We especially thank all the people involved in the field work over the years.

Conflicts of Interest: The authors declare no conflict of interest.

Appendix A. Model Code and Data for Density Estimation

(A)   Model code

The present code describes a 2-pass depletion sampling method with constant sampling effort to estimate local density, under OPENBUGS 3.X software. The model reproduces Carle and Strub's approach [88]. # j: removals

```
# k: number of removals
# m: number of sites sampled
# C0/C1: number of caught fish
# mu: catchability
# n: population size
# area: area in square meters.
# lambda: abundance/area
# density: expressed in individuals per square meters
```

```
model {
  for(i in 1:m){
    logit(mui[i])<-alpha[i]
    density[i]<-N[i]/area[i]
    N[i] ~ dpois( lambda[i])
    log(lambda[i])<- beta[i] *log(area[i]/1000+1)
    t[i,1] <- N[i]
    for ( j in 1:k ) {
      C1[i,j] ~ dbin(mui[i],t[i,j]) ## replace C1 by C2 to obtain estimates for fish > 2years
      t[i,j + 1] <- N[i]—sum(C1[i,1:j]) ## same here
    }
  alpha[i] ~dnorm(a,b)
  beta[i] ~dnorm(c,d)
  }
  a~dunif(−10,10)
  b~dgamma(0.01,0.01)
  c~dunif(0.1,100)
  d~dgamma(0.01,0.01)
}
```

(A) Data for Clarée

```
#example of inits
list(N = c(24,13,19,41,22),a = 1.366,alpha = c(0.5,1.152,0.676,1.008,1.178),b = 0.137,c = 8,d =
1,beta =c( 6,8,9,10,11),meanN = 35.0)
#data
# sampled areas for each site/date
list(area = c(975.645,213.885,173.2,400.325,219.2),
## captures of fish between 1 and 2 years old
C1 = structure(.Data = c(16,7,9,0,14,2,30,4,15,3),.Dim = c(5,2)),
## captures of fish above 2 years old
C2 = structure(.Data = c(0,2,11,4,8,0,0,0,1,0),.Dim = c(5,2)),
m = 5,
k = 2)
```

(A) Data for Val Travers

```
#example of inits
list(N = c(24,13,49,71,22,39),a = 1.366,alpha = c(0.5,1.152,0.676,1.008,1.178,1),b = 0.137,c =
8,d = 1,beta =c( 6,8,9,10,11,10),meanN = 35.0)
#data
# sampled areas for each site/date
list(area = c(44.1,85.89,340.88,838.125,298.8,360.85),
## captures of fish between 1 and 2 years old
C1 = structure(.Data = c(10,0,5,1,30,8,47,14,11,2,27,9),.Dim = c(6,2)),
## captures of fish above 2 years old
C2 = structure(.Data = c(2,2,5,2,18,4,17,2,13,3,5,1),.Dim = c(6,2)),
m = 6,
k = 2)
```

Appendix B. Genotyping Protocols, Genotyping Errors Detection, and Sample Sizes

(A) Genotyping

Multiplexes A and B amplifications were prepared as follows:4 µL of PCR diluted at 1/50 were added to a mix of 4.8 µL formamide and 0.2 µL genescan 500 LIZ size standard. C1 and C2 amplification were pool-plexed: 2 µL of each amplification diluted at 1/50 were added to the mix formamide-genescan 500 LIZ size standard. Preparations were then denatured at 95 °C during 5 min and put on ice during 5 min. Genotyping were realized on a 3100-Avant capillary sequencor and on a 3730XL capillary sequencer (thermofisher) respectively for 2003–2009 and 2012–2018 samples.

Alleles identification were realized with GENEMAPPER software for 2003 and 2009 samples and with STRAND 2.4.110 software for 2012 and 2018 samples. Binning was carried out with GENEMAPPER for 2003–2009 samples and with MsatAllele package on R 3.5.3 for 2012–2018 samples. The data set was then harmonized on the excel file created by R, throughout the time series, thanks to references samples which were analysed several time on the two sequencers used.

A thousand and seven hundreds sventy-seven trout were genotyped. 77 individuals were excluded of statistical analysis because they had more than 5 microsatellites not genotyped: 9 in 2003, 3 in 2009, 51 in 2012 and 14 in 2018.

So 1700 trout were used to identify families and realize genetic assignation: 136 in 2003 (40 from the Clarée system and 96 for the Val Travers system), 185 in 2009 (94 in Clarée, 91 in Val Travers), 960 in 2012 (528 in Clarée, 432 in Val Travers), 419 in 2018 (183 in Clarée, 236 in Val Travers).

(A)   Microsatellite Error Rates

Ninety-one samples were replicated to evaluate error rates due to laboratory and genotyping errors.

Thirteen microsatellites had a low error rate between 0 and 2.78%. StrUBA and Ssa121NVH had an error rate near 4% due to genotyping error (2nd allele undetected) or to an unspecific amplification which blurs the real allele signal. SSOSL438 was excluded of the analysis owing to a high error rate (7.83%). This marker had a high error rate due to an offset caused by different amplification intensities. This caused difficulties in harmonizing allele names during the time series of the data.

Table A1. Error rates for microsatellite markers arranged by multiplex (i.e., arrangement of markers in sets for simultaneous PCR amplification).

	multiplex A					
microsatellite error rates (%)	Ssa197 0	SsaD190 1.26	StrUBA 4.04	T3-13 2.78	SSOSL438 7.83	Ssa179NVH 2.27
	multiplex B					
microsatellite error rates (%)	Str58 0	Ssa103NVH 0	OmyRT5U 0.76	Ss4 0.51		
	multiplex C					
microsatellite error rates (%)	SSOSL85 0.51	SSOSL311 0.51	SsaT47Lee 1.01	Ssa121NVH 4.29	SSOSL417 1.26	Ssa159NVH 0.51

Appendix C. Density Estimates, and Estimates of the Proportion of Transplanted Individuals

Table A2. Density Estimates for the Clarée population. The density estimates are provided as the number of fish per hectare. Three different sites were selected and sampled in 2009, and two of them were revisited for sampling in 2010. Fish older than 2 years old were too rare in 3 samplings to safely estimate density, and were thus not accounted for.

Age Class	Sampling Date	2.5% Quantile	50% Quantile	97.5% Quantile
1< age < 2 years	27/12/2009	235.7	256.2	3659
	27/12/2009	420.8	420.8	514.3
	27/12/2009	923.8	923.8	1097
	23/12/2010	849.3	874.3	974.2
	23/12/2010	821.2	821.2	1049
Average			650.16	
age > 2 years	27/12/2009	607.8	841.6	2010
	27/12/2009	808.3	808.3	1212
Average			824.95	
Total density per hectare			1475.11	

Table A3. Proportion of transplanted individuals in the Clarée population.

Dispersal Scenario	Restricted	Unrestricted
Area considered (in hectare)	1	18
Population size	1475.11	26551.98
Number of transplanted individuals	151	151
Proportion of transplanted individuals	0.10236525	0.00568696

Table A4. Density Estimates for the Val Travers population. The density estimates are provided as the number of fish per hectare. Three different sites were selected and sampled in 2010, and were all revisited for sampling in 2011 and 2012.

Age Class	Sampling Date	2.5% Quantile	50% Quantile	97.5% Quantile
1< age < 2 years	04/01/2010	2268	2268	2721
	05/01/2010	698.6	698.6	1048
	06/01/2010	1115	1203	1437
	04/01/2011	727.8	787.5	918.7
	16/01/2012	435.1	468.5	602.4
	16/01/2012	997.6	1081	1358
Average			1084.43	
age > 2 years	04/01/2010	2721	3628	10,430
	05/01/2010	815	1514	7335
	06/01/2010	1027	1349	3286
	04/01/2011	584.6	680.1	1110
	16/01/2012	468.5	870.1	3514
	16/01/2012	775.9	969.9	8037
Average			1076.62	
Total density per hectare			2161.05	

Table A5. Proportion of transplanted individuals in the Val Travers population.

Dispersal Scenario	Restricted	Unrestricted
Area considered (in hectare)	1	9
Population size	2161.05	19,449.48
Number of transplanted individuals	151	151
Proportion of transplanted individuals	0.069873333	0.007763704

References

1. Williams, G.C. Natural Selection, the Costs of Reproduction, and a Refinement of Lack's Principle. *Am. Nat.* **1966**, *100*, 687–690. [CrossRef]
2. Baker, H.G.; Stebbins, G.L. Genetics of Colonizing Species, Proceedings. In *International Union of Biological Sciences Symposia on General Biology 1964*; Academic Press: Asilomar, CA, USA, 1965.
3. Hendry, A.P.; Wenburg, J.K.; Bentzen, P.; Volk, E.C.; Quinn, T.P. Rapid Evolution of Reproductive Isolation in the Wild: Evidence from Introduced Salmon. *Science* **2000**, *290*, 516–518. [CrossRef] [PubMed]
4. Barrett, S.C. Foundations of Invasion Genetics: The Baker and Stebbins Legacy. *Mol. Ecol.* **2015**, *24*, 1927–1941. [CrossRef] [PubMed]
5. Wright, S. Evolution in Mendelian Populations. *Genetics* **1931**, *16*, 97. [PubMed]
6. Garant, D.; Forde, S.E.; Hendry, A.P. The Multifarious Effects of Dispersal and Gene Flow on Contemporary Adaptation. *Funct. Ecol.* **2007**, *24*, 434–443. [CrossRef]
7. Von Wettberg, E.J.; Marques, E.; Murren, C.J. Local Adaptation or Foreign Advantage? Effective Use of a Single-Test Site Common Garden to Evaluate Adaptation across Ecological Scales. *New Phytol.* **2016**, *211*, 8–10. [CrossRef] [PubMed]
8. Fitzpatrick, S.W.; Reid, B.N. Does Gene Flow Aggravate or Alleviate Maladaptation to Environmental Stress in Small Populations? *Evol. Appl.* **2019**, *12*, 1402–1416. [CrossRef]
9. Hogg, J.T.; Forbes, S.H.; Steele, B.M.; Luikart, G. Genetic Rescue of an Insular Population of Large Mammals. *Proc. R. Soc. B Biol. Sci.* **2006**, *273*, 1491–1499. [CrossRef]
10. Frankham, R. Genetic Rescue of Small Inbred Populations: Meta-Analysis Reveals Large and Consistent Benefits of Gene Flow. *Mol. Ecol.* **2015**, *24*, 2610–2618. [CrossRef]
11. Tomasini, M.; Peischl, S. When Does Gene Flow Facilitate Evolutionary Rescue? *arXiv* **2020**, arXiv:bio/622142. [CrossRef]
12. Fitzpatrick, S.W.; Bradburd, G.S.; Kremer, C.T.; Salerno, P.E.; Angeloni, L.M.; Funk, W.C. Genomic and Fitness Consequences of Genetic Rescue in Wild Populations. *Curr. Biol.* **2020**, *30*, 517–522. [CrossRef] [PubMed]
13. Schiffers, K.; Bourne, E.C.; Lavergne, S.; Thuiller, W.; Travis, J.M. Limited Evolutionary Rescue of Locally Adapted Populations Facing Climate Change. *Philos. Trans. R. Soc. B Biol. Sci.* **2013**, *368*, 20120083. [CrossRef] [PubMed]
14. Bourne, E.C.; Bocedi, G.; Travis, J.M.; Pakeman, R.J.; Brooker, R.W.; Schiffers, K. Between Migration Load and Evolutionary Rescue: Dispersal, Adaptation and the Response of Spatially Structured Populations to Environmental Change. *Proc. R. Soc. B Biol. Sci.* **2014**, *281*, 20132795. [CrossRef] [PubMed]
15. Ceballos, G.; Ehrlich, P.R.; Dirzo, R. Biological Annihilation via the Ongoing Sixth Mass Extinction Signaled by Vertebrate Population Losses and Declines. *Proc. Natl. Acad. Sci. USA* **2017**, *114*, E6089–E6096. [CrossRef] [PubMed]
16. Razgour, O.; Forester, B.; Taggart, J.B.; Bekaert, M.; Juste, J.; Ibáñez, C.; Puechmaille, S.J.; Novella-Fernandez, R.; Alberdi, A.; Manel, S. Considering Adaptive Genetic Variation in Climate Change Vulnerability Assessment Reduces Species Range Loss Projections. *Proc. Natl. Acad. Sci. USA* **2019**, *116*, 10418–10423. [CrossRef]
17. Davis, M.B.; Shaw, R.G. Range Shifts and Adaptive Responses to Quaternary Climate Change. *Science* **2001**, *292*, 673–679. [CrossRef]
18. Taylor, E.B. A Review of Local Adaptation in Salmonidac, with Particular Reference to Pacific and Atlantic Salmon. *Aquaculture* **1991**, *98*, 185–207. [CrossRef]
19. Adkison, M.D. Population Differentiation in Pacific Salmons: Local Adaptation Genetic Drift, or the Environment? *Can. J. Fish. Aquat. Sci.* **1995**, *52*, 2762–2777. [CrossRef]
20. Haugen, T.O.; Vøllestad, L.A. Population Differences in Early Life-History Traits in Grayling. *J. Evol. Biol.* **2000**, *13*, 897–905. [CrossRef]
21. Garcia de Leaniz, C.; Fleming, I.A.; Einum, S.; Verspoor, E.; Jordan, W.C.; Consuegra, S.; Aubin-Horth, N.; Lajus, D.; Letcher, B.H.; Youngson, A.F. A Critical Review of Adaptive Genetic Variation in Atlantic Salmon: Implications for Conservation. *Biol. Rev.* **2007**, *82*, 173–211. [CrossRef]
22. Fraser, D.J.; Weir, L.K.; Bernatchez, L.; Hansen, M.M.; Taylor, E.B. Extent and Scale of Local Adaptation in Salmonid Fishes: Review and Meta-Analysis. *Heredity* **2011**, *106*, 404–420. [CrossRef] [PubMed]
23. Bartels, D.; Salamini, F. Desiccation Tolerance in the Resurrection Plant craterostigma Plantagineum. A Contribution to the Study of Drought Tolerance at the Molecular Level. *Plant Physiol.* **2001**, *127*, 1346–1353. [CrossRef]
24. Beacham, T.D.; Murray, C.B. Sexual Dimorphism in Length of Upper Jaw and Adipose Fin of Immature and Maturing Pacific Salmon (Oncorhynchus). *Aquaculture* **1986**, *58*, 269–276. [CrossRef]

25. Bernatchez, L. *Ecological Theory of Adaptive Radiation: Empirical Assessment from Coregonine Fishes (Salmoniformes)*; Hendry, A.P., Stearns, S.C., Eds.; Oxford University Press: Oxford, UK, 2004; pp. 175–207.

26. Derome, N.; Bernatchez, L. The Transcriptomics of Ecological Convergence between 2 Limnetic Coregonine Fishes (Salmonidae). *Mol. Biol. Evol.* **2006**, *23*, 2370–2378. [CrossRef] [PubMed]

27. Normandeau, E.; Hutchings, J.A.; Fraser, D.J.; Bernatchez, L. Population-Specific Gene Expression Responses to Hybridization between Farm and Wild Atlantic Salmon. *Evol. Appl.* **2009**, *2*, 489–503. [CrossRef] [PubMed]

28. Vandersteen Tymchuk, W.; O'reilly, P.; Bittman, J.; MacDonald, D.; Schulte, P. Conservation Genomics of Atlantic Salmon: Variation in Gene Expression between and within Regions of the Bay of Fundy. *Mol. Ecol.* **2010**, *19*, 1842–1859. [CrossRef] [PubMed]

29. Bernatchez, L.; Renaut, S.; Whiteley, A.R.; Derome, N.; Jeukens, J.; Landry, L.; Lu, G.; Nolte, A.W.; Østbye, K.; Rogers, S.M.; et al. On the Origin of Species: Insights from the Ecological Genomics of Lake Whitefish. *Philos. Trans. R. Soc. B Biol. Sci.* **2010**, *365*, 1783–1800. [CrossRef]

30. Goetz, F.; Rosauer, D.; Sitar, S.; Goetz, G.; Simchick, C.; Roberts, S.; Johnson, R.; Murphy, C.; Bronte, C.R.; Mackenzie, S. A Genetic Basis for the Phenotypic Differentiation between Siscowet and Lean Lake Trout (Salvelinus Namaycush). *Mol. Ecol.* **2010**, *19*, 176–196. [CrossRef]

31. Jensen, L.F.; Hansen, M.M.; Pertoldi, C.; Holdensgaard, G.; Mensberg, K.-L.D.; Loeschcke, V. Local Adaptation in Brown Trout Early Life-History Traits: Implications for Climate Change Adaptability. *Proc. R. Soc. B Biol. Sci.* **2008**, *275*, 2859–2868. [CrossRef]

32. Stelkens, R.B.; Jaffuel, G.; Escher, M.; Wedekind, C. Genetic and Phenotypic Population Divergence on a Microgeographic Scale in Brown Trout. *Mol. Ecol.* **2012**, *21*, 2896–2915. [CrossRef]

33. Westley, P.A.; Ward, E.J.; Fleming, I.A. Fine-scale local adaptation in an invasive freshwater fish has evolved in contemporary times. *Proc. R. Soc. B* **2013**, *280*, 2012–2327. [CrossRef] [PubMed]

34. Yamamoto, S.; Maekawa, K.; Tamate, T.; Koizumi, I.; Hasegawa, K.; Kubota, H. Genetic Evaluation of Translocation in Artificially Isolated Populations of White-Spotted Charr (Salvelinus Leucomaenis). *Fish. Res.* **2006**, *78*, 352–358. [CrossRef]

35. Robinson, Z.L.; Coombs, J.A.; Hudy, M.; Nislow, K.H.; Letcher, B.H.; Whiteley, A.R. Experimental Test of Genetic Rescue in Isolated Populations of Brook Trout. *Mol. Ecol.* **2017**, *26*, 4418–4433. [CrossRef]

36. Lecomte, F.; Beall, E.; Chat, J.; Davaine, P.; Gaudin, P. The Complete History of Salmonid Introductions in the Kerguelen Islands, Southern Ocean. *Polar Biol.* **2013**, *36*, 457–475. [CrossRef]

37. Labonne, J.; Vignon, M.; Prévost, E.; Lecomte, F.; Dodson, J.J.; Kaeuffer, R.; Aymes, J.-C.; Jarry, M.; Gaudin, P.; Davaine, P.; et al. Invasion Dynamics of a Fish-Free Landscape by Brown Trout (Salmo Trutta). *PLoS ONE* **2013**, *8*, e71052. [CrossRef] [PubMed]

38. Verfaillie, D.; Favier, V.; Dumont, M.; Jomelli, V.; Gilbert, A.; Brunstein, D.; Gallée, H.; Rinterknecht, V.; Menegoz, M.; Frenot, Y. Recent Glacier Decline in the Kerguelen Islands (49 S, 69 E) Derived from Modeling, Field Observations, and Satellite Data. *J. Geophys. Res. Earth Surf.* **2015**, *120*, 637–654. [CrossRef]

39. Pitman, K.J.; Moore, J.W.; Sloat, M.R.; Beaudreau, A.H.; Bidlack, A.L.; Brenner, R.E.; Hood, E.W.; Pess, G.R.; Mantua, N.J.; Milner, A.M.; et al. Glacier Retreat and Pacific Salmon. *BioScience* **2020**, *70*, 220–236. [CrossRef]

40. Kirkpatrick, M.; Barton, N.H. Evolution of a Species' Range. *Am. Nat.* **1997**, *150*, 1–23. [CrossRef]

41. Labonne, J.; Kaeuffer, R.; Guéraud, F.; Zhou, M.; Manicki, A.; Hendry, A.P. From the Bare Minimum: Genetics and Selection in Populations Founded by Only a Few Parents. *Evol. Ecol. Res.* **2016**, *17*, 21–34.

42. Turesson, G. The Genotypical Response of the Plant Species to the Habitat. *Hereditas* **1922**, *3*, 211–350. [CrossRef]

43. Hereford, J. A Quantitative Survey of Local Adaptation and Fitness Trade-Offs. *Am. Nat.* **2009**, *173*, 579–588. [CrossRef] [PubMed]

44. Kawecki, T.J.; Ebert, D. Conceptual Issues in Local Adaptation. *Ecol. Lett.* **2004**, *7*, 1225–1241. [CrossRef]

45. Jarry, M.; Beall, E.; Davaine, P.; Guéraud, F.; Gaudin, P.; Aymes, J.-C.; Labonne, J.; Vignon, M. Sea Trout (Salmo Trutta) Growth Patterns during Early Steps of Invasion in the Kerguelen Islands. *Polar Biol.* **2018**, *41*, 925–934. [CrossRef]

46. Gharbi, K.; Gautier, A.; Danzmann, R.G.; Gharbi, S.; Sakamoto, T.; Høyheim, B.; Taggart, J.B.; Cairney, M.; Powell, R.; Krieg, F.; et al. A Linkage Map for Brown Trout (Salmo Trutta): Chromosome Homeologies and Comparative Genome Organization with Other Salmonid Fish. *Genetics* **2006**, *172*, 2405–2419. [CrossRef] [PubMed]

47. Anderson, E.; Thompson, E. A Model-Based Method for Identifying Species Hybrids Using Multilocus Genetic Data. *Genetics* **2002**, *160*, 1217–1229. [PubMed]

48. Wang, J. Estimating Genotyping Errors from Genotype and Reconstructed Pedigree Data. *Methods Ecol. Evol.* **2018**, *9*, 109–120. [CrossRef]

49. Ovidio, M.; Philippart, J.-C.; Baras, É. Biais Methodologique Dans l'estimation Du Domaine Vital et de La Mobilite de S. Trutta Suite a l'adoption de Differentes Frequences de Localisation. *Aquat. Living Resour.* **2001**, *6*, 449–454. [CrossRef]

50. Jonsson, B.; Jonsson, N. Habitats as Template for Life Histories. In *Ecology of Atlantic Salmon and Brown Trout*; Springer: Berlin/Heidelberg, Germany, 2011; pp. 1–21.

51. Aparicio, J.; Ortego, J.; Cordero, P. What Should We Weigh to Estimate Heterozygosity, Alleles or Loci? *Mol. Ecol.* **2006**, *15*, 4659–4665. [CrossRef]

52. Alho, J.S.; Välimäki, K.; Merilä, J. Rhh: An R Extension for Estimating Multilocus Heterozygosity and Heterozygosity–Heterozygosity Correlation. *Mol. Ecol. Resour.* **2010**, *10*, 720–722. [CrossRef]

53. Delph, L.F. The Study of Local Adaptation: A Thriving Field of Research. *J. Hered.* **2017**, *109*, 1–2. [CrossRef]

54. Latreille, A.C.; Pichot, C. Local-Scale Diversity and Adaptation along Elevational Gradients Assessed by Reciprocal Transplant Experiments: Lack of Local Adaptation in Silver Fir Populations. *Ann. For. Sci.* **2017**, *74*, 77. [CrossRef]

55. Labonne, J.; Aymes, J.-C.; Beall, E.; Chat, J.; Dopico-Rodriguez, E.V.; Vasquez, E.G.; Gaudin, P.; Gueraud, F.; Henry, A.; Horreo-Escandon, J.-L.; et al. Ecologie Évolutive de La Colonisation Des Îles Kerguelen Par Les Salmonidés. Scientific report, Université de Pau et des Pays de l'Adour. Available online: https://hal.archives-ouvertes.fr/hal-01210200/document (accessed on 17 November 2020).

56. Pavey, S.A.; Collin, H.; Nosil, P.; Rogers, S.M. The Role of Gene Expression in Ecological Speciation. *Ann. N. Y. Acad. Sci.* **2010**, *1206*, 110. [CrossRef] [PubMed]

57. Quinn, T.P.; Kinnison, M.T.; Unwin, M.J. Evolution of Chinook Salmon (Oncorhynchus Tshawytscha) Populations in New Zealand: Pattern, Rate, and Process. In *Microevolution Rate, Pattern, Process*; Springer: Berlin/Heidelberg, Germany, 2001; pp. 493–513.

58. Aubin-Horth, N.; Bourque, J.-F.; Daigle, G.; Hedger, R.; Dodson, J.J. Longitudinal Gradients in Threshold Sizes for Alternative Male Life History Tactics in a Population of Atlantic Salmon (Salmo Salar). *Can. J. Fish. Aquat. Sci.* **2006**, *63*, 2067–2075. [CrossRef]

59. Miskelly, C.M.; Taylor, G.A.; Gummer, H.; Williams, R. Translocations of Eight Species of Burrow-Nesting Seabirds (Genera Pterodroma, Pelecanoides, Pachyptila and Puffinus: Family Procellariidae). *Biol. Conserv.* **2009**, *142*, 1965–1980. [CrossRef]

60. Sax, D.F.; Stachowicz, J.J.; Brown, J.H.; Bruno, J.F.; Dawson, M.N.; Gaines, S.D.; Grosberg, R.K.; Hastings, A.; Holt, R.D.; Mayfield, M.M.; et al. Ecological and Evolutionary Insights from Species Invasions. *Trends Ecol. Evol.* **2007**, *22*, 465–471. [CrossRef]

61. Charlesworth, D.; Charlesworth, B. Inbreeding Depression and Its Evolutionary Consequences. *Annu. Rev. Ecol. Syst.* **1987**, *18*, 237–268. [CrossRef]

62. Cain, B.; Wandera, A.B.; Shawcross, S.G.; Edwin Harris, W.; Stevens-Wood, B.; Kemp, S.J.; OKITA-OUMA, B.; Watts, P.C. Sex-Biased Inbreeding Effects on Reproductive Success and Home Range Size of the Critically Endangered Black Rhinoceros. *Conserv. Biol.* **2014**, *28*, 594–603. [CrossRef]

63. Forcada, J.; Hoffman, J.I. Climate Change Selects for Heterozygosity in a Declining Fur Seal Population. *Nature* **2014**, *511*, 462–465. [CrossRef]

64. Haanes, H.; Markussen, S.S.; Herfindal, I.; Røed, K.H.; Solberg, E.J.; Heim, M.; Midthjell, L.; Sæther, B.-E. Effects of Inbreeding on Fitness-Related Traits in a Small Isolated Moose Population. *Ecol. Evol.* **2013**, *3*, 4230–4242. [CrossRef]

65. Servedio, M.R. Beyond Reinforcement: The Evolution of Premating Isolation by Direct Selection on Preferences and Postmating, Prezygotic Incompatibilities. *Evolution* **2001**, *55*, 1909–1920. [CrossRef]

66. Svensson, E.I.; Eroukhmanoff, F.; Friberg, M. Effects of Natural and Sexual Selection on Adaptive Population Divergence and Premating Isolation in a Damselfly. *Evolution* **2006**, *60*, 1242–1253. [CrossRef] [PubMed]

67. Dolgin, E.; Whitlock, M.; Agrawal, A. Male Drosophila Melanogaster Have Higher Mating Success When Adapted to Their Thermal Environment. *J. Evol. Biol.* **2006**, *19*, 1894–1900. [CrossRef] [PubMed]

68. Hendry, A.P.; Nosil, P.; Rieseberg, L.H. The speed of ecological speciation. *Funct. Ecol.* **2007**, *21*, 455–464. [CrossRef] [PubMed]

69. Qvarnström, A.; Brommer, J.E.; Gustafsson, L. Testing the Genetics Underlying the Co-Evolution of Mate Choice and Ornament in the Wild. *Nature* **2006**, *441*, 84–86. [CrossRef]

70. Labonne, J.; Hendry, A.P. Natural and Sexual Selection Giveth and Taketh Away Reproductive Barriers: Models of Population Divergence in Guppies. *Am. Nat.* **2010**, *176*, 26–39. [CrossRef]

71. Maan, M.E.; Seehausen, O. Ecology, Sexual Selection and Speciation. *Ecol. Lett.* **2011**, *14*, 591–602. [CrossRef]

72. Wedekind, C.; Seebeck, T.; Bettens, F.; Paepke, A.J. MHC-Dependent Mate Preferences in Humans. *Proc. R. Soc. Lond. B Biol. Sci.* **1995**, *260*, 245–249.

73. Daniel, M.J.; Koffinas, L.; Hughes, K.A. Mating Preference for Novel Phenotypes Can Be Explained by General Neophilia in Female Guppies. *Am. Nat.* **2020**. [CrossRef]

74. Gil, J.; Caudron, A.; Labonne, J. Can Female Preference Drive Intraspecific Diversity Dynamics in Brown Trout (Salmo Trutta, L.)? *Ecol. Freshw. Fish* **2016**, *25*, 352–359. [CrossRef]

75. Forsberg, L.A.; Dannewitz, J.; Petersson, E.; Grahn, M. Influence of Genetic Dissimilarity in the Reproductive Success and Mate Choice of Brown Trout—Females Fishing for Optimal MHC Dissimilarity. *J. Evol. Biol.* **2007**, *20*, 1859–1869. [CrossRef]

76. Garner, S.R.; Bortoluzzi, R.N.; Heath, D.D.; Neff, B.D. Sexual Conflict Inhibits Female Mate Choice for Major Histocompatibility Complex Dissimilarity in Chinook Salmon. *Proc. R. Soc. B Biol. Sci.* **2010**, *277*, 885–894. [CrossRef] [PubMed]

77. Tentelier, C.; Barroso-Gomila, O.; Lepais, O.; Manicki, A.; Romero-Garmendia, I.; Jugo, B. Testing Mate Choice and Overdominance at MH in Natural Families of Atlantic Salmon Salmo Salar. *J. Fish Biol.* **2017**, *90*, 1644–1659. [CrossRef] [PubMed]

78. Ayala, F.J.; Campbell, C.A. Frequency-Dependent Selection. *Annu. Rev. Ecol. Syst.* **1974**, *5*, 115–138. [CrossRef]

79. Hughes, K.A.; DU, L.; RODD, F.H.; REZNICK, D.N. Familiarity Leads to Female Mate Preference for Novel Males in the Guppy, Poecilia Reticulata. *Anim. Behav.* **1999**, *58*, 907–916. [CrossRef]

80. Gauthey, Z.; Hendry, A.; Elosegi, A.; Tentelier, C.; Labonne, J. The Context Dependence of Assortative Mating: A Demonstration with Conspecific Salmonid Populations. *J. Evol. Biol.* **2016**, *29*, 1827–1835. [CrossRef]

81. Schwartz, A.K.; Hendry, A.P. Sexual Selection and the Detection of Ecological Speciation. *Evol. Ecol. Res.* **2006**, *8*, 399–413.

82. Svensson, E.; Gosden, T. Contemporary Evolution of Secondary Sexual Traits in the Wild. *Funct. Ecol.* **2007**, *21*, 422–433. [CrossRef]

83. Hendry, A.P. Ecological Speciation! Or the Lack Thereof? *Can. J. Fish. Aquat. Sci.* **2009**, *66*, 1383–1398. [CrossRef]

84. Candolin, U.; Heuschele, J. Is Sexual Selection Beneficial during Adaptation to Environmental Change? *Trends Ecol. Evol.* **2008**, *23*, 446–452. [CrossRef]
85. Doherty, P.F.; Sorci, G.; Royle, J.A.; Hines, J.E.; Nichols, J.D.; Boulinier, T. Sexual Selection Affects Local Extinction and Turnover in Bird Communities. *Proc. Natl. Acad. Sci. USA* **2003**, *100*, 5858–5862. [CrossRef]
86. Kokko, H.; Brooks, R. Sexy to Die for? Sexual Selection and the Risk of Extinction. In *Annales Zoologici Fennici*; JSTOR: New York, NY, USA, 2003; pp. 207–219.
87. Servedio, M.R.; Boughman, J.W. The Role of Sexual Selection in Local Adaptation and Speciation. *Annu. Rev. Ecol. Evol. Syst.* **2017**, *48*, 85–109. [CrossRef]
88. Carle, F.L.; Strub, M.R. A new method for estimating population size from removal data. *Biometrics* **1978**, *34*, 621–630. [CrossRef]

Article

Associative Overdominance and Negative Epistasis Shape Genome-Wide Ancestry Landscape in Supplemented Fish Populations

Maeva Leitwein *[iD], Hugo Cayuela and Louis Bernatchez

Institut de Biologie Intégrative et des Systèmes (IBIS), Université Laval, Québec, QC G1V 0A6, Canada;
hugo.cayuela51@gmail.com (H.C.); Louis.Bernatchez@bio.ulaval.ca (L.B.)
* Correspondence: maeva.leitwein.pro@gmail.com

Abstract: The interplay between recombination rate, genetic drift and selection modulates variation in genome-wide ancestry. Understanding the selective processes at play is of prime importance toward predicting potential beneficial or negative effects of supplementation with domestic strains (i.e., human-introduced strains). In a system of lacustrine populations supplemented with a single domestic strain, we documented how population genetic diversity and stocking intensity produced lake-specific patterns of domestic ancestry by taking the species' local recombination rate into consideration. We used 552 Brook Charr (*Salvelinus fontinalis*) from 22 small lacustrine populations, genotyped at ~32,400 mapped SNPs. We observed highly variable patterns of domestic ancestry between each of the 22 populations without any consistency in introgression patterns of the domestic ancestry. Our results suggest that such lake-specific ancestry patterns were mainly due to variable associative overdominance (AOD) effects among populations (i.e., potential positive effects due to the masking of possible deleterious alleles in low recombining regions). Signatures of AOD effects were also emphasized by highly variable patterns of genetic diversity among and within lakes, potentially driven by predominant genetic drift in those small isolated populations. Local negative effects such as negative epistasis (i.e., potential genetic incompatibilities between the native and the introduced population) potentially reflecting precursory signs of outbreeding depression were also observed at a chromosomal scale. Consequently, in order to improve conservation practices and management strategies, it became necessary to assess the consequences of supplementation at the population level by taking into account both genetic diversity and stocking intensity when available.

Keywords: introgression; associative-overdominance (AOD); stocking; genomic landscape; evolutionary mechanisms; salmonid

 check for updates

Citation: Leitwein, M.; Cayuela, H.; Bernatchez, L. Associative Overdominance and Negative Epistasis Shape Genome-Wide Ancestry Landscape in Supplemented Fish Populations. *Genes* **2021**, *12*, 524. https://doi.org/10.3390/genes12040524

Academic Editor: Nico M. Van Straalen

Received: 3 March 2021
Accepted: 31 March 2021
Published: 3 April 2021

Publisher's Note: MDPI stays neutral with regard to jurisdictional claims in published maps and institutional affiliations.

1. Introduction

Genetic admixture resulting from introgressive hybridization is a fundamental mechanism influencing populations and species evolution [1–5]. This may result from natural secondary contacts following an allopatric period of geographic isolation [3,6]. Genetic admixture may also result from human activities, including biological invasions, human-mediated translocation [7–10] or the supplementation of wild populations with non-local sources [11–13]. Although most of the introgressed genetic variation is likely neutral, admixture may introduce beneficial genetic variants improving the fitness of individuals in the recipient population [5,14,15]. Conversely, introgressed alleles may be maladaptive and have deleterious effects on fitness, for instance due to genomic incompatibilities [16–19]. Admixture from non-local populations may also have a dilution effect on locally adapted populations [20]. Despite the plethora of studies addressing these issues, few have documented how the interplay between the introduction pressure of foreign alleles (e.g., supplementation history), genetic drift and selection may produce complex admixture landscapes along the genome [21,22].

One important factor to consider for understanding the evolutionary outcomes of admixture is the local variation in recombination rate [13]. In the first generations following hybridization or in low recombining genomic regions, genetic drift and selection (either positive or negative) will act on large genomic tracts [13,22,23]. Hence, two main outcomes are expected: first, if the foreign population has a lower effective size (N_e) and a stronger genetic load (i.e., accumulation of recessive deleterious alleles) than the recipient population, outbreeding depression may occur due to genetic incompatibilities. Consequently, in the case of negative epistasis (i.e., genetic incompatibilities between the recipient and the foreign population), blocks of introgressed foreign ancestry are expected to be purged more quickly in low recombining regions of the genome [24]. Alternatively, if the genetic diversity of the foreign population is higher than that of the recipient population [11], positive effects may occur in first hybrids generation by masking the effect of link recessive deleterious alleles present in the recipient population (i.e., associative overdominance (AOD), [25,26]). Over time, introgressed tracts will shorten more quickly in highly recombining genomic regions [27]. Therefore, both beneficial or negative selective outcomes are expected to occur following genetic admixture [13,23]. The interplay between variation in recombination rate (which modulates the size of the introgressed tracts) and local genetic diversity is also expected to result in a complex genome wide admixture landscape.

To date, the understanding of such processes shaping genome-wide admixture landscapes remain obscure due to limited number of empirical studies that specifically addressed this issue [13,23]. Consequently, predicting the evolutionary outcomes of biological invasions or the supplementation practices (e.g., genetic rescue, stocking practices) remains challenging and important to develop viable conservation practices. In this study, we empirically investigate the evolutionary mechanisms shaping the genome-wide admixture landscape in fish populations supplemented with a foreign hatchery strain. The Brook Charr (*S. fontinalis*) is a socio-economically important fish species supporting a major recreational fishing industry in eastern North America. Consequently, the species has a long history of intense supplementation (i.e., stocking to supply to angling demand). In Eastern Canada, the domestication of local strains and the development of hatcheries has increased to supplement rivers and lakes over the last century [28–30]. Here and further on, we will used the term "domestic" to refer to the hatchery stock that has been used to supplement the populations we studied. Thus, population supplementation performed over the last 50 years in the Province of Québec has led to introgressive hybridization between stocked domestic strains and wild populations resulting in a complex admixture landscape along the genome [31] with an increase in domestic ancestry associated with the increase in stocking intensity [28,32]. Our research group previously evaluated the relationship between the domestic ancestry and the intensity of stocking practices [28], and we subsequently investigated the role of the local recombination rate in shaping the genome-wide ancestry landscape [31]. We observed that the main mechanism determining the domestic ancestry rate was associative overdominance (AOD), indicating that domestic ancestry was mainly favored by masking the effect of linked recessive deleterious alleles [13,31,33]. However, the influence of the genetic diversity and the intensity of stocking effort on the genome-wide landscape of ancestry remained to be investigated. Additionally, the relative contribution of AOD and negative epistasis on the ancestry landscape of the 22 studied populations still remained to be documented at both an intrapopulation level (i.e., among genomes) and an interpopulation level (i.e., similarities between populations). Using an integrative approach, we thus investigated how local population genetic diversity along the genome and the introduction pressure of domestic alleles (i.e., lakes stocking intensity) interplays with AOD and negative epistasis to produce lake-specific genome-wide ancestry patterns.

We predicted three alternative evolutionary scenarios considering the relationships between genome-wide domestic ancestry rate variation and (i) local recombination rate, (ii) local genetic diversity and (iii) lake-specific stocking history (Figure 1A–C). Scenario A corresponds to patterns of associative overdominance, with higher domestic ancestry in low recombining regions and in low-diversity genomic regions [13,31]. We also predicted more domestic ancestry for populations that have experienced higher stocking intensity, given that the probability for hybridization events is expected to increase with the number of fish stocked. Scenarios B and C correspond to negative epistasis predictions, where reduced domestic ancestry is expected in low recombining regions. Scenario B implies the introgression of potential beneficial domestic alleles, whereby we predict locally a lower genetic diversity due to the fixation of the beneficial alleles and a higher introgression rate of domestic ancestry. In Scenario C we predicted lower domestic ancestry in low recombining regions and in genomic regions of lower diversity due to genomic incompatibility. [13,34–36]. By taking advantage of a large dataset of 22 supplemented lacustrine populations of Brook Charr in Québec (previously genotyped using Genotype-By-Sequencing (GBS); [28]), we examined which evolutionary scenarios most probably explain the genome-wide domestic ancestry patterns in those populations. We searched for similarities in patterns of ancestry between populations that may reflect adaptive or maladaptive introgression from the domestic strain. Then, we determined which one of the aforementioned alternative scenarios (i.e., AOD vs. negative epistasis) best explained the observed patterns of domestic ancestry across linkage groups.

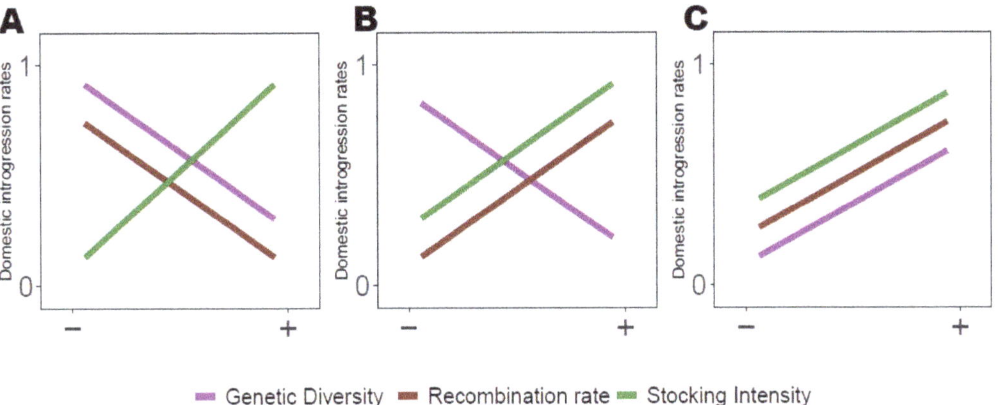

Figure 1. Predicted relationships between domestic ancestry rate, local recombination rate, local genetic diversity and stocking intensity. (**A**) Predicted relationship expected with associative overdominance (AOD), where more domestic ancestry are observed in low-diversity genomic region by masking the effects of potentially recessive deleterious mutation [13]. (**B,C**) Predicted relationship with negative epistasis where loci of incompatibility results in lower domestic ancestry in low recombining genomic regions. (**B**) Predicted relationship in presence of potential local beneficial alleles where lower diversity is expected due to the fixation of the beneficial alleles and higher introgression of domestic ancestry. (**C**) Predicted higher domestic ancestry in low-diversity genomic regions, namely due to the purge of incompatibility alleles [13,34]. X-axis illustrates the range of variation for each of the three variables (i.e., from lower to higher).

2. Materials and Methods

2.1. Study System and Sampling

The dataset used here and produced by Létourneau et al. (2018) comprises the genotypes of 553 Brook Charr collected from 22 isolated lakes from two wildlife reserves (Mastigouche and St-Maurice) in Québec, Canada as well as 37 individuals from the Truite de la Mauricie aquaculture domestic brood stock (Table 1, Figure S1). These lakes have been supplemented with this domestic Brook Charr strain which historically originated from crosses between two populations (Nashua and Baldwin) (details on supplementation history can be found in [3]). The domestic strain has been maintained for more than 100 years, and an average of 6–7 million domestic Brook Charr are released each year into the wild [28,37]. The history of stocking and the domestic strain used for supplementation has been rigorously recorded over time [28,37], and the number of fish stocked per hectare is reported in Table 1. Additional information on the domestic history can be found in Table S1.

Table 1. Description of the 22 sampled Brook Charr lakes in Québec.

Reserve	Lake	Label	N_samples_filters	N_late_hybrids	N_SNPs_mapped	lakes_size	total_ha	Mean Ancestry Rate	Mean Genetic Diversity
Mastigouche	Abénakis	ABE	21	16	31529	5	1915	0.0253	0.259
	Arbout	ARB	28	24	31713	5	380	0.0261	0.259
	Chamberlain	CHA	20	19	33290	18	958	0.0324	0.261
	Cougouar	COU	29	28	33455	8	1669	0.0426	0.252
	Deux-Etapes	DET	28	22	33580	12.3	3647	0.0282	0.246
	Gélinotte	GEL	25	18	31377	5	1652	0.0209	0.271
	Grignon	GRI	23	17	33298	29.6	846	0.0186	0.26
	Jones	JON	26	17	30553	28	468	0.0062	0.25
	Ledoux	LED	26	12	30941	13.7	3956	0.0663	0.236
	Lemay	LEM	28	19	30253	19.1	914	0.0538	0.23
Saint-Maurice	Brulô t	BRU	25	17	35463	8.1	247	0.0557	0.166
	Corbeil	COR	23	16	28071	9.5	526	0.0491	0.261
	Gaspard	GAS	28	14	34844	11.6	259	0.0766	0.168
	Maringouins	MAR	26	17	31102	6.2	1073	0.0734	0.23
	Melchior	MEL	26	20	30720	4.4	455	0.038	0.212
	Milord	MIL	25	16	41035	46.7	1175	0.0843	0.158
	Perdu	PER	24	17	32484	22.1	2293	0.075	0.247
	Porc-Epic	POE	26	17	32100	2.7	1648	0.0829	0.247
	Portage	POR	27	12	34415	46.9	1044	0.0615	0.157
	Tempête	TEM	17	11	32852	12.5	3784	0.046	0.286
	À la truite	TRU	26	21	31741	6.5	1814	0.0597	0.26
	Vierge	VIE	26	12	33747	5.7	208	0.0052	0.176

N_samples_filters: Number of individuals after filtering; N_late_hybrids: Number of individuals identified as late hybrids; N_SNPs_mapped: Number of mapped SNPs; lakes_size: Laked size in ha; total_ha: The total number of fish stocked/ha; Mean ancestry rate: Mean ancestry rate (estimated from ELAI) per lakes; Mean genetic diversity: Mean genetic diversity based on SNPs within 2 Mb sliding windows.

2.2. Genomic Data and Local Ancestry Inference

The SNP calling and ancestry inference of the 22 populations have been performed in [4]. Briefly, after applying quality filters, raw reads were demultiplexed with STACKS version 1.40 [38] and aligned to the closely related Arctic charr (S. alpinus) reference genome [39] with BWA_MEM version 0.7.9 [40]. SNP calling was performed separately for each population, and the domestic strain with STACKS version 1.40. Retained RAD loci had a minimum depth of four reads per locus, present in at least 60% of the populations, a minimum allele frequency of 2% and a maximum of 20% of missing data (see further details in Leitwein et al. [4]). To avoid merging paralogs, each individual locus with more than two alleles were removed with the R package STACKR [41]. The local ancestry inference was performed according to Leitwein et al. [5]. RAD loci were mapped against the Artic charr reference genome and ordered along each of the 42 Brook Charr linkage groups (LGs) by applying the MapComp software to the Brook Charr linkage map used as a reference [42]. MapComp allowed retrieving the mapping position of each marker by controlling for synteny and collinearity between the Artic Charr and the Brook Charr genomes [31]. The local ancestry inference was performed with the program ELAI version 1.01 [43] using the domestic strain as the foreign population and wild fish caught in each lake as the admixed recipient populations. ELAI was run 20 times for each 42 linkage groups (LG) to assess convergence. The number of upper clusters ($-C$) was set to 2 (i.e., assuming that each fish was a mixture of domestic and wild populations), the number of lower clusters ($-c$) to 15 and the number of expectation-maximization steps ($-s$) to 20 [31].

2.3. Describing Genome-Wide Variation of Domestic Ancestry and Genetic Diversity

We used the ancestry estimates provided in Leitwein et al. [4] to perform subsequent analyses. Briefly, the number and length of domestic tracts within each individual genome was retrieved from ELAI [43] and used to assess hybrids classes. The late-generation hybrids that mostly comprise wild-type ancestry and displayed a chromosomal ancestry imbalance (CAI) of < 0.125 [31,44] were retained for subsequent analyses (Table 1). We chose to remove the early hybrids (e.g., F1, F2, BC1 and BC2) as they were not represented by a sufficient number of fish to perform rigorous statistical analyses. For each population, we then assessed the genome-wide domestic ancestry rate for those late hybrid individuals.

We identified regions presenting excess or deficit of domestic ancestry considering local variation in recombination rate. To do so, we built a linear mixed model, where the log-transformed domestic ancestry rate was introduced as the response variable and the standardized (i.e., center-reduced) recombination rate was included as the explanatory variable (details on the recombination rate estimation can be found in Leitwein et al. [4]). The LGs and 2Mbp windows were incorporated as random effects in the model. We then retrieved the model residuals to highlight genomic regions displaying excess or deficit of domestic ancestry. Heatmaps of the domestic ancestry and the residuals of the model were plotted in R with the package "ggplot2" [45]. In order to obtain unbiased estimates of population genetic diversity, we removed all domestic ancestry tracts previously identified in the ancestry inference step with ELAI [43] from individual genomic data. We quantified the nucleotide diversity based on the population VCF (Variant Call Format) files. Then, we estimated the average nucleotide diversity using the R package PopGenome [46] along sliding windows of 2 Mb weighted by the number of observations (i.e., locus) within each window. The genetic diversity heatmap was plotted in R with the package "ggplot2" [45].

2.4. Evaluating the Influence of Genetic Diversity, Stocking Intensity and Recombination Rate on Genomewide Variation in Domestic Ancestry

We examined if and how local genetic diversity, stocking intensity (Table 1) and recombination rate simultaneously affected domestic ancestry at both the genome and linkage group (LG) levels. To estimate the genome-wide recombination rate, we generated a Brook Charr reference genome by anchoring the Brook Charr linkage map from Sutherland et al. [42] to the Artic charr reference genome after controlling for collinearity between those two sister species [31]. Then, MAREYMAP [47] was used to estimate the recombination rate by comparing the physical (pb) and genetic position (cM); the weighted mean recombination rate between the two closest markers was computed for the markers not included in the map [31].

At the genome level, we used linear mixed models where the log-transformed local domestic ancestry rate was included as the response variable. Recombination rate, genetic diversity and stocking intensity were standardized and incorporated as explanatory terms in the models. To consider the non-independency of ancestry estimates across windows, the LGs and the 2-Mb windows were treated as random effects in the models. We used a likelihood ratio test to assess the significance of the tested relationship by comparing the models with and without the explanatory term. We calculated marginal R^2 to quantify the proportion of variance explained by the explanatory variables only.

We then investigated how domestic ancestry was influenced by local genetic diversity, stocking intensity and recombination rate at the linkage group (LG) level. For this, linkage groups were analyzed separately by building linear mixed models which included as fixed effects the best-supported combination of variables (i.e., recombination, genetic diversity and stocking intensity, see Results Section 3.2) evaluated at the genome level. The 2-Mb sliding windows were incorporated as random effect. We evaluated the three possible scenarios presented in Figure 1 across the 42 LG by reporting the slope coefficient of the fixed effects and their 95% confidence intervals. We also calculated p-values to evaluate the significance of the three explanatory variables across linkage groups. To correct for multiple testing, we applied the conservative Bonferroni correction and only values with $p < 0.001$ were considered significant (i.e., $\alpha_{corrected} = 0.05/42$ LGs) [48]. All analyses were performed using the R package "LME4" [45].

3. Results

3.1. Genome-Wide Variation of Domestic Ancestry and Genetic Diversity

The genome-wide level of domestic ancestry rate was estimated for each population using the 405 late-hybrid individuals and an average of 32,442 ± 1799 mapped SNPs per population (Table 1). The domestic ancestry was highly variable among populations with mean rate ranging from 0.0052 (VIE population) to 0.0843 (MIL population) (Table 1). Domestic ancestry was also highly variable among individuals within population, ranging from 0.0000 to 0.6000 (Figure 2). When controlling for recombination rate, we did not observe any shared pattern of deficit or excess of introgression among populations. Consequently, each population presented unique pattern of introgression. Our analyses also showed that some genomic regions showed strong excess of domestic ancestry compared to the rest of the genome and those regions varied among populations. For instance, the GEL population displayed an excess of domestic ancestry for LG 40 and 42 whereas the GRI population showed an ancestry excess in LG 41 and LG 42 (Figure S2).

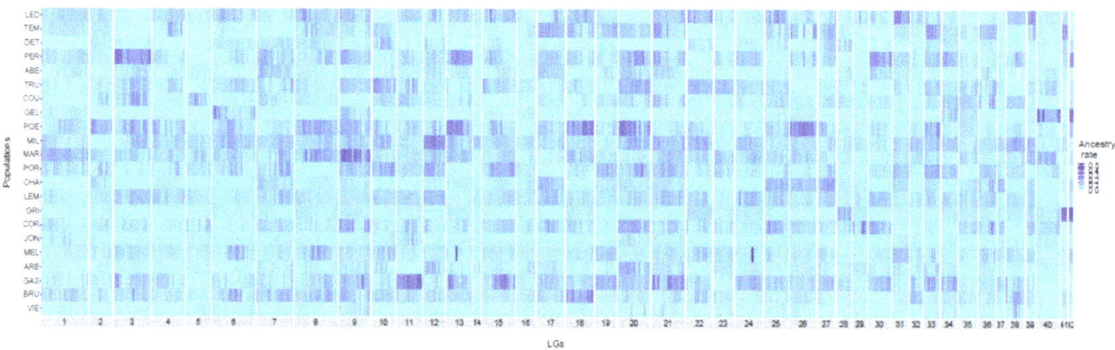

Figure 2. Heatmap of the proportion of domestic ancestry rate along the 42 Brook Charr linkage groups (LGs) for each lake. Lakes are ordered as a function of stocking intensity (less stocked lakes at the bottom). Labels are detailed in Table 1.

The level of genetic diversity estimated within 2MB sliding windows was highly variable both among and within populations. The mean population genetic diversity ranged from 0.157 (POR population) to 0.285 (TEM population). The MIL, BRU, GAS and VIE populations also displayed relatively low genetic diversity, with a mean of 0.158, 0.166, 0.168 and 0.176, respectively (Table 1 and Figure S3). Within each population, genetic diversity was also highly variable among individuals ranging from 0.0652 to 0.510 (Table 1 and Figure S3). While some genomic regions displayed particularly high level of genetic diversity, we did not observe any linkage group with a mean diversity significantly lower or higher in comparison to the whole genome (Table 1 and Figure S3).

3.2. Influence of Genetic Diversity, Stocking Intensity and Recombination Rate on Domestic Ancestry

At the genome-wide level, we observed more domestic ancestry in low recombining and low-diversity genomic regions (i.e., a pattern of associated overdominance (AOD) Figure 1, scenario A). The global R^2 and the marginal R^2 of the full model (i.e., where the log-transformed domestic ancestry rate was introduced as the response variable and the standardized (i.e., center reduced) recombination rate was included as the explanatory variable) were 0.084 and 0.015, respectively (Figure 3). The likelihood ratio test was significant for genetic diversity, which displayed the strongest effect ($\chi^2_{Diversity}$ = 5020.6, $p < 2.2 \times 10^{-16}$, coefficient slope = -4.95×10^{-3} (CI95%: -5.08×10^{-3}, -4.8×10^{-3}); Figure 3A), stocking intensity ($\chi^2_{Stocking}$ = 4143.9, $p < 2.2 \times 10^{-16}$; coefficient slope = 4.4×10^{-3}; (CI95%: 4.3×10^{-3}, -4.57×10^{-3}); Figure 3B), as well as recombination rate ($\chi^2_{Recombination}$ = 1797.7, $p < 2.2 \times 10^{-16}$; coefficient slope = -3.41×10^{-3} (CI95%: -3.57×10^{-3}, -3.26×10^{-3});

Figure 3C). We also found that domestic ancestry was negatively correlated with genetic diversity (slope coefficient $\beta_{Diversity}$ = -4.950×10^{-3}, $p < 2.2 \times 10^{-16}$; Figure 3A) and recombination rate ($\beta_{Recombination}$ = -3.418×10^{-3}, $p < 2.2 \times 10^{-16}$; Figure 3C). The domestic ancestry was positively associated with stocking intensity ($\beta_{Stocking}$ = 4.441×10^{-3}, $p < 2.2 \times 10^{-16}$) (i.e., Figure 3B).

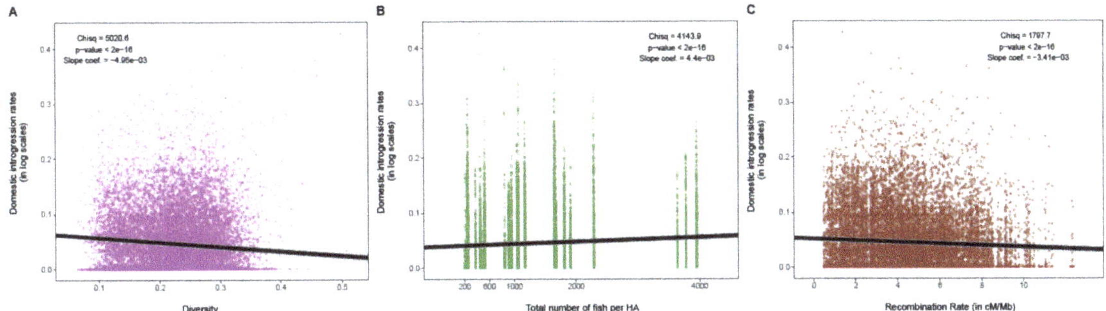

Figure 3. Whole genome relationship between the domestic ancestry and (**A**) genetic diversity, (**B**) stocking intensity and (**C**) local recombination rate, assessed with a generalized linear mixed model (GLM; R^2c = 0.084, R^2m = 0.015, AIC = -1952637, χ^2 = 4143.9, $p < 2.2 \times 10^{-16}$ (**B**); $\beta_{Recombination}$ = -3.418×10^{-3}, $\beta_{Diversity}$ = -4.950×10^{-3}, $\beta_{Stocking}$ = 4.441×10^{-3}, $p < 2.2 \times 10^{-16}$). Grey lines represent the 95% confidence interval.

At the linkage group level, we observed contrasted patterns of relationships between domestic ancestry, local recombination rate, local genetic diversity and stocking intensity among LGs (Figure 4). Ten LGs showed a pattern of associative overdominance with a negative correlation between the domestic ancestry rate and recombination rate and diversity. Among these, seven (LGs 8, 15, 18, 19, 22, 34 and 38; Figure 4) showed a positive correlation with stocking intensity (i.e., Figure 1, scenario A), and three LGs (LGs 9, 11 and 24; Figure 4) showed a higher domestic ancestry for the lakes harboring the lowest stocking intensity. Six LGs (LGs 1, 21, 25, 31, 36 and 39; Figure 4) displayed a pattern of negative epistasis supported by a positive correlation between domestic ancestry rate and recombination rate (Figure 1B,C). Among these, four LGs (LGs 1, 21, 31 and 39; Figure 4) displayed a negative correlation between domestic ancestry rate and genetic diversity (Figure 1, scenario B), and two LGs (LGs 25 and 36) displayed a positive correlation between domestic ancestry rate and all three explanatory variables (Figure 1C). Two LGs did not fit any of the three predicted scenarios (Figure 1), with LG16 displaying same pattern as for negative epistasis associated with the occurrence of beneficial alleles but with more domestic ancestry when the intensity of stocking was lower (Figure 4). LG 17 had more domestic ancestry within high genetic diversity and low recombining genomic regions (Figure 4).

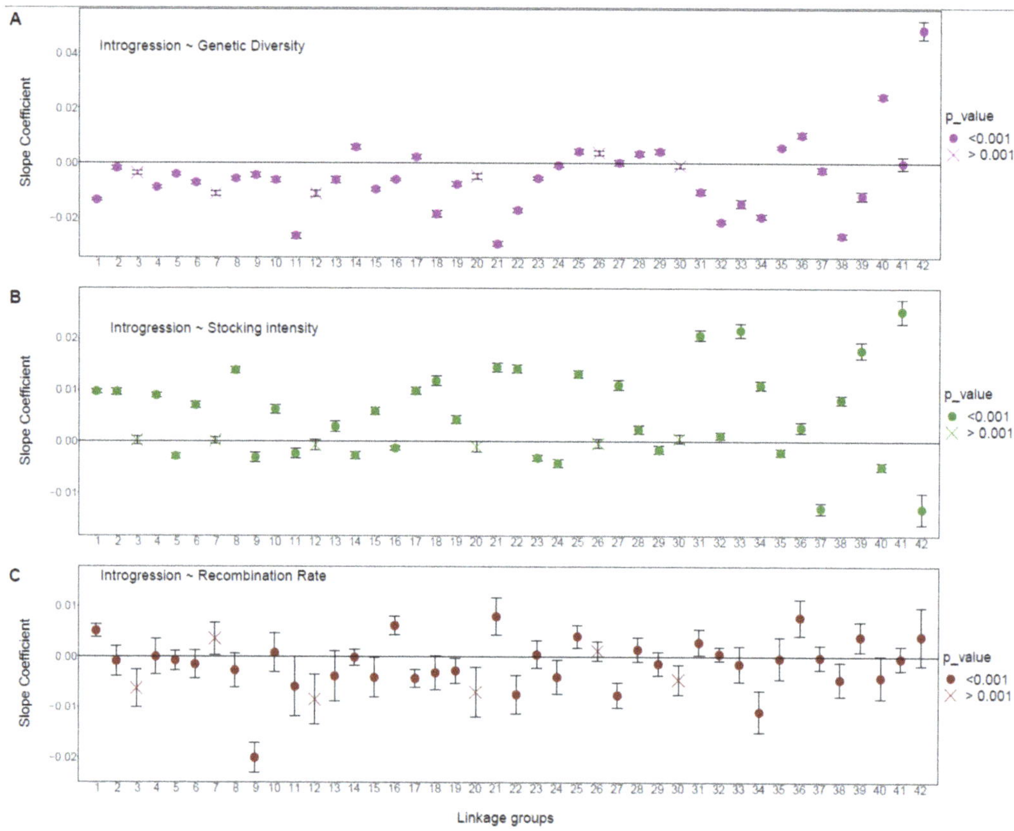

Figure 4. Relationship between the domestic ancestry rate and (**A**) genetic diversity, (**B**) stocking intensity and (**C**) local recombination rate for each linkage group. Slope coefficients of the GLM model are reported for each 42 Brook Charr linkage group. Dots represent the significance of each model with circles and crosses representing $p < 0.001$ and $p > 0.001$, along with the 95% confidence interval, respectively.

4. Discussion

Our study suggests that genetic admixture resulting from supplementation of Brook Charr populations with a domestic strain during the last century produced a lake-specific pattern of ancestry modulated by the interplay between local recombination rate, local genetic diversity and stocking intensity. As previously shown by Létourneau et al. (2018) [28] based on single SNP information, but here based on haplotype information, domestic ancestry increased with stocking intensity. Based on a correlative approach using linear mixed models, our results suggest that associative overdominance (AOD) is the main evolutionary mechanism shaping the genome-wide ancestry landscape among the 22 lakes analyzed. At the linkage group scale, however, our results suggest that negative epistasis effects may also play a role in shaping the local pattern of domestic ancestry, with more domestic ancestry in highly recombining regions. As such, our study suggest that supplementation of small isolated lacustrine populations could result in some form of genetic rescue [15], although potential negative effects possibly reflecting genetic incompatibilities may also occur locally along the genome.

4.1. Associative Overdominance as the Main Mechanism Shaping Domestic Ancestry Landscape

Evidence for associative overdominance was supported by prevailing domestic ancestry in low recombining and low diversity genomic regions which could result in the masking of potentially deleterious alleles. AOD effects are particularly expected when the foreign strain used for supplementation harbors a higher level of genetic diversity than populations being supplemented [15,49]. In our study system, the domestic strain does display higher genetic diversity than the wild lacustrine populations [50], which could indeed favor AOD effects. Additionally, populations with small N_e such as wild Brook Charr found in small isolated lakes tend to accumulate more deleterious alleles and suffer from inbreeding depression [20,51]. The lacustrine Brook Charr populations studied here are isolated, display small N_e [52] and tend to accumulate more deleterious mutations compared to more connected populations [53]. An increasing number of studies has documented that in some circumstances, supplementation of wild populations can result in genetic rescue which translated in an increase of both census and effective population size by mitigating fitness loss associated with inbreeding and reduced genetic polymorphism [20,54–56]. To our knowledge, however, only a few studies have provided empirical evidence for the underlying role of associative overdominance [57] and/or the variable consequences of admixture over time following the onset of supplementation [58].

Although AOD could be the main mechanism shaping domestic ancestry patterns at the genome level, the strength of its effect was variable among LGs. The domestic ancestry landscape was dominated by AOD in 10 LGs (Figure 4), where we observed more domestic ancestry in genomic regions characterized by a relatively low recombining rate and low diversity. As both local recombination rate and genetic diversity were variable along the genome, it is not surprising that some linkage groups displayed more AOD signals than others. Additionally, genomic regions tending to accumulate more recessive deleterious mutations will be more subject to AOD effects [59]. To our knowledge, only a few studies have attempted to document genome wide variation of AOD effects. In particular, chromosomal variation of AOD effects has been observed between autosomal and X-chromosomes of *Drosophila melanogaster* [60], whereby autosomal chromosomes maintain higher genetic diversity due to AOD. Variation in AOD effects was also documented in humans, whereby 22 genomic regions displayed unusual peaks of diversity in low recombining regions associated with AOD effects [61].

Overall, while highly significant from a statistical standpoint, the effect size of the correlations between foreign ancestry, recombination, diversity and stocking were small, which can most likely be imputable to the diluting effect of the statistical signals caused by the large amount of data. Given that the effective population sizes (Ne) of lacustrine Brook Charr populations is generally very small (Median Ne of 35 CI 32:28 in [53] and 53.6 CI 5.8:1069.4 in [52]), genetic drift is expected to be pronounced. Consequently, it may have contributed to increase the proportion of unexplained genetic variance. Moreover, the relatively recent stocking history (a mean duration of stocking of 17 years over all 22 populations, Table S1) and the time at maturity of three years for the Brook Charr [31,37] likely decreased our ability to detect a strong genomic signature of AOD and negative epistasis. Indeed, it has been shown with simulated data that the ability to detect genetic signals of adaptive and maladaptive introgression increases with the number of generations after the introduction of the foreign genotypes [23].

4.2. Negative Epistasis also Contributes to the Dynamics of Domestic Ancestry Landscape

Negative epistasis also interplayed with AOD in modulating the dynamics of the genome-wide variation in domestic ancestry rate, with more domestic ancestry observed in high recombining regions of six LGs. In theory, hybridization between two species and even divergent populations of the same species may lead to "Dobzhansky–Muller" incompatibilities [16,19,62–64]. In particular, individuals bred in captivity may adapt to captive conditions and can then develop genetic incompatibilities with their wild counterparts, leading to outbreeding depression when they are released into the wild [65,66], in addition to potentially diluting local adaptation. Indeed, outbreeding depression has previously been documented in wild populations following supplementation with captive populations [67], and these cases have received considerable attention from conservation biologists over the last three decades [20,68–72]. Here, it is therefore not surprising to detect in some genomic regions the molecular signature of negative epistasis since the domestic strain has been bred in captivity for more than 100 years (or at least 30 generations). Along with signals of negative epistasic effects, we were also expecting to observe a higher level of domestic ancestry in genomic regions harboring the highest level of genetic diversity, given that regions with high recombination rate generally tend to show high genetic diversity [73]. However, and contrary to expectations, we observed more domestic ancestry in low diversity genomic regions for four LGs. This pattern could hypothetically be explained by the presence of beneficial domestic alleles as in the case of fixation of a beneficial allele through selective sweeps, which may result in locally reduced neutral variation around such beneficial alleles [74].

4.3. Perspectives for Conservation

Assessing the evolutionary mechanisms at play and their consequences in the context of population supplementation can help to better understand the tradeoff between costs and benefits of human-driven hybridization that may result from supplementation practices [13]. Stocking programs in a fishery context are conducted to increase the census size of fish populations and the number of fish that can be harvested but often neglect the evolutionary consequences of introducing foreign genotypes in the supplemented populations. Ultimately, being able to more accurately predict whether the introduction of foreign alleles will improve (i.e., genetic rescue) or decrease (i.e., outbreeding depression) the mean fitness of the supplemented populations could inform management decisions and assist in designing conservation programs [20,58,67,69]. Here, however, our results suggest that the outcomes of supplementation are largely unpredictable as they showed that each population displayed a unique pattern of ancestry landscape following stocking operations using the same domestic strain. After controlling for local variation in recombination rate along the genome, our analyses provided little evidence for shared patterns of domestic ancestry among populations. Such population-specific ancestry patterns could be partly explained by the highly variable pattern of genome-wide genetic diversity among populations, a likely consequence of pronounced genetic drift being enhanced by small effective population sizes of these populations [52] and very limited gene flow between them [53]. Furthermore, local adaptation to contrasted environmental conditions prevailing in the different lakes [53] likely modulates the variation in genetic diversity along the genome via soft selective sweeps. This may favor or disfavor the foreign alleles introduced and could therefore contribute to generate the population-specific ancestry landscape observed in our study and in previous ones [32,50]. Admittedly, however, in the absence of phenotypic and/or fitness information predating and postdating supplementation, the resulting demographic consequences of the opposite effects of AOD and negative epistasis we have documented here remain hypothetical and deserve further investigation. Nevertheless, our results suggest that supplementation could possibly be acting as a form of genetic rescue in those small isolated populations characterized by very small N_e and a low level of genetic diversity. Yet, this potentially positive asset must be considered cautiously as the use of the same domestic strain to supplement multiple populations can lead to an homogenization of the population genetic structure [50] and to the dilution of local adaptation via the disruption of co-adapted gene networks [32,75]. Moreover, our results also suggest that, alongside the main positive effects of AOD, negative epistasis may also determine local variation of domestic ancestry, which may in turn be at the source of outbreeding depression, ultimately impacting the outcomes of supplementations [76,77]. Finally, while AOD effects decrease with time after hybridization, maladaptive effects could arise later on, especially in populations with small N_e [78].

5. Conclusions

To conclude, our study emphasizes the need for conservation biologists to examine how the complex interplay between the introduction pressure of exogenic genetic makeup, genetic diversity and recombination rate along the genome modulates introgression landscapes of recipient populations. As such, this study offers new avenues for the study of the mechanisms shaping the evolution of exploited populations and regulating the dynamics of hybridization, either occurring naturally (e.g., hybrid zones) or caused by humans.

Supplementary Materials: The following are available online at https://www.mdpi.com/article/10.3390/genes12040524/s1, Figure S1: Geographical localization of sampled lakes within two wildlife reserves (Mastigouche and St Maurice) in the province of Quebec, Canada. Lakes labels are reported in Table S1; Figure S2: Heatmap of the residuals of the linear mixed model of the domestic ancestry rate controlled by the recombination controlled by the recombination models along the 42 Brook Charr linkage groups (LGs) for each lake. Lakes are ordered as a function of stocking intensity (less stocked at the bottom), labels are detailed in Table S1. Figure S3: Heatmap of the variation in genetic diversity estimated on SNPs within 2Mb sliding windows along the 42 Brook Charr linkage groups (LGs) for each lake. Lakes are ordered as a function of stocking intensity (less stocked at the bottom), labels are described in Table S1; Table S1: Description of the 22 sampled Brook Charr lakes in Québec, Canada along with the stocking variables: lakes_size: size of the lake in hectare; total_ha: the total number of fish stocked per hectare; mean_year: the number of years since the mean year of stocking; nb_stock_ev: The total number of stocking events; mean_stock_fish: The mean number of fish stocked per stocking event; FIRST_STOCK: First year of stocking events; LAST_STOCK: Last year of stocking events and DURATION: Number of years of stocking events.

Author Contributions: M.L., H.C. and L.B. conceived the study. M.L. and H.C. designed the analyses and performed the analyses. M.L. wrote the manuscript. H.C. helped with the manuscript structure, and all co-authors critically revised the manuscript and approved the final version to be published. All authors have read and agreed to the published version of the manuscript.

Funding: This research was funded by the MFFP, the SEPAQ, the Canadian Research Chair in Genomics and Conservation of Aquatic Resources, Ressources Aquatiques Québec (RAQ), as well as by a Strategic Project Grant from the Natural Science and Engineering Research Council of Canada (NSERC) to L. Bernatchez, D. Garant and P. Sirois.

Institutional Review Board Statement: Not applicable.

Informed Consent Statement: Not applicable.

Data Availability Statement: The data presented in this study are openly available in Létourneau et al. [3] at Dryad Digital Repository: https://doi.org/10.5061/dryad.s5qt3.

Acknowledgments: We thank Martin Laporte and Eric Normandeau for their advices and statistical help. We thank the biologists and technicians of the Société d'Etablissement de Plain Air du Québec (SEPAQ) and the Ministère des Forêts, de la Faune et des Parcs du Québec (MFFP), in particular Amélie Gilbert and Isabel Thibault for their implication in the project and/or their field assistance and sampling. We are also grateful to R. Waples and one anonymous reviewer for their constructive comments on a previous version of this manuscript.

Conflicts of Interest: The authors declare no conflict of interest.

References

1. Anderson, E. *Introgressive Hybridization*; John Wiley and Sons, Inc.: New York, NY, USA, 1949.
2. Dowling, T.E.; Secor, A.C.L. The role of hybridization and introgression in the diversification of animals. *Annu. Rev. Ecol. Syst.* **1997**, *28*, 593–619. [CrossRef]
3. Barton, N.H. The role of hybridization in evolution. *Mol. Ecol.* **2008**, *10*, 551–568. [CrossRef]
4. Harrison, R.G.; Larson, E.L. Hybridization, Introgression, and the Nature of Species Boundaries. *J. Hered.* **2014**, *105*, 795–809. [CrossRef]
5. Racimo, F.; Sankararaman, S.; Nielsen, R.; Huerta-Sánchez, E. Evidence for archaic adaptive introgression in humans. *Nat. Rev. Genet.* **2015**, *16*, 359–371. [CrossRef] [PubMed]
6. Lindtke, D.; Buerkle, C.A. The genetic architecture of hybrid incompatibilities and their effect on barriers to introgression in secondary contact. *Evol.* **2015**, *69*, 1987–2004. [CrossRef] [PubMed]

7.   Rhymer, J.M.; Simberloff, D. Extinction by hybridization and introgression. *Annu. Rev. Ecol. Syst.* **1996**, *27*, 83–109. [CrossRef]

8.   Currat, M.; Ruedi, M.; Petit, R.J.; Excoffier, L. The hidden side of invasions: Massive introgression by local genes. *Evol. Int. J. Org. Evol.* **2008**, *62*, 1908–1920. [CrossRef]

9.   Knytl, M.; Kalous, L.; Symonova, R.; Rylková, K.; Ráb, P. Chromosome Studies of European Cyprinid Fishes: Cross-Species Painting Reveals Natural Allotetraploid Origin of a Carassius Female with 206 Chromosomes. *Cytogenet. Genome Res.* **2013**, *139*, 276–283. [CrossRef] [PubMed]

10.   Knytl, M.; Kalous, L.; Rylková, K.; Choleva, L.; Merilä, J.; Ráb, P. Morphologically indistinguishable hybrid Carassius female with 156 chromosomes: A threat for the threatened crucian carp, C. carassius, L. *PLoS ONE* **2018**, *13*, e0190924. [CrossRef]

11.   Laikre, L.; Schwartz, M.K.; Waples, R.S.; Ryman, N.; GeM Working Group. Compromising Genetic Diversity in the Wild: Unmonitored Large-Scale Release of Plants and Animals. *Trends Ecol. Evol.* **2010**, *25*, 520–529. [CrossRef]

12.   Hagen, I.J.; Jensen, A.J.; Bolstad, G.H.; Diserud, O.H.; Hindar, K.; Lo, H.; Karlsson, S. Supplementary stocking selects for domesticated genotypes. *Nat. Commun.* **2019**, *10*, 199. [CrossRef]

13.   Leitwein, M.; Duranton, M.; Rougemont, Q.; Gagnaire, P.-A.; Bernatchez, L. Using Haplotype Information for Conservation Genomics. *Trends Ecol. Evol.* **2020**, *35*, 245–258. [CrossRef]

14.   Hedrick, P.W. Adaptive introgression in animals: Examples and comparison to new mutation and standing variation as sources of adaptive variation. *Mol. Ecol.* **2013**, *22*, 4606–4618. [CrossRef] [PubMed]

15.   Frankham, R. Genetic rescue of small inbred populations: Meta-analysis reveals large and consistent benefits of gene flow. *Mol. Ecol.* **2015**, *24*, 2610–2618. [CrossRef] [PubMed]

16.   Orr, H.A. The population genetics of speciation: The evolution of hybrid incompatibilities. *Genetics* **1995**, *139*, 1805–1813. [CrossRef] [PubMed]

17.   Maheshwari, S.; Barbash, D.A. The Genetics of Hybrid Incompatibilities. *Annu. Rev. Genet.* **2011**, *45*, 331–355. [CrossRef] [PubMed]

18.   Dion-Côté, A.-M.; Renaut, S.; Normandeau, E.; Bernatchez, L. RNA-seq Reveals Transcriptomic Shock Involving Transposable Elements Reactivation in Hybrids of Young Lake Whitefish Species. *Mol. Biol. Evol.* **2014**, *31*, 1188–1199. [CrossRef] [PubMed]

19.   Laporte, M.; Le Luyer, J.; Rougeux, C.; Dion-Côté, A.-M.; Krick, M.; Bernatchez, L. DNA methylation reprogramming, TE derepression, and postzygotic isolation of nascent animal species. *Sci. Adv.* **2019**, *5*, eaaw1644. [CrossRef]

20.   Allendorf, F.W.; Hohenlohe, P.A.; Luikart, G. Genomics and the future of conservation genetics. *Nat. Rev. Genet.* **2010**, *11*, 697–709. [CrossRef] [PubMed]

21.   Sankararaman, S.; Mallick, S.; Dannemann, M.; Prüfer, K.; Kelso, J.; Pääbo, S.; Patterson, N.; Reich, D. The genomic landscape of Neanderthal ancestry in present-day humans. *Nat. Cell Biol.* **2014**, *507*, 354–357. [CrossRef]

22.   Schumer, M.; Xu, C.; Powell, D.L.; Durvasula, A.; Skov, L.; Holland, C.; Blazier, J.C.; Sankararaman, S.; Andolfatto, P.; Rosenthal, G.G.; et al. Natural selection interacts with recombination to shape the evolution of hybrid genomes. *Science* **2018**, *360*, 656–660. [CrossRef]

23.   Martin, S.H.; Jiggins, C.D. Interpreting the genomic landscape of introgression. *Curr. Opin. Genet. Dev.* **2017**, *47*, 69–74. [CrossRef]

24.   Veller, C.; Edelman, N.B.; Muralidhar, P.; Nowak, M.A. Recombination, Variance in Genetic Relatedness, and Selection against Introgressed DNA. *bioRxiv* **2019**, 846147. [CrossRef]

25.   Lippman, Z.B.; Zamir, D. Heterosis: Revisiting the magic. *Trends Genet.* **2007**, *23*, 60–66. [CrossRef]

26.   Chen, Z.J. Molecular mechanisms of polyploidy and hybrid vigor. *Trends Plant Sci.* **2010**, *15*, 57–71. [CrossRef]

27.   Medina, P.; Thornlow, B.; Nielsen, R.; Corbett-Detig, R. Estimating the Timing of Multiple Admixture Pulses During Local Ancestry Inference. *Genetics* **2018**, *210*, 1089–1107. [CrossRef]

28.   Létourneau, J.; Ferchaud, A.; Le Luyer, J.; Laporte, M.; Garant, D.; Bernatchez, L. Predicting the genetic impact of stocking in Brook Charr (Salvelinus fontinalis) by combining RAD sequencing and modeling of explanatory variables. *Evol. Appl.* **2017**, *11*, 577–592. [CrossRef]

29.   White, S.L.; Miller, W.L.; Dowell, S.A.; Bartron, M.L.; Wagner, T. Limited hatchery introgression into wild brook trout (Salvelinus fontinalis) populations despite reoccurring stocking. *Evol. Appl.* **2018**, *11*, 1567–1581. [CrossRef] [PubMed]

30.   Lehnert, S.J.; Baillie, S.M.; Macmillan, J.; Paterson, I.G.; Buhariwalla, C.F.; Bradbury, I.R.; Bentzen, P. Multiple decades of stocking has resulted in limited hatchery introgression in wild brook trout (Salvelinus fontinalis) populations of Nova Scotia. *Evol. Appl.* **2020**, *13*, 1069–1089. [CrossRef] [PubMed]

31.   Leitwein, M.; Cayuela, H.; Ferchaud, A.; Normandeau, É.; Gagnaire, P.; Bernatchez, L. The role of recombination on genome-wide patterns of local ancestry exemplified by supplemented brook charr populations. *Mol. Ecol.* **2019**, *28*, 4755–4769. [CrossRef] [PubMed]

32.   Lamaze, F.C.; Sauvage, C.; Marie, A.; Garant, D.; Bernatchez, L. Dynamics of introgressive hybridization assessed by SNP population genomics of coding genes in stocked brook charr (Salvelinus fontinalis). *Mol. Ecol.* **2012**, *21*, 2877–2895. [CrossRef] [PubMed]

33.   Kim, B.Y.; Huber, C.D.; Lohmueller, K.E. Deleterious variation shapes the genomic landscape of introgression. *PLoS Genet.* **2018**, *14*, e1007741. [CrossRef]

34.   Schumer, M.; Rosenthal, G.G.; Andolfatto, P. How common is homoploid hybrid speciation? *Evolution* **2014**, *68*, 1553–1560. [CrossRef]

35. Duranton, M.; Allal, F.; Fraïsse, C.; Bierne, N.; Bonhomme, F.; Gagnaire, P.-A. The origin and remolding of genomic islands of differentiation in the European sea bass. *Nat. Commun.* **2018**, *9*, 2518. [CrossRef] [PubMed]

36. Martin, S.H.; Davey, J.W.; Salazar, C.; Jiggins, C.D. Recombination rate variation shapes barriers to introgression across butterfly genomes. *PLoS Biol.* **2019**, *17*, e2006288. [CrossRef] [PubMed]

37. Ministère du Développement Durable, de l'Environnement, de la Faune et des Parcs. *Outil D'aide À L'ensemencement Des Plans D'eau—Omble de Fontaine (Salvelinus Fontinalis). Direction de La Faune Aquatique: Québec, Canada*; Direction de La Faune Aquatique: Québec, Canada; Direction Générale de l'expertise Sur La Faune et Ses Habitats: Québec, Canada, 2013; pp. 1–12.

38. Catchen, J.; Hohenlohe, P.A.; Bassham, S.; Amores, A.; Cresko, W.A. Stacks: An analysis tool set for population genomics. *Mol. Ecol.* **2013**, *22*, 3124–3140. [CrossRef]

39. Christensen, K.A.; Rondeau, E.B.; Minkley, D.R.; Leong, J.S.; Nugent, C.M.; Danzmann, R.G.; Ferguson, M.M.; Stadnik, A.; Devlin, R.H.; Muzzerall, R.; et al. The Arctic charr (Salvelinus alpinus) genome and transcriptome assembly. *PLoS ONE* **2018**, *13*, e0204076. [CrossRef]

40. Li, H.; Durbin, R. Fast and accurate long-read alignment with Burrows–Wheeler transform. *Bioinform.* **2010**, *26*, 589–595. [CrossRef] [PubMed]

41. Gosselin, T.; Bernatchez, L. Stackr: GBS/RAD Data Exploration, Manipulation and Visualization Using R. 2016. Available online: Https://Github.Com/Thierrygosselin/Stackr (accessed on 22 March 2021).

42. Sutherland, B.J.G.; Gosselin, T.; Normandeau, E.; Lamothe, M.; Isabel, N.; Audet, C.; Bernatchez, L. Salmonid chromosome evolution as revealed by a novel method for comparing RADseq linkage maps. *Genome Biol. Evol.* **2016**, *8*, 3600–3617. [CrossRef]

43. Guan, Y. Detecting Structure of Haplotypes and Local Ancestry. *Genetics* **2014**, *196*, 625–642. [CrossRef]

44. Leitwein, M.; Gagnaire, P.-A.; Desmarais, E.; Berrebi, P.; Guinand, B. Genomic consequences of a recent three-way admixture in supplemented wild brown trout populations revealed by local ancestry tracts. *Mol. Ecol.* **2018**, *27*, 3466–3483. [CrossRef] [PubMed]

45. Brasil República. Decreto—Lei n° 227, de 28 de fevereiro de 1967. Dá nova redação ao Decreto-lei n° 1.985, de 29 de janeiro de 1940 (Código de Minas). 1967. Available online: http://www.planalto.gov.br/ccivil_03/Decreto-Lei/Del0227.htm (accessed on 19 October 2020).

46. Pfeifer, B.; Wittelsbürger, U.; Ramos-Onsins, S.E.; Lercher, M.J. PopGenome: An Efficient Swiss Army Knife for Population Genomic Analyses in R. *Mol. Biol. Evol.* **2014**, *31*, 1929–1936. [CrossRef] [PubMed]

47. Rezvoy, C.; Charif, D.; Guéguen, L.; Marais, G.A. MareyMap: An R-based tool with graphical interface for estimating recombination rates. *Bioinformatics* **2007**, *23*, 2188–2189. [CrossRef] [PubMed]

48. Etymologia: Bonferroni Correction. *Emerg. Infect. Dis.* **2015**, *21*, 289. [CrossRef]

49. Todesco, M.; Pascual, M.A.; Owens, G.L.; Ostevik, K.L.; Moyers, B.T.; Hübner, S.; Heredia, S.M.; Hahn, M.A.; Caseys, C.; Bock, D.G.; et al. Hybridization and extinction. *Evol. Appl.* **2016**, *9*, 892–908. [CrossRef]

50. Marie, A.D.; Bernatchez, L.; Garant, D. Loss of genetic integrity correlates with stocking intensity in brook charr (Salvelinus fontinalis). *Mol. Ecol.* **2010**, *19*, 2025–2037. [CrossRef]

51. Luikart, G.; Ryman, N.; Tallmon, D.A.; Schwartz, M.K.; Allendorf, F.W. Estimation of census and effective population sizes: The increasing usefulness of DNA-based approaches. *Conserv. Genet.* **2010**, *11*, 355–373. [CrossRef]

52. Gossieaux, P.; Bernatchez, L.; Sirois, P.; Garant, D. Impacts of stocking and its intensity on effective population size in Brook Charr (Salvelinus fontinalis) populations. *Conserv. Genet.* **2019**, *20*, 729–742. [CrossRef]

53. Ferchaud, A.; Leitwein, M.; Laporte, M.; Boivin-Delisle, D.; Bougas, B.; Hernandez, C.; Normandeau, É.; Thibault, I.; Bernatchez, L. Adaptive and maladaptive genetic diversity in small populations: Insights from the Brook Charr (Salvelinus fontinalis) case study. *Mol. Ecol.* **2020**, *29*, 3429–3445. [CrossRef]

54. Richards, C.M. Inbreeding Depression and Genetic Rescue in a Plant Metapopulation. *Am. Nat.* **2000**, *155*, 383–394. [CrossRef]

55. Uller, T.; Leimu, R. Founder events predict changes in genetic diversity during human-mediated range expansions. *Glob. Chang. Biol.* **2011**, *17*, 3478–3485. [CrossRef]

56. Hedrick, P.W.; Garcia-Dorado, A. Understanding Inbreeding Depression, Purging, and Genetic Rescue. *Trends Ecol. Evol.* **2016**, *31*, 940–952. [CrossRef] [PubMed]

57. Hedrick, P.W. What is the evidence for heterozygote advantage selection? *Trends Ecol. Evol.* **2012**, *27*, 698–704. [CrossRef] [PubMed]

58. Harris, K.; Zhang, Y.; Nielsen, R. Genetic rescue and the maintenance of native ancestry. *Conserv. Genet.* **2019**, *20*, 59–64. [CrossRef]

59. Charlesworth, D.; Willis, J.H. The genetics of inbreeding depression. *Nat. Rev. Genet.* **2009**, *10*, 783–796. [CrossRef]

60. Schou, M.F.; Loeschcke, V.; Bechsgaard, J.; Schlötterer, C.; Kristensen, T.N. Unexpected high genetic diversity in small populations suggests maintenance by associative overdominance. *Mol. Ecol.* **2017**, *26*, 6510–6523. [CrossRef] [PubMed]

61. Gilbert, K.J.; Pouyet, F.; Excoffier, L.; Peischl, S. Transition from Background Selection to Associative Overdominance Promotes Diversity in Regions of Low Recombination. *Curr. Biol.* **2020**, *30*, 101–107. [CrossRef]

62. Turelli, M.; Orr, H.A. Dominance, epistasis and the genetics of postzygotic isolation. *Genetics* **2000**, *154*, 1663–1679.

63. Orr, H.A.; Turelli, M. The evolution of postzygotic isolation: Accumulating Dobzhansky-Muller incompatibilities. *Evol. Int. J. Org. Evol.* **2001**, *55*, 1085–1094. [CrossRef]

64. Renaut, S.; Bernatchez, L. Transcriptome-wide signature of hybrid breakdown associated with intrinsic reproductive isolation in lake whitefish species pairs (Coregonus spp. Salmonidae). *Heredity* **2010**, *106*, 1003–1011. [CrossRef]

65. Frankham, R. Genetic adaptation to captivity in species conservation programs. *Mol. Ecol.* **2008**, *17*, 325–333. [CrossRef]
66. Bosse, M.; Megens, H.; Derks, M.F.L.; De Cara, M.; Ángeles, R.; Groenen, M.A.M. Deleterious alleles in the context of domestication, inbreeding, and selection. *Evol. Appl.* **2018**, *12*, 6–17. [CrossRef]
67. Frankham, R.; Ballou, J.D.; Eldridge, M.D.B.; Lacy, R.C.; Ralls, K.; Dudash, M.R.; Fenster, C.B. Predicting the Probability of Outbreeding Depression. *Conserv. Biol.* **2011**, *25*, 465–475. [CrossRef]
68. Marshall, T.C.; Spalton, J.A. Simultaneous inbreeding and outbreeding depression in reintroduced Arabian oryx. *Anim. Conserv.* **2000**, *3*, 241–248. [CrossRef]
69. Frankham, R. Where are we in conservation genetics and where do we need to go? *Conserv. Genet.* **2009**, *11*, 661–663. [CrossRef]
70. Huff, D.D.; Miller, L.M.; Chizinski, C.J.; Vondracek, B. Mixed-source reintroductions lead to outbreeding depression in second-generation descendents of a native North American fish. *Mol. Ecol.* **2011**, *20*, 4246–4258. [CrossRef] [PubMed]
71. Kaulfuß, F.; Reisch, C. Reintroduction of the endangered and endemic plant species Cochlearia bavarica—Implications from conservation genetics. *Ecol. Evol.* **2017**, *7*, 11100–11112. [CrossRef] [PubMed]
72. Mable, B.K. Conservation of adaptive potential and functional diversity: Integrating old and new approaches. *Conserv. Genet.* **2019**, *20*, 89–100. [CrossRef]
73. Spencer, C.C.A.; Deloukas, P.; Hunt, S.; Mullikin, J.; Myers, S.; Silverman, B.; Donnelly, P.; Bentley, D.; McVean, G. The Influence of Recombination on Human Genetic Diversity. *PLoS Genet.* **2006**, *2*, e148. [CrossRef]
74. Burke, M.K. How does adaptation sweep through the genome? Insights from long-term selection experiments. *Proc. R. Soc. B Biol. Sci.* **2012**, *279*, 5029–5038.
75. Bougas, B.; Granier, S.; Audet, C.; Bernatchez, L. The Transcriptional Landscape of Cross-Specific Hybrids and Its Possible Link with Growth in Brook Charr (Salvelinus fontinalis Mitchill). *Genetics* **2010**, *186*, 97–107. [CrossRef]
76. Frankham, R. Challenges and opportunities of genetic approaches to biological conservation. *Biol. Conserv.* **2010**, *143*, 1919–1927. [CrossRef]
77. Cook, C.N.; Sgrò, C.M. Poor understanding of evolutionary theory is a barrier to effective conservation management. *Conserv. Lett.* **2019**, *12*, 12619. [CrossRef]
78. Harris, K.; Nielsen, R. The Genetic Cost of Neanderthal Introgression. *Genetics* **2016**, *203*, 881–891. [CrossRef] [PubMed]

genes

Article

Patterns of Epigenetic Diversity in Two Sympatric Fish Species: Genetic vs. Environmental Determinants

Laura Fargeot [1],*, Géraldine Loot [2,3], Jérôme G. Prunier [1], Olivier Rey [4], Charlotte Veyssière [2] and Simon Blanchet [1,2,*]

[1] Centre National de la Recherche Scientifique (CNRS), Université Paul Sabatier (UPS), Station d'Ecologie Théorique et Expérimentale, UMR 5321, F-09200 Moulis, France; jerome.prunier@gmail.com
[2] CNRS, UPS, École Nationale de Formation Agronomique (ENFA), UMR 5174 EDB (Laboratoire Évolution & Diversité Biologique), 118 route de Narbonne, F-31062 Toulouse CEDEX 4, France; geraldine.loot@univ-tlse3.fr (G.L.); veyssiere.charlotte@gmail.com (C.V.)
[3] Université Paul Sabatier (UPS), Institut Universitaire de France (IUF), F-75231 Paris CEDEX 05, France
[4] CNRS, Interaction Hôtes-Parasites-Environnements (IHPE), UMR 5244, F-66860 Perpignan, France; olivier.rey@univ-perp.fr
* Correspondence: laura.fargeot@sete.cnrs.fr (L.F.); simon.blanchet@sete.cnrs.fr (S.B.); Tel.: +33-561040361 (S.B.)

Abstract: Epigenetic components are hypothesized to be sensitive to the environment, which should permit species to adapt to environmental changes. In wild populations, epigenetic variation should therefore be mainly driven by environmental variation. Here, we tested whether epigenetic variation (DNA methylation) observed in wild populations is related to their genetic background, and/or to the local environment. Focusing on two sympatric freshwater fish species (*Gobio occitaniae* and *Phoxinus phoxinus*), we tested the relationships between epigenetic differentiation, genetic differentiation (using microsatellite and single nucleotide polymorphism (SNP) markers), and environmental distances between sites. We identify positive relationships between pairwise genetic and epigenetic distances in both species. Moreover, epigenetic marks better discriminated populations than genetic markers, especially in *G. occitaniae*. In *G. occitaniae*, both pairwise epigenetic and genetic distances were significantly associated to environmental distances between sites. Nonetheless, when controlling for genetic differentiation, the link between epigenetic differentiation and environmental distances was not significant anymore, indicating a noncausal relationship. Our results suggest that fish epigenetic variation is mainly genetically determined and that the environment weakly contributed to epigenetic variation. We advocate the need to control for the genetic background of populations when inferring causal links between epigenetic variation and environmental heterogeneity in wild populations.

Keywords: genetic structure; empirical comparative study; DNA methylation; nongenetic heredity; population genomics; freshwater

1. Introduction

Describing and understanding spatial patterns of intraspecific diversity in natural populations constitutes the basis for predicting the evolutionary dynamics of populations. So far, most studies have focused on spatial patterns of intraspecific genetic diversity [1,2], using neutral and/or nonneutral (or adaptive) molecular markers. Neutral markers are influenced by mutation, drift and gene flow, and are not directly associated to individual fitness. Nonneutral markers are influenced not only by the same processes but also by natural selection associated to the surrounding environment; they are hence associated to the fitness and adaptation of organisms [3]. Recently, there has been an increasing interest in documenting the distribution of intraspecific epigenetic diversity in wild populations because it may also have a role in adaptive potential of organisms [4]. Epigenetic variation is a major potential source of adaptive variation since it can be directly sensitive to the environment and transmitted across generations [5–7]. In particular, epigenetic variation may allow for the rapid adaptation of populations to changing environments, at a pace actually higher than adaptation by natural selection on standing genetic variation [5,8–11]. In this context, an important question concerns the spatial covariation that may exist between genetic and epigenetic diversity patterns in natural populations, i.e., whether genetic and epigenetic variants follow similar spatial patterns across landscapes or not. Answering this question allows testing whether these two markers carry distinct/complementary pieces of information, and speculating about their relative roles in the adaptive potential of organisms across spatial scales. This question is not trivial, as the inherent characteristics of genetic and epigenetic marks can lead to opposite predictions regarding the spatial covariation of epigenetic and genetic diversity patterns.

On the one side, some epigenetic marks are directly sensitive to external environmental cues (epimutations can be triggered by the surrounding environment at a lifetime scale), which may lead to uncorrelated genetic and epigenetic diversity patterns since genetic mutations are not directly sensitive to the environment. Indeed, environmental constraints (e.g., contaminants, diet, social stimuli, etc.) experienced by individuals along their life can alter the distribution of epigenetic marks across the whole genome [12–15]. Some of these marks induced by the environment can be transmitted across generations (inherited), and in that case, they are comparable to nonneutral genetic variance, except that (i) the mutation rate of epigenetic markers is higher and (ii) these marks are less stable over the long time [7,16–18]. Consequently, the epigenome (all of individual epigenetic marks on the DNA sequence) could theoretically carry a footprint of the contemporary (biotic and abiotic) environment in which the last few generations have lived. Epigenetic marks are hence expected to transmit (environmental) information that is not necessarily transmitted by (and that is hence complementary to) genetic marks [19]. In this situation, we can predict that spatial patterns of epigenetic diversity should deviate from those documented for neutral and nonneutral genetic markers. In particular, epigenetic diversity patterns should be strongly linked to environmental heterogeneity, whereas this should be less the case for nonneutral genetic diversity patterns, and obviously not the case for neutral genetic diversity.

On the other side, there is mounting evidence that alternative mechanisms can generate correlated patterns of epigenetic and (neutral) genetic diversity across natural landscapes. A first alternative hypothesis rests upon the assumption that epigenetic marks depend (either completely or partially) on genetic variation rather than on environmental variation [20–22], since an individual transmitting a given genetic allele during mitosis also transmits the epigenetic information carried by this allele, i.e., the epiallele [20]. Therefore, there would be a physical link between alleles and epialleles. Moreover, different types of genetically encoded molecules are required to modulate the expression of genes, such as RNA or proteins [23,24]. These molecules are involved in the establishment and stability of histone tail modifications or DNA methylation across generations [16,25]. Consequently, the establishment and stability of epigenetic marks is allowed by genetic information. A second alternative hypothesis hence states that the same neutral processes (drift, mutation, and gene flow) can influence both genetic and epigenetic markers in a similar direction [26–28]. Indeed, epi-mutations have been reported to occur naturally in wild populations [29] and age-related methylation drift is known to reflect imperfect maintenance of epigenetic marks through cell renewal [30]. These two hypotheses both suggest that, under certain circumstances, epigenetic and neutral genetic diversity patterns could actually strongly covary spatially across natural landscapes.

Documenting and understanding the joint distribution of genetic and epigenetic marks in natural populations is essential to tease apart the potential role of epigenetic and genetic backgrounds for the adaptive potential of populations, and a few studies have paved the way toward such an objective [26]. Up to now, these studies have led to contrasting and context-dependent results. For instance, a significant correlation was detected between genetic and epigenetic differentiation among natural population pairs of *Hordeum brevi-subulatum* [31], whereas it was not the case in *Vitex negundo var. heterophylla* [32]. These contrasted patterns have also been observed in animal species [33,34], although most previous studies have focused on plant populations. At a first glance, it therefore seems that no generalities can be drawn about the spatial covariation between epigenetic and genetic diversity patterns in natural populations. We argue that an insightful way to tackle this question is to study species living in sympatry within a single landscape. Indeed, by "controlling" for the common environment, it would become possible to test whether the epigenetic response of populations to the same local environment is species-dependent or predictable across species, and to test the causal link between genetic and epigenetic marks. To date, these kinds of empirical comparative study are scarce [35,36], while they may be very helpful for generalizing findings across species.

The general objective of this study was to generate novel insights into the spatial patterns of genetic and epigenetic diversity of wild populations, and to empirically test the link between epigenetic population structure and the local environment. We conducted an empirical "comparative" study involving two sympatric freshwater fish species (*Gobio occitaniae* and *Phoxinus phoxinus*) in a common riverscape to gain insights into the links between genetic diversity, epigenetic diversity, and the environment. By focusing on the same set of sites for the two species, we tested the correlation between genetic and epigenetic diversity structure within each species by controlling for environmental covariation, and we compared this correlation between species. Furthermore, we used both supposedly neutral (microsatellites) and nonneutral (single nucleotide polymorphism, SNP) genetic markers, to gain further insights into the processes sustaining patterns of epigenetic diversity in wild populations. Because the two fish species belong to the same trophic level and display similar life history traits (similar generation time, for instance), we expected similar patterns for the two species. In particular, assuming that epigenetic marks are under partial genetic control [11,20,37], we predicted a positive and significant correlation between pairwise genetic and epigenetic differentiation for the two species, irrespectively of the type (neutral or not) of genetic marker. An absence of significant correlation would indirectly indicate that epigenetic diversity is controlled by other factors, i.e., the environment. We further tested the correlation between pairwise epigenetic (and genetic) differentiation and environmental distances between sites to quantify to which extent epigenetic diversity was determined by the local environment. For the two species, we expected that populations living in strongly distinct habitats would be highly differentiated epigenetically, which should not be observed for neutral (microsatellite) genetic differentiation, and only partially observed in nonneutral (SNP) markers. Given their different ecological requirements, we finally expected that the environmental component of epigenetic differentiation, if any, would not be driven by the same environmental factors in the two species.

2. Materials and Methods

2.1. Sampling and Site Selection

2.1.1. Biological Models

The two focal fish species belong to the cyprinidae family: *Gobio occitaniae* (the Occitan gudgeon) and *Phoxinus phoxinus* (the European minnow). These two species are phylogenetically related, they belong to the same trophic level, they have a similar generation time (2–3 years), and they face similar selective pressures as they coexist in sympatry in many areas [38]. Nonetheless, they slightly differ ecologically since *G. occitaniae* is ubiquitous over a large part of the whole upstream-downstream gradient in rivers, whereas *P. phoxinus* is more specialized and lives preferentially in upstream areas. In addition, and despite the fact that they are both insectivorous, *G. occitaniae* feeds preferentially on the bottom, whereas *P. phoxinus* feeds mainly in the water column. Finally, *G. occitaniae* is larger in body length than *P. phoxinus* (mean body length at adult size is ~80–150 and ~50–90 mm, respectively).

2.1.2. Study Area and Sampling

Based on a priori knowledge [39], we sampled the two fish species in 13 sites from the Garonne River basin (South-Western France, Figure 1). Sites varied according to key abiotic factors to optimize the likelihood of detecting unique epigenetic marks in these populations. In particular, we selected sites varying according to two key environmental variables directly affecting fish fitness and populations [39–41]: mean annual temperature (ranging from 16.4 to 23.3 °C, Table A1, see also Figure 1) and oxygen saturation (ranging from 77.5% to 114.7%, Table A1). Electric fishing was conducted during summer 2014 and performed under the authorization of "Arrêté Préfectoraux" delivered by the "Direction Départementale des Territoires" of each administrative department (Ariège, Aveyron, Haute-Garonne, Hautes-Pyrénées, Lot and Tarn et Garonne). We sampled 24 fish per species in each site, leading to a total of 312 individuals per species (*n* = 624). Fish were treated in accordance to the European Communities Council Directive (2010/63/EU) regarding the use of animals in Research, French Law for Animal Protection R214-87 to R214-137. Although DNA methylation diversity can show tissue-specific differences within an individual [42–44], we favored a non-lethal approach and hence a small piece of pelvic fin was sampled on each individual. It is noteworthy that, in fish, the shape and color of fins can be linked to abiotic environmental conditions [45,46]. All individuals were anaesthetized using benzocaine before fin clips. Each fin tissue was preserved in 70% ethanol for further genetic and epigenetic analyses. All individuals were released in their respective sampling site.

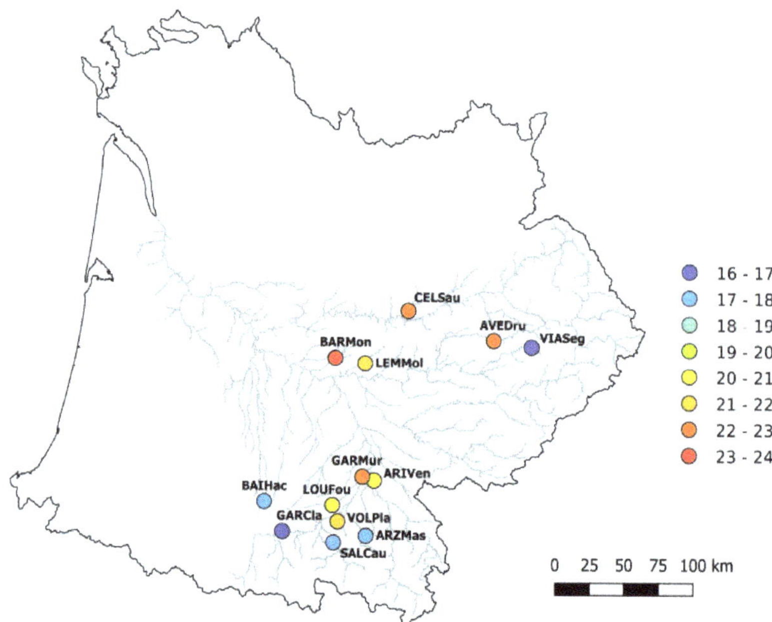

Figure 1. Distribution of the 13 sampling sites in the Garonne river basin. Names and localization are highlighted in bold. Twenty-four individuals per site and per species (*Gobio occitaniae* and *Phoxinus phoxinus*) have been sampled. Color of circles indicates mean water temperature (°C).

2.2. Environmental Data

All sites were characterized for 14 variables related to physicochemical characteristics (overall water quality) and river topography so as to test for association between epigenetic (and genetic) markers and environment (see Table A1). Topographical variables included river flow ($m^3 \cdot s^{-1}$), river width (m), river slope (%), and altitude (m) and were retrieved from the French Theoretical Hydrological Network (RHT) [47]. Physicochemical characteristics included concentrations in nitrate, nitrite, orthophosphate, and oxygen ($mg \cdot L^{-1}$), biological oxygen demand (BOD, $mg \cdot L^{-1}$), water conductivity ($mS.cm^{-1}$), pH, suspended matter (SM, $mg \cdot L^{-1}$), oxygen saturation (%), and temperature (°C). They were obtained from the databases of the Water Information System of the Adour Garonne basin (SIEAG "Système d'Information sur l'Eau du Bassin Adour Garonne"; http://adour-garonne. eaufrance.fr). Here, we used values measured in July from 2013 to 2015, to take into account interannual variability and potential measurement errors. Values were averaged (for each parameter) across this period. July was chosen as the reference month since this is a period in which environmental constraints are likely to be strong on fish fauna (low water level, hyperthermia, hypoxia, etc.) and because it is the most informed month in the SIEAG database. All these variables are known to affect dynamics of fish populations and properly characterize the environmental conditions encountered by fish [38,48].

A principal component analysis (PCA) was performed on the 14 environmental variables using the R package "ade4" [49], to synthetize data into orthogonal variables. The three first axes represented 71.96% of the total variance (Table 1 for details), and were hence retained as synthetic environmental variables. The first axis, defined by a strong contribution of (in decreasing order) oxygen concentration, water conductivity, nitrite concentration, oxygen saturation, and nitrate concentration (Table 1), stands for a eutrophication gradient. Sites with positive values along this axis were characterized by a low concentration of oxygen, a high conductivity and high concentrations in nitrate and nitrite (i.e., the more eutrophic sites). The second axis, defined by a strong contribution of river flow, river width, and pH (Table 1), stands for an upstream–downstream gradient. Sites with positive values along this axis were characterized by a large river bed (high water flow) and high pH values. The third axis is defined by a strong contribution of orthophosphate concentration, slope, altitude, and suspended matter (Table 1). Sites with positive values along this axis were characterized by high altitude sites with a steep slope and high values of nutrient and suspended matter.

Table 1. Characteristics of the three first principal components from the principal component analysis (PCA) ran on the 14 environmental variables and used to characterize each of the 13 sampling sites. The part of the total environmental variance (%) and the contribution of each variable to each component are shown. The variables that contributed significantly to the axis are shown in bold. BOD = biological oxygen demand; SM = suspended matter.

–	Component 1	Component 2	Component 3
Part of total variance (%)	37.03	20.87	14.06
River flow	−0.273	**0.888**	−0.056
River width	−0.295	**0.864**	0.037
Slope	−0.434	−0.407	**0.614**
Altitude	−0.538	−0.442	**0.596**
Conductivity	**0.878**	0.258	0.212
BOD	0.555	−0.020	0.045
MS	0.262	0.060	**0.529**
Nitrate	**0.748**	0.166	0.344
Nitrite	**0.855**	0.199	0.266
Orthophosphate	0.354	0.303	**0.757**
Oxygen	**−0.903**	0.267	0.191
pH	−0.341	**0.719**	0.075
Oxygen saturation	**−0.848**	0.304	0.238
Temperature	0.577	0.339	−0.208
Global characteristic	Oligotrophic water– Eutrophic water	Small river–Large river	Low altitude and nutrient– High altitude and nutrient

2.3. Genetic and Epigenetic Data

2.3.1. Genetic Data

The DNA of all individuals (n = 624) was extracted using a salt-extraction protocol [50]. Individual genetic data consisted in both microsatellite (supposedly neutral) and SNP markers (potentially nonneutral).

Microsatellites data (13 and 17 loci for *G. occitaniae* and *P. phoxinus*, respectively) were obtained from a previous study [39]. Details on accession numbers, conditions for polymerase chain reactions (PCRs), and preliminary analyses (e.g., search for null alleles or possible linkage disequilibrium between loci) are provided in Fourtune et al. [39].

SNP markers (1892 and 1244 in *G. occitaniae* and *P. phoxinus*, respectively) were obtained from the restriction site-associated DNA (RAD) sequencing of pooled DNAs at the site and species levels [51], using laboratory and bioinformatic procedures described in Prunier et al. [52]. As the DNA of individuals was pooled at the site level, we were unable to retrieve individual genotypes (contrary to microsatellite markers) and we therefore used the frequencies of alleles (from each SNP) at the population level as raw genomic data for the SNPs.

2.3.2. Epigenetic Data

Individuals were then genotyped using Methylation-Sensitive-AFLP (MS-AFLP). MS-AFLP allows identifying "genome-wide" methylation patterns. This is a modified version of standard AFLP (Amplified Fragment Length Polymorphism) technique [53] that is well suited for nonmodel species (without a reference genome) and useful to assess epigenetic diversity for large sample sizes (>200) [37]. MS-AFLP relies on two separate double digestions with EcoRI (rare cutter, on 5'G|AATTC restriction site) and either one of two methylation-sensitive restriction enzymes (HpaII and MspI, frequent cutters, on 5'CC|GG restriction site). Because HpaII and MspI have different cytosine methylation sensitivities, comparison of the two digestion fragment profiles (EcoRI/MspI and EcoR1/HpaII) leads to the distinction of four methylation conditions for each DNA fragment: Condition I = fragments are present in both profiles, indicating an unmethylated state; Condition II = fragments are present only in EcoRI/MspI profile indicating an hemimethylation of internal cytosine (^{HMe}CG-sites) or a full methylation of (both) internal cytosines (^{Me}CG-sites); Condition III = fragments are present only in EcoRI/HpaII profile indicating an hemimethylation of external cytosine (^{HMe}CCG-sites); Condition IV = fragments are absent in both profiles, indicating an uninformative state [54]. This last case can have multiple origins such as full methylation on external cytosine (^{Me}CCG-), hemimethylation of both cytosines (^{HMe}C^{HMe}CG-sites), full-methylation of both cytosines (^{Me}C^{Me}CG-sites), or more rarely genetic mutation leading to polymorphism of the restriction site.

2.3.3. MS-AFLP Protocol

The first step consists in two separate double digest reactions of 3 µL of extraction product (30–40 ng·µL^{-1}) with 0.5 µL of FastDigest EcoRI (1 FDU·µL^{-1}) and 0.5 µL of either FastDigest MspI (1 FDU·µL^{-1}) or FastDigest HpaII (1 FDU·µL^{-1}) isoschizomers. DNA was digested at 37 °C for 15 min. Double-stranded adaptors (see Table A2 for details) [32,55,56] were then ligated onto the sticky end of all the digestion products (10 µL) with 0.3 µL of a T4 DNA ligase (5 U/µL, Thermo Scientific) and 1 µL of each adaptor (EcoRI adaptors 2.5 µM; MspI/HpaII adaptor 0.25 µM) at 25 °C for 1 h. After a step of enzyme killing, the product was subjected to two rounds of increasingly selective PCR amplification (PCR1 and PCR2). Preselective amplification (PCR1, see Appendix C for details) was performed in a total volume of 25 µL using 5 µL of 5× buffer, 1.5 µL of dNTP (10 mM), 2 µL of each preselective primer (10 µM, see Table A2 for sequences), 0.3 µL of Taq DNA polymerase (5 U/µL Thermo Scientific®), and 2 µL of ligation product. Preamplified products were then diluted to 1:50 in sterile water. Selective amplification (PCR2, see Appendix C for details) was then performed under the same conditions (reagents and total volume) than the preselective amplification, except that three specific selective primers couples were used (see Table A2 for sequences). Primers for selective PCRs were chosen among a set of 24 and 23 primers for *G. occitaniae* and *P. phoxinus*, respectively, that we previously tested for optimal conditions (number of loci amplified per primer, not shown). Amplified products were then diluted to 1:15 in sterile water and 2.2 μL of this mix was added in 7.8 μL of a mix composed of 800 μL of Hi-Di formamide (Applied Biosystems®) and 15 μL of ROX500 (Applied Biosystems®) prior to analyzing and sizing the fragments. Fragment analysis was performed on an ABI PRISM 3730 capillary sequencer (Applied Biosystems®, Foster City, CA, USA) at the Génopôle Toulouse Midi-Pyrénées.

Fragment profiles were analyzed with GENEMAPPER 5.0® and we scored fragments (loci) between 150 and 500 bp to avoid homoplasy [57]. Binning of fragments was performed using a peak height threshold at 750 Relative Fluorescence Units (RFU) to exclude all ambiguous peaks. Manual verification permitted to eliminate false positive such as peaks just above or below the threshold set, fluorescence blobs, or peaks too close one from the other to be correctly resolved by automated analysis. Absence and presence of data at each locus were then converted into Conditions I, II, II, or IV as explained above [54]. All loci that contained Condition IV (i.e., uninformative state) for more than 95% of the individuals were excluded from further analyses. This resulted in a total of 251 polymorphic loci for *G. occitaniae* and 274 polymorphic loci for *P. phoxinus*, respectively (see the number of loci per primer in Table A2). We considered each of the four conditions as carrying unique information, and we therefore ran statistical analyses directly on a four-state data matrix, which permitted us to keep all the information contained in the dataset.

2.4. Statistical Analyses

To test the part of the molecular variance that was explained by the between-population component, an analysis of molecular variance (AMOVA; Excoffier et al. 1992) was performed on either genetic or epigenetic markers and for each species separately ("poppr.amova" function from the poppr R package). Regarding genetic markers, only microsatellites markers were considered here, because we did not have the within-population component (individual genotypes) in our SNPs dataset (see above). If epigenetic marks are more sensitive to the environment, they should be more discriminant and the between-population component should be higher for epigenetic markers than for genetic markers.

We then estimated measures of genetic (for both marker types separately) and epigenetic differentiation (for each species separately) by calculating the Gst" index of differentiation between each pair of populations. We preferred this metric of differentiation over other metrics (e.g., Fst, Gst, Jost's D, etc.) as it has been shown to be robust to variations in mutation rates and sample sizes [58,59].

To test whether pairwise epigenetic differentiation was dependent upon genetic differentiation (i.e., whether epigenetic differentiation was genetically determined), a simple Mantel test was first performed between pairwise genetic and epigenetic distances for each species separately and for each genetic marker type separately ("mantel.rtest" function from the ade4 R package). Simple Mantel tests were also used to assess the significance of the correlation between pairwise differentiation measured from microsatellite markers and differentiation measured from SNP markers.

To test whether epigenetic differentiation between populations resulted from environmental differences among sites (i.e., whether epigenetic differentiation was environmentally determined), simple Mantel tests were also performed between either genetic or epigenetic pairwise distances and each of the three environmental distance matrices computed from retained principal components (Euclidian distances) and a geographical distance matrix based on riparian distance between sites (to control for a potential confounding spatial effect and to test for patterns of isolation-by-distance). To further investigate the relationship between epigenetic pairwise differentiation and environmental or geographical distance matrices, multiple regressions on distance matrices (MRM, "MRM" function from the ecodist R package) were then performed. MRM is an extension of partial Mantel test allowing to test the relationship between a response matrix and any number of explanatory matrices, where each matrix contains distance or similarities (Smouse et al. 1986). For each species, the pairwise matrix of epigenetic differentiation was the response variable, and explanatory variables where the three environmental distance matrices, the geographical distance matrix, and the pairwise matrix of genetic differentiation based on SNP markers to account for a possible genetic determinism of epigenetic marks. For the sake of simplicity, we did not include the pairwise matrix of genetic differentiation based on microsatellites, although results were very similar whether we integrated it or not in the models (not shown).

3. Results

3.1. Genetic and Epigenetic Discrimination

Molecular analysis of variance (AMOVA) revealed that a significant part of the genetic (microsatellite markers) and epigenetic variance was attributed to the between-population component, for both species (p-value < 0.001; permutation tests with 1000 repetitions). For *G. occitaniae*, the part of the total variance explained by the between-population component was twice as high for epigenetic markers as it was for genetic markers (20.15% and 10.34%, respectively, see Table 2). For *P. phoxinus*, a similar trend was observed although less pronounced (19.59% and 16.75%, respectively, see Table 2). This suggests that, in both species, epigenetic markers were more powerful to discriminate among populations than genetic markers.

Table 2. Outputs of analyses of molecular variance (AMOVA) aiming at testing the part of the molecular variance that was explained by the between-population component (the within-population component is not shown here). Results are presented for the two fish species (*G. occitaniae* and *P. phoxinus*) and the two molecular marker types (genetic and epigenetic markers) separately. For the genetic marker, only microsatellite markers have been considered in this analysis (see the text for details). The percentages of the total variance ("Variation") explained by the between-population component (and the associated Phi-st values) are presented, as well as the respective p-values based on permutation tests with 1000 repetitions.

	Degrees of Freedom	Sum of Squares	Variance Components	Variation (%)	*Phi-st*	*p*-value
G. occitaniae						
Genetic markers	12	217.182	0.481	10.34	0.103	<0.001
Epigenetic markers	12	5726.943	17.210	20.15	0.202	<0.001
P. phoxinus						
Genetic markers	12	484.297	1.160	16.75	0.168	<0.001
Epigenetic markers	12	6369.97	19.478	19.59	0.196	<0.001

3.2. Simple Associations between Epigenetic, Genetic, and Environmental Distances

Simple Mantel tests demonstrated that there was a significant correlation between pairwise genetic and epigenetic distance matrices in *G. occitaniae* for both microsatellite and SNP markers ($r = 0.363$, p-value < 0.05 and $r = 0.531$, p-value < 0.001, for microsatellites and SNPs, respectively; Figure 2a and Table 3). In *P. phoxinus*, although the same tendency was observed, the correlation was not significant ($r = 0.287$, p-value = 0.089 and $r = 0.294$, p-value = 0.121 for microsatellites and SNPs, respectively; Figure 2b and Table 3). Moreover, most Gst" values measured using epigenetic markers were above the 1:1 line, indicating that the mean pairwise differentiation among populations was higher when using epigenetic markers than when using genetic markers (Figure 2a,b). As expected, relationships between pairwise genetic distances measured using microsatellites in the one hand and SNPs on the other hand were strong and highly significant (Table 3).

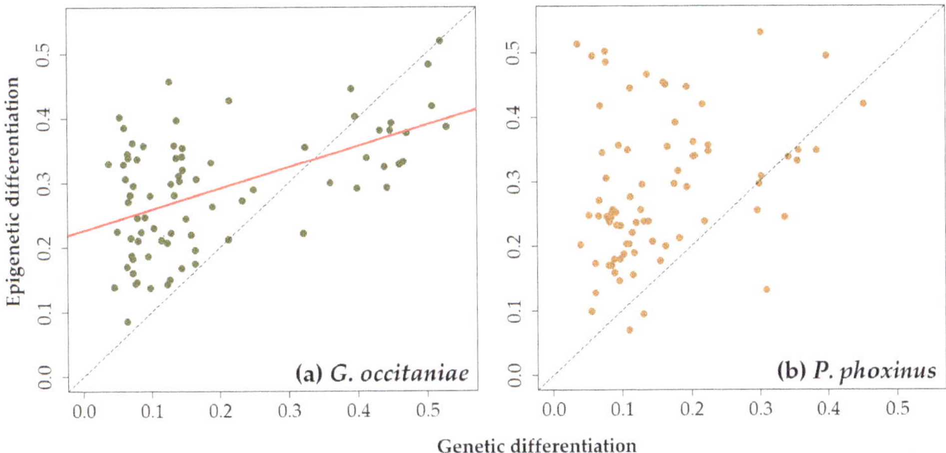

Figure 2. Biplots illustrating the relationships between pairwise genetic (based on SNP markers) and epigenetic differentiation (based on MS-AFLP markers) in (**a**) *G. occitaniae* (the red line indicates a significant relationship based on a simple Mantel test) and (**b**) *P. phoxinus* (no significant relationship was detected based on a simple Mantel test). Each dot represents a pairwise distance between two sites. The dashed line indicates the 1:1 line. Similar trends were observed with microsatellite markers but are not shown here.

Table 3. Summary of simple Mantel tests testing the relationships between genetic and epigenetic differentiation between pairs of populations in *G. occitaniae* and *P. Phoxinus*. Results are presented for both genetic (microsatellites and SNPs) and epigenetic (MS-AFLP) markers. Mantel statistics are presented (above diagonal), as well as the associated *p*-values based on 1000 permutations (below diagonal). Significant relationships are bolded.

	Microsatellites	**SNP**	**MS-AFLP**
G. occitaniae			
Microsatellites	–	**0.616**	0.363
SNP	**0.009**	–	0.531
MS-AFLP	**0.011**	**<0.001**	–
P. phoxinus			
Microsatellites	–	**0.894**	0.287
SNP	**<0.001**	–	0.294
MS-AFLP	0.089	0.121	–

In *G. occitaniae*, there was a significant relationship between epigenetic pairwise distances and environmental distances computed from the first principal component (Figure 3a and Table 4). A similar relationship was observed between pairwise genetic differentiation measured using SNPs and distance between sites measured from the same PCA axis, whereas such a relationship was not significant when considering microsatellite markers (Figure 3b and Table 4). In *G. occitaniae*, environmental distances measured from other PCA axes were not correlated to epigenetic or genetic pairwise matrices of differentiation (Table 4). In *P. phoxinus*, none of the relationships between epigenetic or genetic pairwise differentiation and environmental and geographic pairwise distances were significant (Table 4).

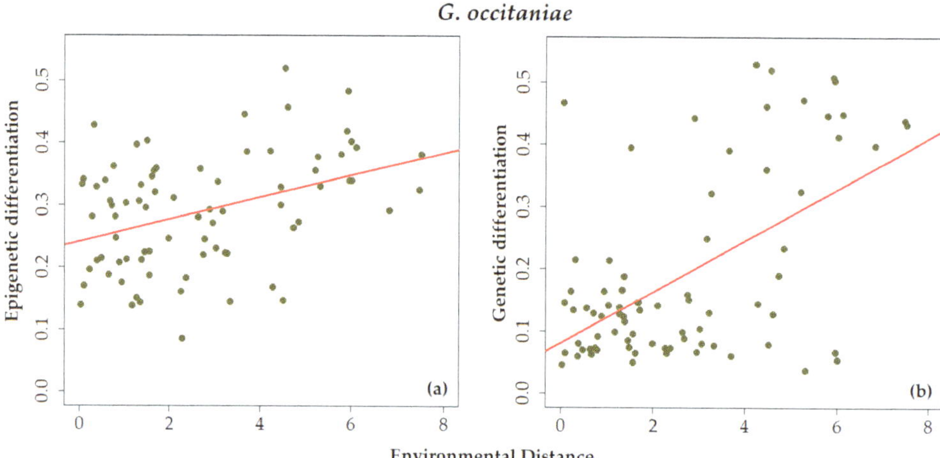

Figure 3. Biplot illustrating the relationship between (**a**) epigenetic differentiation between pairs of populations (based on MS-AFLP) and (**b**) genetic differentiation between pairs of populations (based on SNPs markers) and environmental distances (along a eutrophication gradient) between pairs of sites for *G. occitaniae* (the red line indicates a significant relationship based on simple Mantel test). Each dot represents a pairwise distance between two sites.

Table 4. Summary of simple Mantel tests testing the relationships between epigenetic and genetic pairwise differentiation, environmental (PCA components 1 to 3), and geographical (riparian) pairwise distances in *G. occitaniae* and *P. phoxinus*. Results are presented for epigenetic (MS-AFLP) and two genetic (microsatellite and SNPs) markers. Mantel statistics (r) are presented, as well as the associated *p*-values based on permutation tests with 1000 repetitions. Significant relationships are bolded.

	Component 1		Component 2		Component 3		Riparian Distance	
	r	*p*-value	r	*p*-value	r	*p*-value	r	*p*-value
G. occitaniae								
MS-AFLP	**0.395**	**<0.01**	0.033	0.397	−0.027	0.570	−0.060	0.670
SNP	**0.562**	**<0.05**	−0.142	0.670	−0.104	0.661	−0.117	0.666
Microsatellites	0.161	0.195	−0.066	0.538	−0.162	0.765	0.095	0.333
P. phoxinus								
MS-AFLP	−0.154	0.844	−0.114	0.669	−0.143	0.797	−0.080	0.639
SNP	−0.217	0.872	−0.072	0.539	−0.035	0.479	0.107	0.328
Microsatellites	−0.11	0.663	−0.180	0.747	−0.001	0.436	0.186	0.228

3.3. Multiple Associations between Epigenetic, Genetic, and Environmental Distances

Multiple regressions on distance matrices (MRM) revealed that, in both species, there was no relationship between epigenetic pairwise differentiation and environmental and geographic pairwise distances (Table 5). MRM showed that there was a significant relationship between genetic and epigenetic differentiation (Table 5) in *G. occitaniae*, but not in *P. phoxinus*.

Table 5. Summary of multiple regression on distance matrices (MRM) testing the relationships between epigenetic differentiation, genetic differentiation (based on SNP markers), environmental (PCA components 1 to 3), and geographical (riparian) distances between sites in *G. occitaniae* and *P. phoxinus*. Parameters associated to each explanatory variable are shown, together with their *p*–values (1000 permutations). Significant relationships are bolded.

	G. occitaniae		*P. phoxinus*	
	Coefficients	*p*-value	Coefficients	*p*-value
Intercept	0.197	0.945	0.312	0.362
Component 1	0.007	0.317	−0.008	0.370
Component 2	0.006	0.473	−0.006	0.672
Component 3	<0.001	0.925	−0.009	0.559
Riparian Dist.	<0.001	0.817	−0.001	0.534
Genetics (SNP)	**0.290**	**0.003**	0.307	0.220

4. Discussion

Although patterns of genetic and epigenetic structure in natural populations have already been investigated in plants [26], the relative role of genetic and epigenetic variation in driving the adaptive potential of animal populations remains unclear. Here, we compared the epigenetic structure of two sympatric freshwater fish species along the same environmental gradient, and tested the relationship between genetic diversity, epigenetic diversity, and the environment. Our results suggest that epigenetic diversity is mostly influenced by the genetic background of organisms and weakly influenced by environmental variation.

We found a tendency toward a positive correlation between pairwise genetic and epigenetic distance matrices in both species, although the correlation was significant only in one of the two species (*G. occitaniae*). This suggests that epigenetic diversity might be partly genetically controlled and/or that similar processes operate in the same manner on these two molecular markers (genetic and epigenetic). Indeed, consistent with our hypothesis, there was spatial congruency between pairwise genetic and epigenetic differentiation in *G. occitaniae*, irrespective of the type (microsatellites or SNP) of genetic marker used to assess genetic differentiation. A similar tendency, yet not significant, was observed in *P. phoxinus*, indicating that the strength of the positive association between genetic and epigenetic differentiation might slightly differ among species from the same ecological guild. A very few comparative studies in wild populations have been performed so far, and it is hence extremely difficult to draw general conclusions. In two plants species (*Spartina alterniflora* and *Borrichia frutescens*) sharing the same habitats, a positive correlation between pairwise genetic and epigenetic differentiation have been highlighted in one species (*S. alterniflora*) and not in the other (*B. frutescens*) [35]. This latter study and our findings suggest that patterns of covariation between genetic and epigenetic diversity in wild populations is likely to be species-dependent and hard to predict.

In *G. occitaniae*, the association between genetic and epigenetic differentiation was particularly strong when genetic differentiation was calculated using SNP markers. SNP markers (or some of them) are supposedly nonneutral and hence are significantly affected by natural selection, whereas microsatellites are supposedly neutral and hence mainly affected by drift and dispersal. Beyond these characteristics, epigenetic markers considered here, MS-AFLP, are likely to be more similar to SNP than to microsatellite markers in terms of the amount of evolutionary information they reveal about wild populations of fish species. Indeed, microsatellite markers are known to have a faster mutation rate and thus a higher level of polymorphism [60] than MS-AFLP and SNP markers. Consequently, the impact of neutral processes such as mutation, genetic drift, or gene flow are probably more similar between MS-AFLP and SNP markers than between MS-AFLP and microsatellite markers. In order to test the association between genetic and epigenetic differentiation, comparison between different genetic markers have previously been done in a few studies [26,31,61–65]. These studies also found different patterns according to the type of genetic markers that was used to estimate genetic differentiation, which confirms that the different regions of the genome are not affected by neutral processes in the same manner [66–68] and strongly suggests that estimating genetic differentiation based on a single marker type can potentially lead to imprecise conclusions, especially, if this latter differs in mutation rate and polymorphism level compared to epigenetic marks.

We found a significant link between epigenetic variation and environmental heterogeneity in *G. occitaniae*, but not in *P. phoxinus*. Nonetheless, when we controlled for the underlying genetic structure of populations, this link was no longer significant, suggesting a non-causal (spurious) association [69]. This indicates that, in *G. occitaniae*, the association between epigenetic diversity and environment actually occurred because of an actual causal relationship between genetic diversity and environment and a covariation at the genome level between epigenetic and genetic marks. On the contrary in *P. phoxinus*, none of these associations between environment, genetic diversity, and epigenetic diversity was uncovered. The strong and significant association between the environment and genetic diversity observed in *G. occitaniae* and the absence of such association in *P. phoxinus* is in agreement with previous findings that phenotypic differentiation between sites is strongly associated with the environment in *G. occitaniae* but not in *P. phoxinus* [39]. In particular, oxygen saturation was strongly associated with phenotypic divergence among *G. occitaniae*, and complementary analyses (Commonality Analysis, not shown) also revealed that oxygen concentration was the most impacting environmental variable associated to genetic (SNPs) differentiation. Overall, these findings are consistent with the hypothesis of natural selection, triggered by environmental stress—here oxygen—and modulated by genetic marks, resulting in a local phenotypic adaptation of *G. occitaniae* [11,39,70]. To sum up, the epigenetic-environment association found in *G. occitaniae* was actually spurious, which stresses the importance of controlling for the genetic background of populations to infer the causal link between epigenetic variations and environmental heterogeneity in wild populations.

In both species, our results highlighted that epigenetic marks were more powerful to discriminate populations than the two genetic markers. Indeed, in both species, the AMOVA revealed that the variance measured among populations was higher when using epigenetic markers than when using microsatellite markers. Moreover, using the same metric of differentiation (Gst"), we found that the mean pairwise differentiation among populations was higher when using epigenetic markers than when using genetic markers (either microsatellites or SNPs, see Figure 2; most Gst" values measured using epigenetic marks are above the 1:1 line). In other words, environmentally -and geographically- distant populations were more different epigenetically than genetically. This result suggests that although a part of the epigenetic marks seems to be genetically controlled, epigenetic diversity also contained information that seems independent from genetic variation and that allows discriminating populations further. This strong discriminative power of epigenetic marks is unlikely to be mainly driven by the sensitivity of epigenetic marks to the environment as we found little evidence that, in these species, the epigenetic structure of populations was causally linked to the environment (see above). Rather, we can speculate that the inherent characteristics of epigenetic marks (in particular, higher mutation rates) and their sensitivity to neutral processes (drift and dispersal) make them extremely relevant as natural markers for population discrimination. This strong discriminative ability might be highly relevant for species conservation, for instance, to identify ecologically and evolutionary isolated populations (and hence Evolutionary Significant Units) [71–73] and/or infer connectivity among populations [4].

Finally, we want to address some methodological limitations to our work. First, we worked on fin tissue to favor a non-lethal approach, supported by the fact that the shape of the fin and its coloration can be linked to abiotic environmental conditions [45,46]. In this context, the fin appears to be a good compromise between both scientific and ethical concerns. However, given that DNA methylation diversity can show tissue-specific differences within an individual [42–44], this choice is not trivial. The fin is likely not the tissue that responds the most, at the molecular level, to environmental conditions, and is consequently probably less linked to fitness than other tissues. Our results might have been different with other tissues like muscle [33,74–76], blood [34,77,78], liver [79,80], or gill tissue [81]. Some authors made different choices to avoid tissue-specific differences such as gonads [56] or even whole organism when it is relatively small in body size [36,82–84]. Second, we used a MS-AFLP protocol, which is currently the most widely used approach for inferring epigenetic diversity in wild populations [37,78,81,85]. Although MS-AFLP has several advantages, such as being efficient and economical to assess epigenetic diversity for large sample sizes, this method only provides anonymous and dominant markers leading to a fragment analysis that is subject to homoplasy (i.e., two fragments of the same size but with different sequence) [57]. Consequently, DNA methylation marks are difficult to link to the functional context or to compare directly with genetic data [78,86]. Promising approaches based on reduced representation bisulfite sequencing (RRBS) approach and next-generation sequencing (NGS) might partly solve these issues [87,88], particularly by allowing the identification of the specific loci potentially implied in the response to the environment. RRBS should be explicitly compared to MS-AFLP to isolate further the potential limits of MS-AFLP approaches and to gain novel insights into the loci underlying adaptation to the local environment [89].

5. Conclusions

To conclude, our study provided an attempt to link epigenetic variation in wild populations to the surrounding environment, a work that has been almost always carried out in plants and much more rarely in animals [20,90]. In our empirical comparative study, we showed that, contrary to expectations, there was no link between epigenetic variation and environmental constraints in *G. occitaniae* and *P. phoxinus*. This suggests that epigenetic diversity might be poorly associated to adaptation in these two species. Nonetheless, in both cases, epigenetic variation seems to be genetically determined, indicating a genetic control of epigenetic variation, as suspected in previous works [20–22]. Interestingly, epigenetic differentiation was linked (or show a tendency to be linked) to microsatellites (i.e., neutral) genetic differentiation, reinforcing the idea of an impact of the same neutral processes on genetic and epigenetic variation [26–28]. This implies that, in the species we investigated, epigenetic variation is more likely driven by neutral than nonneutral processes. Nonetheless, epigenetic marks are still more efficient than genetic markers to discriminate populations and can hence provide a tool to improve conservation strategies of endangered populations [4]. Future works should hence consider the dual use of genetic and epigenetic marks to inform conservation strategies, such as the delimitation of significant units of conservation or the quantification of biological connectivity in fragmented landscapes.

Author Contributions: Conceptualization, L.F., O.R., and S.B.; methodology, L.F., C.V., and G.L.; software, L.F., J.G.P., and G.L.; validation, J.G.P., G.L., and S.B.; formal analysis, L.F., J.G.P., and S.B.; investigation, L.F. and S.B.; resources, S.B. and G.L.; data curation, L.F., C.V., and G.L.; writing—original draft preparation, L.F.; writing—review and editing, S.B.; visualization, L.F. and S.B.; supervision, S.B.; project administration, S.B.; funding acquisition, S.B. and G.L. All authors have read and agreed to the published version of the manuscript.

Funding: This research was funded by the Agence Nationale de la Recherche, grant numbers ANR-10LABX and ANR 18-CE02-0006.

Institutional Review Board Statement: Not applicable.

Informed Consent Statement: Not applicable.

Data Availability Statement: The data presented in this study are openly available in (FigShare) at (https://doi.org/10.6084/m9.figshare.13580534.v1), reference number (13580534).

Acknowledgments: We thank Héloïse Bossuat, Lisa Fourtune, Christelle Leung, and all other colleagues who help at different stages of the work.

Conflicts of Interest: The authors declare no conflict of interest. The funders had no role in the design of the study; in the collection, analyses, or interpretation of data; in the writing of the manuscript; or in the decision to publish the results.

Appendix A

Table A1. Physical and chemical characteristics of the 13 sites in which 24 individuals per species have been sampled. These sites have been selected from previous studies [39] to maximize environmental variation among sites. Details on the measurement of each parameter is explained in the main text and in Fourtune et al. [39]. Latitude and longitude are expressed in Lambert93.

Site Code	River Name	Latitude	Longitude	River Flow (m^3/s)	River Width (m)	River Slope (%)	Altitude (m)	Nitrates $(mg \cdot L^{-1})$	Nitrites $(mg \cdot L^{-1})$	Ortho-Phosphates $(mg \cdot L^{-1})$	Oxygen $(mg \cdot L^{-1})$	BOD $(mg \cdot L^{-1})$	Water Conductivity $(mS \cdot cm^{-1})$	pH	SM $(mg \cdot L^{-1})$	Oxygen Saturation (%)	Temperature ($^\circ$C)
ARIVen	Ariege	573460.37	6261035.16	59	60.00	1	160	23.03	0.09	0.27	8.80	1.03	796.67	8.17	52.40	92.00	20.33
ARZMas	Arize	567474.38	6221932.39	4	6.35	2	284	3.70	0.02	0.03	9.37	0.67	343.00	8.17	4.63	100.33	17.23
AVEDru	Aveyron	659451.23	6359878.75	9	14.85	8	481	5.96	0.06	0.39	9.02	1.20	489.67	8.32	13.33	114.67	22.80
BAIHac	Baise	493842.07	6246291.53	0	6.00	6	279	2.67	0.02	0.04	9.20	0.60	157.67	8.00	39.33	99.00	17.67
BARMon	Barguelonne	545090.06	6347536.21	2	6.05	1	85	12.43	0.04	0.01	7.57	2.03	531.33	8.03	10.33	88.33	23.33
CELSau	Cele	597974.87	6380867.77	19	17.25	1	143	8.40	0.03	0.11	8.68	0.53	249.67	7.92	3.20	100.57	22.27
GARCla	Garonne	506893.70	6225179.17	53	67.00	4	388	1.73	0.01	0.02	10.33	1.17	165.33	8.07	18.37	109.67	16.63
GARMur	Garonne	564902.27	6263764.69	109	83.50	1	155	2.33	0.03	0.02	9.65	0.67	214.71	8.38	6.70	112.00	22.06
LEMMol	Lemboulas	566796.13	6343683.15	1	3.95	2	101	9.81	0.12	0.11	6.95	1.57	689.33	7.85	14.67	77.53	21.73
LOUFou	Louge	543137.54	6243604.74	1	9.25	1	249	4.90	0.03	0.05	8.83	0.80	264.50	8.08	34.70	99.50	20.08
SALCau	Salat	543868.20	6217247.36	34	32.50	5	355	1.33	0.01	0.03	10.30	1.13	165.83	8.33	3.23	107.50	17.48
VIASeg	Viaur	687141.24	6355312.09	1	5.70	6	750	10.32	0.02	0.03	9.11	0.93	140.00	7.74	22.67	100.33	16.40
VOLPla	Volp	546927.08	6232038.71	1	10.15	5	251	2.70	0.01	0.02	8.57	0.53	435.67	8.10	2.93	100.67	21.93

Appendix B

Table A2. Sequences of adaptors and primers used in the Methylation-Sensitive-Amplified Fragment Length Polymorphism (MS-AFLP) protocol (and their dyes and the number of loci per primer). Forward (F) and Reverse (R) strand. The same preselective primer was used for both species and three different selective primers were used for each species.

Name	Sequence (5′-3′)	Dye	Number of loci
EcoRI adaptors	F: CTCGTAGACTGCGTACC R: AATTGGTACGCAGTCTAC	–	–
MspI/HpaII adaptors	F: GATCATGAGTCCTGCT R: CGAGCAGGACTCATGA	–	–
Preselective primers	F: GACTGCGTACCAATTC+**A** R: ATCATGAGTCCTGCTCGG+**A**	–	–
Selective primers *G. occitaniae*	F: GACTGCGTACCAATTC+**AGA** R: ATCATGAGTCCTGCTCGG+**ATC**	HEX	66
	F: GACTGCGTACCAATTC+**AAT** R: ATCATGAGTCCTGCTCGG+**AAT**	AT550	69
	F: GACTGCGTACCAATTC+**AAT** R: ATCATGAGTCCTGCTCGG+**ATC**	AT550	116
Selective primers *P. phoxinus*	F: GACTGCGTACCAATTC+**AAC** R: ATCATGAGTCCTGCTCGG+**AAA**	FAM	71
	F: GACTGCGTACCAATTC+**AAC** R: ATCATGAGTCCTGCTCGG+**AAT**	FAM	105
	F: GACTGCGTACCAATTC+**AGA** R: ATCATGAGTCCTGCTCGG+**ACT**	HEX	98

Appendix C. —PCR Conditions

PCR1: Reaction started with a first elongation step at 72 °C for 2 min (to ensure that ligation will not vanish during the first denaturation step), followed by 25 cycles of amplification: a denaturation step at 92 °C for 30 s, a hybridization step at 56 °C for 1 min, and an elongation step at 72 °C for 1 min and finished with a final extension step at 72 °C for 5 min.

PCR2: Reaction started with a first denaturation step at 94 °C for 30 s, followed by 15 cycles of a decreasingly selective amplification: a denaturation step at 94 °C for 30 s, a decreasingly selective hybridization step from 65 to 56 °C (dropping the temperature at each cycle), and an elongation step at 72 °C for 2 min. This is followed by 30 cycles of more classic amplification: 94 °C for 30 s, 56 °C for 30 s, and 72 °C for 2 min and finished with a final long extension step at 60 °C for 30 min.

References

1. Manel, S.; Schwartz, M.K.; Luikart, G.; Taberlet, P. Landscape genetics: Combining landscape ecology and population genetics. *Trends Ecol. Evol.* **2003**, *18*, 189–197. [CrossRef]
2. Sexton, J.P.; Hangartner, S.B.; Hoffmann, A.A. Genetic isolation by environment or distance: Which pattern of gene flow is most common? *Evolution* **2014**, *68*, 1–15. [CrossRef] [PubMed]
3. Holderegger, R.; Kamm, U.; Gugerli, F. Adaptive vs. neutral genetic diversity: Implications for landscape genetics. *Landsc. Ecol.* **2006**, *21*, 797–807. [CrossRef]
4. Rey, O.; Eizaguirre, C.; Angers, B.; Baltazar-Soares, M.; Sagonas, K.; Prunier, J.G.; Blanchet, S. Linking epigenetics and biological conservation: Towards a *conservation epigenetics* perspective. *Funct. Ecol.* **2020**, *34*, 414–427. [CrossRef]
5. Rey, O.; Danchin, E.; Mirouze, M.; Loot, C.; Blanchet, S. Adaptation to Global Change: A Transposable Element–Epigenetics Perspective. *Trends Ecol. Evol.* **2016**, *31*, 514–526. [CrossRef] [PubMed]

6. Richards, C.L.; Alonso, C.; Becker, C.; Bossdorf, O.; Bucher, E.; Colomé-Tatché, M.; Durka, W.; Engelhardt, J.; Gaspar, B.; Gogol-Döring, A.; et al. Ecological plant epigenetics: Evidence from model and non-model species, and the way forward. *Ecol. Lett.* **2017**, *20*. [CrossRef] [PubMed]

7. Danchin, E.; Pocheville, A.; Rey, O.; Pujol, B.; Blanchet, S. Epigenetically facilitated mutational assimilation: Epigenetics as a hub within the inclusive evolutionary synthesis: Epigenetics as a hub for genetic assimilation. *Biol. Rev.* **2019**, *94*, 259–282. [CrossRef]

8. Greer, E.L.; Maures, T.J.; Ucar, D.; Hauswirth, A.G.; Mancini, E.; Lim, J.P.; Benayoun, B.A.; Shi, Y.; Brunet, A. Transgenerational epigenetic inheritance of longevity in Caenorhabditis elegans. *Nature* **2011**, *479*, 365–371. [CrossRef]

9. Nicotra, A.B.; Segal, D.L.; Hoyle, G.L.; Schrey, A.W.; Verhoeven, K.J.F.; Richards, C.L. Adaptive plasticity and epigenetic variation in response to warming in an Alpine plant. *Ecol. Evol.* **2015**, *5*, 634–647. [CrossRef]

10. Verhoeven, K.J.F.; vonHoldt, B.M.; Sork, V.L. Epigenetics in ecology and evolution: What we know and what we need to know. *Mol. Ecol.* **2016**, *25*, 1631–1638. [CrossRef]

11. Hu, J.; Barrett, R.D.H. Epigenetics in natural animal populations. *J. Evol. Biol.* **2017**, *30*, 1612–1632. [CrossRef] [PubMed]

12. Feil, R.; Fraga, M.F. Epigenetics and the environment: Emerging patterns and implications. *Nat. Rev. Genet.* **2012**, *13*, 97–109. [CrossRef] [PubMed]

13. Nilsson, E.; Larsen, G.; Manikkam, M.; Guerrero-Bosagna, C.; Savenkova, M.I.; Skinner, M.K. Environmentally Induced Epigenetic Transgenerational Inheritance of Ovarian Disease. *PLoS ONE* **2012**, *7*, e36129. [CrossRef] [PubMed]

14. Alvarado, S.; Fernald, R.D.; Storey, K.B.; Szyf, M. The Dynamic Nature of DNA Methylation: A Role in Response to Social and Seasonal Variation. *Integr. Comp. Biol.* **2014**, *54*, 68–76. [CrossRef] [PubMed]

15. Lillycrop, K.A.; Burdge, G.C. Maternal diet as a modifier of offspring epigenetics. *J. Dev. Orig. Health Dis.* **2015**, *6*, 88–95. [CrossRef]

16. Berger, S.L.; Kouzarides, T.; Shiekhattar, R.; Shilatifard, A. An operational definition of epigenetics. *Genes Dev.* **2009**, *23*, 781–783. [CrossRef]

17. Jablonka, E.; Raz, G. Transgenerational Epigenetic Inheritance: Prevalence, Mechanisms, and Implications for the Study of Heredity and Evolution. *Q. Rev. Biol.* **2009**, *84*, 131–176. [CrossRef]

18. Bollati, V.; Baccarelli, A. Environmental epigenetics. *Heredity* **2010**, *105*, 105–112. [CrossRef]

19. Danchin, É.; Charmantier, A.; Champagne, F.A.; Mesoudi, A.; Pujol, B.; Blanchet, S. Beyond DNA: Integrating inclusive inheritance into an extended theory of evolution. *Nat. Rev. Genet.* **2011**, *12*, 475–486. [CrossRef]

20. Richards, E.J. Inherited epigenetic variation—Revisiting soft inheritance. *Nat. Rev. Genet.* **2006**, *7*, 395–401. [CrossRef]

21. Dubin, M.J.; Zhang, P.; Meng, D.; Remigereau, M.-S.; Osborne, E.J.; Paolo Casale, F.; Drewe, P.; Kahles, A.; Jean, G.; Vilhjálmsson, B.; et al. DNA methylation in Arabidopsis has a genetic basis and shows evidence of local adaptation. *eLife* **2015**, *4*, e05255. [CrossRef] [PubMed]

22. Taudt, A.; Colomé-Tatché, M.; Johannes, F. Genetic sources of population epigenomic variation. *Nat. Rev. Genet.* **2016**, *17*, 319–332. [CrossRef] [PubMed]

23. Heo, J.B.; Lee, Y.-S.; Sung, S. Epigenetic regulation by long noncoding RNAs in plants. *Chromosome Res.* **2013**, *21*, 685–693. [CrossRef] [PubMed]

24. Harvey, Z.H.; Chen, Y.; Jarosz, D.F. Protein-Based Inheritance: Epigenetics beyond the Chromosome. *Mol. Cell* **2018**, *69*, 195–202. [CrossRef] [PubMed]

25. Norouzitallab, P.; Baruah, K.; Vanrompay, D.; Bossier, P. Can epigenetics translate environmental cues into phenotypes? *Sci. Total Environ.* **2019**, *647*, 1281–1293. [CrossRef]

26. Herrera, C.M.; Medrano, M.; Bazaga, P. Comparative spatial genetics and epigenetics of plant populations: Heuristic value and a proof of concept. *Mol. Ecol.* **2016**, *25*, 1653–1664. [CrossRef] [PubMed]

27. Trucchi, E.; Mazzarella, A.B.; Gilfillan, G.D.; Lorenzo, M.T.; Schönswetter, P.; Paun, O. BsRADseq: Screening DNA methylation in natural populations of non-model species. *Mol. Ecol.* **2016**, *25*, 1697–1713. [CrossRef]

28. Richards, C.L.; Bossdorf, O.; Verhoeven, K.J.F. Understanding natural epigenetic variation: Commentary. *New Phytol.* **2010**, *187*, 562–564. [CrossRef]

29. Cubas, P.; Vincent, C.; Coen, E. An epigenetic mutation responsible for natural variation in floral symmetry. *Nature* **1999**, *401*, 157–161. [CrossRef]

30. Issa, J.-P. Aging and epigenetic drift: A vicious cycle. *J. Clin. Investig.* **2014**, *124*, 24–29. [CrossRef]

31. Shan, X.H.; Li, Y.D.; Liu, X.M.; Wu, Y.; Zhang, M.Z.; Guo, W.L.; Liu, B.; Yuan, Y.P. Comparative analyses of genetic/epigenetic diversities and structures in a wild barley species (Hordeum brevisubulatum) using MSAP, SSAP and AFLP. *Genet. Mol. Res.* **2012**, *11*, 2749–2759. [CrossRef] [PubMed]

32. Lele, L.; Ning, D.; Cuiping, P.; Xiao, G.; Weihua, G. Genetic and epigenetic variations associated with adaptation to heterogeneous habitat conditions in a deciduous shrub. *Ecol. Evol.* **2018**, *8*, 2594–2606. [CrossRef] [PubMed]

33. Liu, S.; Sun, K.; Jiang, T.; Ho, J.P.; Liu, B.; Feng, J. Natural epigenetic variation in the female great roundleaf bat (Hipposideros armiger) populations. *Mol. Genet. Genom.* **2012**, *287*, 643–650. [CrossRef] [PubMed]

34. Sheldon, E.L.; Schrey, A.; Andrew, S.C.; Ragsdale, A.; Griffith, S.C. Epigenetic and genetic variation among three separate introductions of the house sparrow (*Passer domesticus*) into Australia. *R. Soc. Open Sci.* **2018**, *5*, 172185. [CrossRef] [PubMed]

35. Foust, C.M.; Preite, V.; Schrey, A.W.; Alvarez, M.; Robertson, M.H.; Verhoeven, K.J.F.; Richards, C.L. Genetic and epigenetic differences associated with environmental gradients in replicate populations of two salt marsh perennials. *Mol. Ecol.* **2016**, *25*, 1639–1652. [CrossRef] [PubMed]
36. Ni, P.; Li, S.; Lin, Y.; Xiong, W.; Huang, X.; Zhan, A. Methylation divergence of invasive *Ciona* ascidians: Significant population structure and local environmental influence. *Ecol. Evol.* **2018**, *8*, 10272–10287. [CrossRef]
37. Bossdorf, O.; Richards, C.L.; Pigliucci, M. Epigenetics for ecologists. *Ecol. Lett.* **2007**, *11*, 106–115. [CrossRef]
38. Fourtune, L.; Paz-Vinas, I.; Loot, G.; Prunier, J.G.; Blanchet, S. Lessons from the fish: A multi-species analysis reveals common processes underlying similar species-genetic diversity correlations. *Freshw. Biol.* **2016**, *61*, 1830–1845. [CrossRef]
39. Fourtune, L.; Prunier, J.G.; Mathieu-Begne, E.; Canto, N.; Veyssiere, C.; Loot, G.; Blanchet, S. Intraspecific genetic and phenotypic diversity: Parallel processes and correlated patterns? *BioRxiv* **2018**, arXiv:288357. [CrossRef]
40. Buisson, L.; Blanc, L.; Grenouillet, G. Modelling stream fish species distribution in a river network: The relative effects of temperature versus physical factors. *Ecol. Freshw. Fish* **2008**, *17*, 244–257. [CrossRef]
41. Crispo, E.; Chapman, L.J. Population genetic structure across dissolved oxygen regimes in an African cichlid fish: POPULATION STRUCTURE ACROSS OXYGEN REGIMES. *Mol. Ecol.* **2008**, *17*, 2134–2148. [CrossRef] [PubMed]
42. Gama-Sosa, M.A.; Midgett, R.M.; Slagel, V.A.; Githens, S.; Kuo, K.C.; Gehrke, C.W.; Ehrlich, M. Tissue-specific differences in DNA methylation in various mammals. *Biochim. Biophys. Acta Gene Struct. Expr.* **1983**, *740*, 212–219. [CrossRef]
43. Sun, Y.; Hou, R.; Fu, X.; Sun, C.; Wang, S.; Wang, C.; Li, N.; Zhang, L.; Bao, Z. Genome-Wide Analysis of DNA Methylation in Five Tissues of Zhikong Scallop, Chlamys farreri. *PLoS ONE* **2014**, *9*, e86232. [CrossRef] [PubMed]
44. Yang, C.; Zhang, Y.; Liu, W.; Lu, X.; Li, C. Genome-wide analysis of DNA methylation in five tissues of sika deer (Cervus nippon). *Gene* **2018**, *645*, 48–54. [CrossRef]
45. Blanchet, S.; Páez, D.J.; Bernatchez, L.; Dodson, J.J. An integrated comparison of captive-bred and wild Atlantic salmon (Salmo salar): Implications for supportive breeding programs. *Biol. Conserv.* **2008**, *141*, 1989–1999. [CrossRef]
46. Paez, D.J.; Hedger, R.; Bernatchez, L.; Dodson, J.J. The morphological plastic response to water current velocity varies with age and sexual state in juvenile Atlantic salmon, *Salmo salar*. *Freshw. Biol.* **2008**, *53*, 1544–1554. [CrossRef]
47. Pella, H.; Lejot, J.; Lamouroux, N.; Snelder, T. Le réseau hydrographique théorique (RHT) français et ses attributs environnementaux. *Géomorphologie Relief Process. Environ.* **2012**, *18*, 317–336. [CrossRef]
48. Prunier, J.G.; Dubut, V.; Loot, G.; Tudesque, L.; Blanchet, S. The relative contribution of river network structure and anthropogenic stressors to spatial patterns of genetic diversity in two freshwater fishes: A multiple-stressors approach. *Freshw. Biol.* **2018**, *63*, 6–21. [CrossRef]
49. Dray, S.; Dufour, A.-B. The ade4 Package: Implementing the Duality Diagram for Ecologists. *J. Stat. Soft.* **2007**, *22*. [CrossRef]
50. Aljanabi, S. Universal and rapid salt-extraction of high quality genomic DNA for PCR- based techniques. *Nucleic Acids Res.* **1997**, *25*, 4692–4693. [CrossRef]
51. Schlötterer, C.; Tobler, R.; Kofler, R.; Nolte, V. Sequencing pools of individuals—mining genome-wide polymorphism data without big funding. *Nat. Rev. Genet.* **2014**, *15*, 749–763. [CrossRef] [PubMed]
52. Prunier, J.G.; Chevalier, M.; Raffard, A.; Loot, G.; Poulet, N.; Blanchet, S. Contemporary loss of genetic diversity in wild fish populations reduces biomass stability over time. *BioRxiv* **2019**, arXiv:884734. [CrossRef]
53. Vos, P.; Hogers, R.; Bleeker, M.; Reijans, M.; van de Lee, T.; Hornes, M.; Friters, A.; Pot, J.; Paleman, J.; Kuiper, M.; et al. AFLP: A new technique for DNA fingerprinting. *Nucleic Acids Res.* **1995**, *23*, 4407–4414. [CrossRef] [PubMed]
54. Schulz, B.; Eckstein, R.L.; Durka, W. Scoring and analysis of methylation-sensitive amplification polymorphisms for epigenetic population studies. *Mol. Ecol. Res.* **2013**, *13*, 642–653. [CrossRef] [PubMed]
55. Xiong, L.Z.; Zhang, Q. Patterns of cytosine methylation in an elite rice hybrid and its parental lines, detected by a methylation-sensitive amplification polymorphism technique. *Mol. Gen. Genet. MGG* **1999**, *261*, 439–446. [CrossRef]
56. Smith, T.A.; Martin, M.D.; Nguyen, M.; Mendelson, T.C. Epigenetic divergence as a potential first step in darter speciation. *Mol. Ecol.* **2016**, *25*, 1883–1894. [CrossRef]
57. Caballero, A.; Quesada, H.; Rolán-Alvarez, E. Impact of Amplified Fragment Length Polymorphism Size Homoplasy on the Estimation of Population Genetic Diversity and the Detection of Selective Loci. *Genetics* **2008**, *179*, 539–554. [CrossRef]
58. Meirmans, P.G.; Hedrick, P.W. Assessing population structure: FST and related measures: Invited technical review. *Mol. Ecol. Res.* **2011**, *11*, 5–18. [CrossRef]
59. Prunier, J.G.; Poesy, C.; Dubut, V.; Veyssière, C.; Loot, G.; Poulet, N.; Blanchet, S. Quantifying the individual impact of artificial barriers in freshwaters: A standardized and absolute genetic index of fragmentation. *Evol. Appl.* **2020**, *13*, 2566–2581. [CrossRef]
60. Ellegren, H. Microsatellites: Simple sequences with complex evolution. *Nat. Rev. Genet.* **2004**, *5*, 435–445. [CrossRef]
61. Gutiérrez-Velázquez, M.V.; Almaraz-Abarca, N.; Herrera-Arrieta, Y.; Ávila-Reyes, J.A.; González-Valdez, L.S.; Torres-Ricario, R.; Uribe-Soto, J.N.; Monreal-García, H.M. Comparison of the phenolic contents and epigenetic and genetic variability of wild and cultivated watercress (Rorippa nasturtium var. aquaticum L.). *Electron. J. Biotechnol.* **2018**, *34*, 9–16. [CrossRef]
62. Avramidou, E.V.; Ganopoulos, I.V.; Doulis, A.G.; Tsaftaris, A.S.; Aravanopoulos, F.A. Beyond population genetics: Natural epigenetic variation in wild cherry (Prunus avium). *Tree Genet. Genomes* **2015**, *11*, 95. [CrossRef]
63. Herrera, C.M.; Medrano, M.; Bazaga, P. Comparative epigenetic and genetic spatial structure of the perennial herb *Helleborus foetidus*: Isolation by environment, isolation by distance, and functional trait divergence. *Am. J. Bot.* **2017**, *104*, 1195–1204. [CrossRef] [PubMed]

64. Roy, N.; Choi, J.-Y.; Lim, M.-J.; Lee, S.-I.; Choi, H.-J.; Kim, N.-S. Genetic and epigenetic diversity among dent, waxy, and sweet corns. *Genes Genom.* **2015**, *37*, 865–874. [CrossRef]

65. Guo, W.; Hussain, N.; Wu, R.; Liu, B. High hypomethylation and epigenetic variation in fragmented populations of wild barley (Hordeum brevisubulatum). *Pak. J. Bot.* **2018**, *50*, 1379–1386.

66. Duchemin, W.; Anselmetti, Y.; Patterson, M.; Ponty, Y.; Bérard, S.; Chauve, C.; Scornavacca, C.; Daubin, V.; Tannier, E. DeCoSTAR: Reconstructing the Ancestral Organization of Genes or Genomes Using Reconciled Phylogenies. *Genome Biol. Evol.* **2017**, *9*, 1312–1319. [CrossRef]

67. DeFaveri, J.; Viitaniemi, H.; Leder, E.; Merilä, J. Characterizing genic and nongenic molecular markers: Comparison of microsatellites and SNPs. *Mol. Ecol. Resour.* **2013**, *13*, 377–392. [CrossRef]

68. Nosil, P.; Funk, D.J.; Ortiz-Barrientos, D. Divergent selection and heterogeneous genomic divergence. *Mol. Ecol.* **2009**, *18*, 375–402. [CrossRef]

69. Fourtune, L.; Prunier, J.G.; Paz-Vinas, I.; Loot, G.; Veyssière, C.; Blanchet, S. Inferring Causalities in Landscape Genetics: An Extension of Wright's Causal Modeling to Distance Matrices. *Am. Nat.* **2018**, *191*, 491–508. [CrossRef]

70. Smith, G.; Ritchie, M.G. How might epigenetics contribute to ecological speciation? *Curr. Zool.* **2013**, *59*, 686–696. [CrossRef]

71. McMahon, B.J.; Teeling, E.C.; Ho, J. How and why should we implement genomics into conservation? *Evol. Appl.* **2014**, *7*, 999–1007. [CrossRef] [PubMed]

72. Shafer, A.B.A.; Wolf, J.B.W.; Alves, P.C.; Bergström, L.; Bruford, M.W.; Brännström, I.; Colling, G.; Dalén, L.; De Meester, L.; Ekblom, R.; et al. Genomics and the challenging translation into conservation practice. *Trends Ecol. Evol.* **2015**, *30*, 78–87. [CrossRef] [PubMed]

73. Eizaguirre, C.; Baltazar-Soares, M. Evolutionary conservation-evaluating the adaptive potential of species. *Evol. Appl.* **2014**, *7*, 963–967. [CrossRef]

74. Massicotte, R.; Angers, B. General-Purpose Genotype or How Epigenetics Extend the Flexibility of a Genotype. *Genet. Res. Int.* **2012**, *2012*, 1–7. [CrossRef] [PubMed]

75. Jiang, Q.; Li, Q.; Yu, H.; Kong, L.-F. Genetic and epigenetic variation in mass selection populations of Pacific oyster Crassostrea gigas. *Genes Genom.* **2013**, *35*, 641–647. [CrossRef]

76. Zhang, X.; Li, Q.; Kong, L.; Yu, H. Epigenetic variation of wild populations of the Pacific oyster Crassostrea gigas determined by methylation-sensitive amplified polymorphism analysis. *Fish. Sci.* **2018**, *84*, 61–70. [CrossRef]

77. Liebl, A.L.; Schrey, A.W.; Richards, C.L.; Martin, L.B. Patterns of DNA Methylation Throughout a Range Expansion of an Introduced Songbird. *Integr. Comp. Biol.* **2013**, *53*, 351–358. [CrossRef]

78. Schrey, A.W.; Alvarez, M.; Foust, C.M.; Kilvitis, H.J.; Lee, J.D.; Liebl, A.L.; Martin, L.B.; Richards, C.L.; Robertson, M. Ecological Epigenetics: Beyond MS-AFLP. *Integr. Comp. Biol.* **2013**, *53*, 340–350. [CrossRef]

79. Wenzel, M.A.; Piertney, S.B. Fine-scale population epigenetic structure in relation to gastrointestinal parasite load in red grouse (*Lagopus lagopus scotica*). *Mol. Ecol.* **2014**, *23*, 4256–4273. [CrossRef]

80. Wogan, G.O.U.; Yuan, M.L.; Mahler, D.L.; Wang, I.J. Genome-wide epigenetic isolation by environment in a widespread *Anolis* lizard. *Mol. Ecol.* **2020**, *29*, 40–55. [CrossRef]

81. Johnson, K.M.; Kelly, M.W. Population epigenetic divergence exceeds genetic divergence in the Eastern oyster *Crassostrea virginica* in the Northern Gulf of Mexico. *Evol. Appl.* **2020**, *13*, 945–959. [CrossRef] [PubMed]

82. Baldanzi, S.; Watson, R.; McQuaid, C.D.; Gouws, G.; Porri, F. Epigenetic variation among natural populations of the South African sandhopper Talorchestia capensis. *Evol. Ecol.* **2017**, *31*, 77–91. [CrossRef]

83. Watson, R.G.A.; Baldanzi, S.; Pérez-Figueroa, A.; Gouws, G.; Porri, F. Morphological and epigenetic variation in mussels from contrasting environments. *Mar. Biol.* **2018**, *165*, 50. [CrossRef]

84. Hawes, N.A.; Amadoru, A.; Tremblay, L.A.; Pochon, X.; Dunphy, B.; Fidler, A.E.; Smith, K.F. Epigenetic patterns associated with an ascidian invasion: A comparison of closely related clades in their native and introduced ranges. *Sci. Rep.* **2019**, *9*, 14275. [CrossRef] [PubMed]

85. Banerjee, A.K.; Guo, W.; Huang, Y. Genetic and epigenetic regulation of phenotypic variation in invasive plants—linking research trends towards a unified framework. *NB* **2019**, *49*, 77–103. [CrossRef]

86. Robertson, M.; Richards, C. Opportunities and challenges of next-generation sequencing applications in ecological epigenetics. *Mol. Ecol.* **2015**, *24*, 3799–3801. [CrossRef] [PubMed]

87. Schield, D.R.; Walsh, M.R.; Card, D.C.; Andrew, A.L.; Adams, R.H.; Castoe, T.A. EpiRADseq: Scalable analysis of genomewide patterns of methylation using next-generation sequencing. *Methods Ecol. Evol.* **2016**, *7*, 60–69. [CrossRef]

88. van Gurp, T.P.; Wagemaker, N.C.A.M.; Wouters, B.; Vergeer, P.; Ouborg, J.N.J.; Verhoeven, K.J.F. epiGBS: Reference-free reduced representation bisulfite sequencing. *Nat. Methods* **2016**, *13*, 322–324. [CrossRef]

89. Paun, O.; Verhoeven, K.J.F.; Richards, C.L. Opportunities and Limitations of Reduced Representation Bisulfite Sequencing in Plant Ecological Epigenomics. *New Phytol.* **2019**, *221*, 738–742. [CrossRef]

90. Hirsch, S.; Baumberger, R.; Grossniklaus, U. Epigenetic Variation, Inheritance, and Selection in Plant Populations. *Cold Spring Harb. Symp. Quant. Biol.* **2012**, *77*, 97–104. [CrossRef]

Article

Phenotypic Response to Light Versus Shade Associated with DNA Methylation Changes in Snapdragon Plants (*Antirrhinum majus*)

Pierick Mouginot [1,†], Nelia Luviano Aparicio [2,†], Delphine Gourcilleau [3,†], Mathieu Latutrie [1], Sara Marin [1], Jean-Louis Hemptinne [3], Christoph Grunau [2] and Benoit Pujol [1,*]

[1] PSL Université Paris: EPHE-UPVD-CNRS, USR 3278 CRIOBE, Université de Perpignan, 52 Avenue Paul Alduy, CEDEX 9, 66860 Perpignan, France; pierick.mouginot@univ-perp.fr (P.M.); mathieu.latutrie@univ-perp.fr (M.L.); marin.sa31@gmail.com (S.M.)

[2] Université Montpellier, CNRS, IFREMER, UPVD, Interactions Hôtes Pathogènes Environnements (IHPE), 66860 Perpignan, France; nelia.luviano@univ-perp.fr (N.L.A.); christoph.grunau@univ-perp.fr (C.G.)

[3] Laboratoire Évolution & Diversité Biologique (EDB, UMR 5174), Université Fédérale de Toulouse Midi-Pyrénées, CNRS, IRD, UPS, 118 route de Narbonne, Bat 4R1, CEDEX 9, 31062 Toulouse, France; delphine.gourcilleau@gmail.com (D.G.); jean-louis.hemptinne@educagri.fr (J.-L.H.)

* Correspondence: benoit.pujol@univ-perp.fr

† These authors share first authorship of this work.

Citation: Mouginot, P.; Luviano Aparicio, N.; Gourcilleau, D.; Latutrie, M.; Marin, S.; Hemptinne, J.-L.; Grunau, C.; Pujol, B. Phenotypic Response to Light Versus Shade Associated with DNA Methylation Changes in Snapdragon Plants (*Antirrhinum majus*). *Genes* **2021**, *12*, 227. https://doi.org/10.3390/genes12020227

Academic Editors: Delphine Legrand and Simon Blanchet
Received: 30 December 2020
Accepted: 27 January 2021
Published: 4 February 2021

Publisher's Note: MDPI stays neutral with regard to jurisdictional claims in published maps and institutional affiliations.

Abstract: The phenotypic plasticity of plants in response to change in their light environment, and in particularly, to shade is a schoolbook example of ecologically relevant phenotypic plasticity with evolutionary adaptive implications. Epigenetic variation is known to potentially underlie plant phenotypic plasticity. Yet, little is known about its role in ecologically and evolutionary relevant mechanisms shaping the diversity of plant populations in nature. Here we used a reference-free reduced representation bisulfite sequencing method for non-model organisms (epiGBS) to investigate changes in DNA methylation patterns across the genome in snapdragon plants (*Antirrhinum majus* L.). We exposed plants to sunlight versus artificially induced shade in four highly inbred lines to exclude genetic confounding effects. Our results showed that phenotypic plasticity in response to light versus shade shaped vegetative traits. They also showed that DNA methylation patterns were modified under light versus shade, with a trend towards global effects over the genome but with large effects found on a restricted portion. We also detected the existence of a correlation between phenotypic and epigenetic variation that neither supported nor rejected its potential role in plasticity. While our findings imply epigenetic changes in response to light versus shade environments in snapdragon plants, whether these changes are directly involved in the phenotypic plastic response of plants remains to be investigated. Our approach contributed to this new finding but illustrates the limits in terms of sample size and statistical power of population epigenetic approaches in non-model organisms. Pushing this boundary will be necessary before the relationship between environmentally induced epigenetic changes and phenotypic plasticity is clarified for ecologically relevant mechanisms with evolutionary implications.

Keywords: phenotypic plasticity; epigenetics; epiGBS; stem elongation; shade avoidance

1. Introduction

Snapdragon plants (*Antirrhinum majus* L.) undergo developmental changes resulting in different morphologies after exposure to shade [1]. This is one if not the most common example of phenotypic plasticity in plants where changes in internode length (stem elongation), apical dominance (reduced branching), and photosynthetic efficiency (increased Specific Leaf Area or SLA) are observed following shade exposure [1–3]. When it allows plants to avoid the presence of neighbouring vegetation, it is part of the widely documented shade avoidance syndrome of plants [2]. This phenotypic plastic response can be adaptive in the presence of competition for light, e.g., by elongating its stem and reaching sunlight and pollinators in a crowded ecosystem [4,5]. The ecological and adaptive significance and the physiological and genetic mechanisms underlying the phenotypic response of plants to shade are well documented [6,7]. However, little is known about the hypothesis that molecular epigenetic variation might underlie this ecologically relevant plastic response of natural populations (but see [8–10]).

1.1. Calling for Ecologically Relevant Tests of the Epigenetic Basis of Phenotypic Plasticity

Epigenetic changes can be involved with phenotypic plastic responses at the molecular level [11,12]. For example, the chromatin organisation and structure can change in relation to DNA methylation or histone post-translational modifications, which can affect gene expression and release transposable elements (TE) [13]. There is growing evidence for epigenetic variation associated with trait variation and phenotypic plasticity [14]. For example, phenotypic plasticity in response to temperature changes—either heat or cold treatments—was found to be associated with epigenetic modifications [15,16]. The role of epigenetic variation as an interface between ecological and genetic mechanisms is increasingly put forward in evolutionary biology studies [17]. More empirical work is needed to assess the ecological significance of epigenetic variation to understand its role in the evolution of natural populations. It is therefore necessary to test whether ecologically relevant phenotypic plastic responses are associated with epigenetic changes. Here we tested the hypothesis that the phenotypic plasticity in response to shade observed in snapdragon plants [1,8] was associated with epigenetic modifications by using an epigenomic approach.

1.2. Separating Genetic and Epigenetic Effects

Ecological and evolutionary epigenetics is a young domain of research that is constantly ongoing technical developments. One issue with epigenetic approaches of phenotypic variation is that the effect of DNA methylation changes can only be assessed in the absence of confounded genetic variation. This constraint challenges the use of epigenetics in studies at the scale of populations. Although statistical approaches are available to estimate simultaneously the genetic and epigenetic variation of phenotypic traits [18,19], they demand a quantity of data that is not adapted for small experiments on epigenomic variation. We, therefore, chose to use a technical solution to this issue. We used highly inbred lines of snapdragon plants in which genomes are nearly if not totally fixed in a homozygous state by successive generations of self-fertilization. We submitted plants from each snapdragon line to regular sunlight or shade, which allowed us to exclude or extremely reduce confounded genetic effects within lines and replicate the experiment across genetically different backgrounds.

1.3. Snapdragon Plants: The Road So Far

Previous work using High Performance Liquid Chromatography suggested that global methylation contents might change under different light treatments, and called for investigating DNA methylation patterns at the genomic level [8]. Several approaches can be used to characterize DNA methylation, such as Whole-Genome Bisulfite Sequencing (WGBS), Bisulfite converted restriction site associated DNA sequencing or bsRADseq, Epi RADseq, methylated DNA Immuno Precipitation, or meDIP. We chose epiGenome Bisulfite Sequencing or epiGBS [20]. This approach characterizes a reduced representation of the genome and therefore aims at detecting global patterns of DNA methylation changes spread across the genome. It is not aimed at identifying a specific gene or genomic region. Although the use of epigenomic methods is still restricted to small sample sizes, which impedes the study of multiple populations, the epiGBS approach allowed us to study enough samples to estimate the effect of ecological factors in snapdragon plant inbred lineages.

We aimed to assess whether the phenotypic response of snapdragon plants to light versus shade was associated with changes in DNA methylation patterns at the genomic level.

We first assessed phenotypic differences associated with light versus shade by exposing plants grown in experimental to regular sunlight or artificially generated shade. Second, we tested whether the light versus shade treatment had an effect on global methylation patterns across the genome by sampling regions of the genome. Finally, we tested whether DNA methylation changes were consistently associated with phenotypic differences when we had found a significant effect of the light treatment on DNA methylation patterns.

2. Materials and Methods

2.1. Study System

We used four inbred lines of *Antirrhinum majus* L. (snapdragon plants) that were produced following successive generations of self-fertilisation. Snapdragon plants harbour hermaphroditic flowers that are usually self-incompatible [21]. It is a short-lived perennial plant characterized by zygomorphic flowers with genomic development and selection that is widely documented [22]. Its natural populations are highly genetically diverse [23,24] and geographically distributed across a large range of environmental conditions, in particular in terms of vegetation cover [25]. Snapdragon plants are locally adapted to their abiotic environment [26], and have been shown to react in terms of growth and development to light quality and intensity [1,27,28]. It is therefore an ecologically relevant study system to investigate the epigenomic basis of phenotypic plasticity in response to shade. Phenotypic plasticity in response to shade was already shown in experiments based on natural populations [1]. Here we chose to study highly inbred lines of snapdragon plants to exclude confounding genetic effects. We used lines from different origins to replicate our approach in different genetic backgrounds. These lines were originally made for horticultural and developmental genomics research programs. Three of them were provided by the John Innes Centre (Norwich Research Park), namely Ji75, Ji98, and Si50. The fourth line, namely E165, was obtained from the Technical University of Cartagena (Instituto de Biotecnología Vegetal, Pr Marcos Egea Gutiérrez-Cortines).

2.2. Experiment

The plant experiment was conducted outside under semi-controlled environmental conditions in the experimental garden facility of the ENSFEA agronomic school of Castanet-Tolosan, France (see photo in the supplementary materials). Seeds were sown on 23 April 2018 in racks filled with mixture compost (50% BP2 Kompact 294, 50% TS3 Argile 404; Klasmann, Bourgoin Jallieu, France). Soon after germination when all seedlings harboured their first two or four leaves (4 June 2018), seedlings were transplanted in individual 9 × 9 cm pots filled with the same mixture compost. Every pot included one plant and was randomly assigned a location in the experimental garden. Half of the plants were randomly chosen and exposed to a shade treatment by using individual shading cages covered with net producing 70% shade. Plants were watered manually with the same amount of water twice a week. A total of 200 plants (50 per inbred line) were used in this experiment. In each inbred line, 25 out of the 50 plants were exposed to shade. The impact of shade nets on light intensity was characterized in a previous study by using multiple spectrophotometer acquisitions. They let pass through around one third of the photosynthetic active radiation (PAR) and two-thirds of the red to far-red ratio (R/FR) [1].

2.3. Adult Plant Phenotypic Measurements

One month after the young seedlings were exposed to shade (2 July 2018), adult plants were measured to test for the effect of shade on phenotypic measurements. Phenotypic measurements included height in cm, number of branches, presence or absence of floral buds, number of internodes, and stem diameter in mm. Internode length was calculated as the average stem length in cm per internode (plant height/number of internodes). Five fully developed leaves were collected, scanned, dried (3 days at 45 °C), and weighed. The area of leaves was measured using the ImageJ software [29] and the specific leaf area (SLA) was calculated as leaves surface (m^2)/leaves mass (kg).

2.4. Second Round of Stem Growth for Plants Sampled for Tissue

After the first round of measurements presented above, stems of adult plants were cut at the first internode level to allow the growth of a new stem, still under the same light or shade treatment. This allowed us to sample tissue from meristems that were young enough not to be close to the stage of producing terminal flowers but for plants that had been exposed to shade or light for more than a month. Shoot apices were chosen because it is the place where new tissues start their differentiated growth and development. This is also where plants perceive external signals that drive phenotypic responses linked to growth or development. The shoot apex and two leaves were harvested on 48 plants, representing six plants by line and by treatment for each type of tissue. We also recorded phenotypic measurements to allow their comparison between treatments to be directly related to epigenetic data. These plants were measured on four different dates to allow their comparisons at the same developmental stage rather than age: three to four developed internodes (on 25 and 30 July and 1 and 6 August respectively 23, 28, 30, and 37 days after cutting). The same phenotypic measurements were taken as for adult plants during the first round of measurements, with an exception made for the presence or absence of a floral bud (none were present). The six phenotypic traits that were analysed were therefore: plant height, internode length, stem diameter, number of leaves, number of branches, and SLA.

2.5. Epigenetic Analysis

Tissue samples were frozen in liquid nitrogen at the moment of sampling and conserved at −80 °C until epigenetic analyses. We chose to sample shoot apexes because it is the part of the plant wherein all new tissues start their growth and develop. When an external signal is perceived by a plant and transformed into a phenotypic response that will drive the modification of the main stem and the organs located onto it (e.g., leaves), the perception and initiation of the response are expected to take place in the stem apical meristem. We also chose to sample leaves in order to explore the epigenetic variation expected to be associated with SLA plasticity. We chose to collect these two tissues because methylation was previously shown to be tissue-specific and so may vary differently between tissues responding to the same environmental treatment [30,31].

Shoot apices and leaves were ground to powder by using Tissue Lyser II (Qiagen, Hilden, Germany) which disrupts biological samples through high-speed shaking in plastic tubes with stainless steels. Total DNA was extracted by using Biosprint 15 DNA Plant kit (Qiagen, Hilden, Germany) which is a rapid and economical automated method that allowed purification of total DNA from plant tissue.

DNA methylation was studied by using the epiGBS method as the *Antirrhinum majus* reference genome was not available at the time [20]. In a nutshell, epiGBS is a reference-free reduced representation bisulfite sequencing method. This method uses genotyping by sequencing of bisulfite-converted DNA followed by reliable de novo reference construction, mapping, variant calling, and distinction of single-nucleotide polymorphisms (SNPs) versus methylation variation (protocol details can be found in the supplementary materials). All library preparations have been realized by Niels Wagemaker (department of Experimental Plant Ecology, Radboud University, Nijmegen) according to their published protocol [20].

2.6. Bioinformatics

We used a bioinformatics pipeline integrated in the snake-make workflow called epiGBS2 [32] to remove PCR duplicates and demultiplex samples. The pipeline is available at: https://github.com/nioo-knaw/epiGBS2. The filtered and demultiplexed reads from epiGBS2 pipeline were used in another pipeline adapted from previous work [33], using the Galaxy project server as applied in [34]. Adapter removing was done using TrimGalore! V06.5 [35]. Single-end reads were aligned to the snapdragon plant genome version 3.0 [36] with BSMAP Mapper [37]. Mapped reads were merged and used as input in BSMAP

Methylation Caller to get a tabular file with cytosine and thymine counts that were used as input to calculate coverage and Frequency of C and T for subsequent analysis.

CpG methylated sites with coverage of at least eight reads per position found in all samples were filtered with the package Methylkit [38]. After BSMAP methylation calling, bedgraph files were used to filter the sites in a CHG and CHH methylation context where H can be A, C, or T. Only the methylation sites covered by eight or more reads were retained for the Principal Component Analysis.

2.7. Statistical Analysis

We summarized the epiGBS data on DNA methylation changes by using PCA for each methylation protocol and line. PCA was conducted with the package factoextra, FactoMineR, emmeans, and missMDA (scripts available at the end of the supplementary materials). We retrieved PCA coordinates per individual, the relative and absolute contributions of the components (also named dimensions) to the global variance, and the contribution of variables (cytosines positions) to the components (also named dimensions) by DNA methylation context (CpG, CHG, and CHH) and by tissue (apex and leaves) for the subsequent analyses.

We assessed the effect of the light versus shade treatment on (1) the phenotypic traits (plant height, internodes length, stem diameter, number of flowers, number of ramifications, and SLA), and (2) methylation patterns summarised by PCA dimensions with Mann–Whitney U-tests for each line, tissue, and methylation protocol. We extracted the effect size and its 95% confidence interval for each test, which allowed us to assess and compare the effect of the light versus shade treatment among lines, tissues, methylation protocols, and PCA dimensions. Where methylation differences due to the light treatment were found, we assessed the correlation between phenotypic trait values and variation in the methylation patterns represented by PCA dimensions with a Spearman correlation test. Each test was conducted on 12 individuals and replicated in the four snapdragon plant inbred lines, 11 to 12 PCA dimensions, two tissues (apex and leaves), and three methylation protocols (CHG, CHH, and CpG). Effect size estimates with confidence intervals that did not include zero were considered as significantly different from zero.

We estimated effect sizes and their 95% confidence interval from Mann–Whitney tests using the R packages "rcompanion" [39] and "coin" [40]. We estimated the Spearman correlation coefficients and their 95% confidence interval using the R package "RVAideMemoire". All analyses were performed in R software version 3.6.3 [41].

We performed power analyses of the Mann–Whitney tests depending on the effect size r and power analyses of the correlations depending on the Spearman correlation coefficient r_S (see 'power analysis' in the supplementary materials for more details and Figures S1 and S2).

3. Results

3.1. Phenotypic Response to Light Versus Shade Treatment

Our analysis revealed a strong effect (effect size $r > 0.5$) of the light versus shade treatment on all phenotypic traits considered except for height (Figure 1, Table S1). Snapdragon plants exposed to regular natural light had more branches, shorter internodes, a larger basal stem diameter, more leaves, and were shorter than their counterparts exposed to shade. We found no difference between inbred lines in the strength of their response to the light treatment, as illustrated by nearly complete overlap between the 95% CIs of the light treatment effect between lines (Table S1). One must note that we had limited statistical power to detect the significance of small size effects (Figure S1).

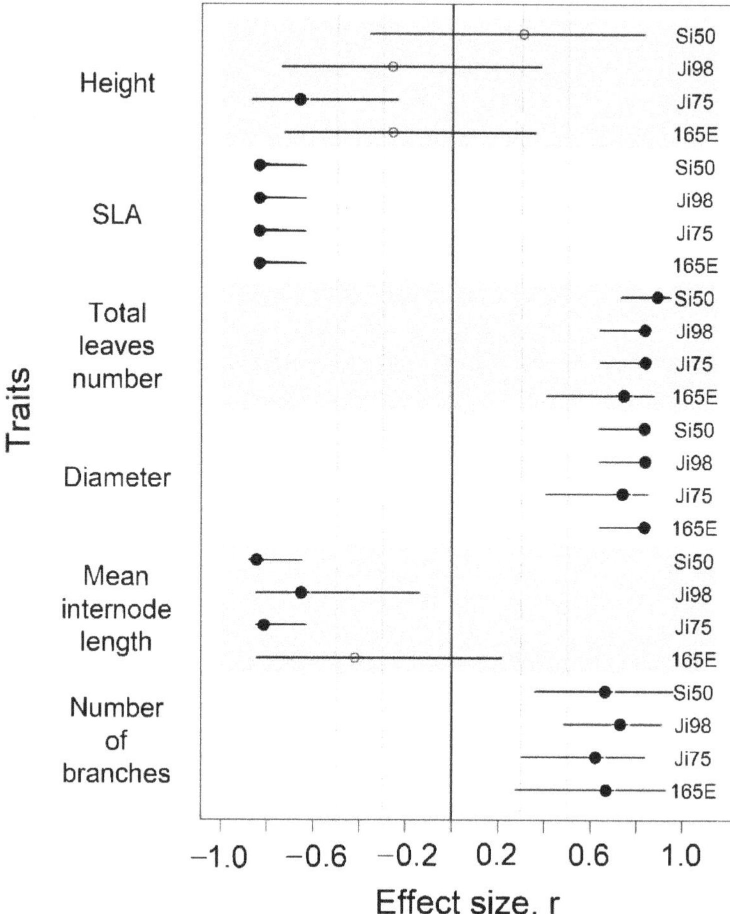

Figure 1. Effect sizes of light treatment on phenotypic traits represented for each snapdragon inbred line flanked by their 95% confidence interval. Line identities are noted in the column on the right side of the graph. Dotted lines represent $r = 0.3$ and $r = 0.5$. Estimates for which the 95% CI does not include zero are represented by black circles while others are represented by empty circles.

3.2. PCA Summary of DNA Methylation Data

The different PCA dimensions explained very similar percentages of DNA methylation data variation for each PCA across the 11 dimensions summarizing the CHH and CHG data variation, and across the 12 dimensions summarizing the CpG data variation, both for apex and leaf tissue (Table S2). Caution must be taken when interpreting the 12th dimension because it explained only c. 10 to the minus 29 power % of the variation. Since DNA methylation data variation could not be summarized to a very low number of dimensions (Table S2), we kept all the dimensions of each PCA and considered them equivalent in the statistical analyses used to test for associations between phenotypic traits measurements and DNA methylation changes.

3.3. DNA Methylation Association with Light Versus Shade Treatment and Phenotypic Variation

The analysis of the effect of the light versus shade treatment on DNA methylation revealed variation within lines between light and shade treatments. Caution must be taken when considering this variation and one should not speculate as to its statistical significance because 95% CI generally overlapped the zero. However, one to three PCA dimensions reflected a large difference between light and shade treatments ($r \geq 0.5$) that can be unambiguously considered as significant in apex tissue (Figure 2, Table S3).

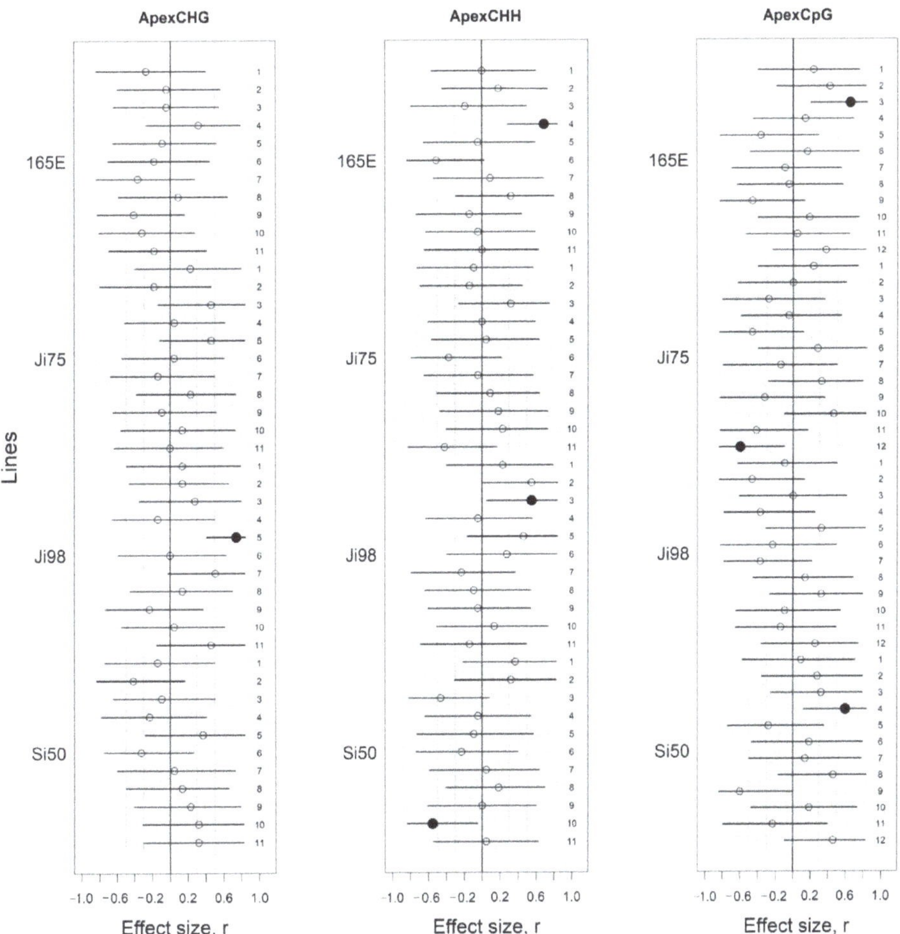

Figure 2. Effect sizes of methylation differences between light versus shade treatments presented for each methylation protocol (CHG, CHH, CpG) applied on apex tissue. Effect sizes are presented for each line and flanked by their 95% confidence interval. Dotted lines represent $r = 0.3$ and $r = 0.5$. Estimates for which the 95% CI does not include zero are represented by black circles while others are empty. Numbers on the right column show the PCA dimension of the methylation protocol.

Equivalent results were found in leaf tissue (Figure 3, Table S4).

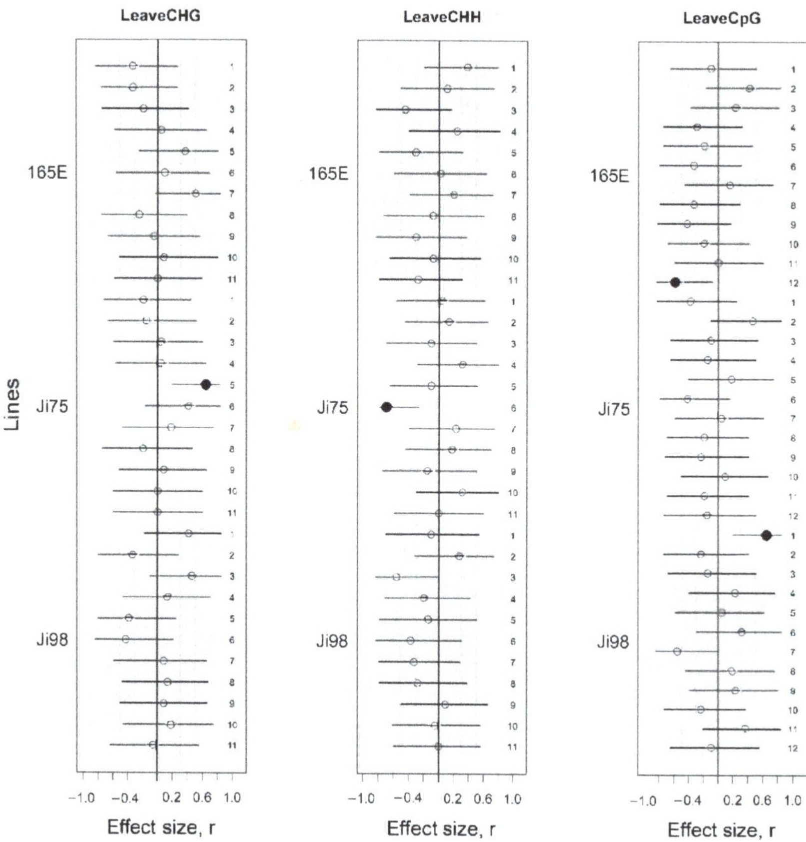

Figure 3. Effect sizes of methylation differences between light versus shade treatments presented for each methylation protocol (CHG, CHH, CpG) applied on leaf tissue. Effect sizes are presented for each line and flanked by their 95% confidence interval. Dotted lines represent $r = 0.3$ and $r = 0.5$. Estimates for which the 95% CI does not include zero are represented by black circles while others are empty. Numbers on the right column show the PCA dimension of the methylation protocol.

3.4. Association between Phenotypic Differences and DNA Methylation Changes

Among the 11 cases presented above that showed wide and significant methylation pattern differences associated with the light versus shade treatment, phenotypic variation was not always found to correlate with DNA methylation variation represented by PCA coordinates. Interestingly, no significant correlation was found between height and PCA coordinates in apex tissue (Figure 4A, Table S5). In the analysis based on leaf tissue samples, it was the number of leaves that did not show any link with DNA methylation variation (Figure 4B, Table S6). Caution must be taken when interpreting these relationships as they characterize the correlation between trait and epigenetic variations but do not take into account the treatment effect.

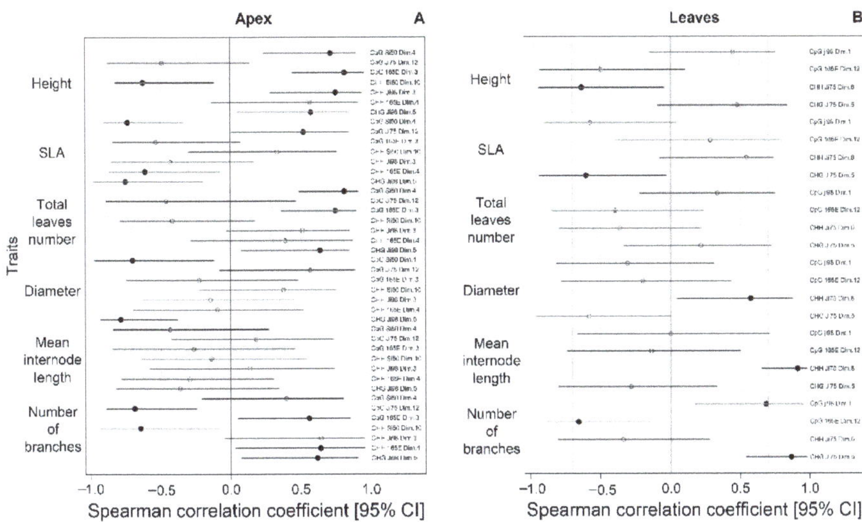

Figure 4. Correlations between each phenotypic trait and methylation PCA coordinate presented only for cases where DNA methylation differences due to the light versus shade treatment were found in apical (**A**) and leaf tissue (**B**). Spearman correlation coefficients (r_s) presented are flanked by their 95% confidence interval. Dotted lines represent $r_s = 0.3$, $r_s = 0.5$ and $r_s = 0.7$. Estimates for which the 95% CI does not include zero are represented by black circles while others are empty. Right column shows the methylation protocol, the line, and the PCA dimension for which the correlation coefficient is presented.

4. Discussion

4.1. Phenotypic Plasticity

Our results confirmed snapdragon plant phenotypic plasticity to light versus shade in highly inbred lines. It is interesting to note that this finding on phenotypic plasticity was replicated in a similar experimental setting for wild snapdragon plant populations [1]. Such phenotypic plastic response is typical of the response described in the presence of shade avoidance syndrome [2,3]. For a similar height, snapdragon plants exposed to shade had flatter or thinner leaves (increased SLA), which is usually associated with a higher growth rate in favorable environments. They were also characterized by increased stem elongation, which is an increase in the mean internode distance and one if not the most documented example of plasticity in plants. Plants grew bigger under shade as illustrated by their higher number of branches, larger basal stem diameter, and greater number of leaves. Although increased internode length and SLA are commonly reported in response to shade, branching is usually reduced because of apical dominance [2]. Our results, together with results found in myrtle plants (*Myrtus communis*) are starting a pool of examples of branching increased by shade [42]. Collectively, these findings and previous findings in wild snapdragon populations and inbred lines support the hypothesis of a strong phenotypic plasticity in response to shade in snapdragon plants. They also suggest that this plasticity was conserved in snapdragon horticultural lines. Interestingly, the magnitude of the plastic response was comparable between lines.

4.2. Epigenetic Response to Light Versus Shade

Previous studies on snapdragon plants found that their ~400 to 500 Mb genome largely harbored methylations to a non-negligible extent (15%), which is comparable to several plant species and seems to vary in relation to the genome size in Angiosperms [8,43]. Our results in highly inbred lines showed that methylation patterns on the snapdragon genome can change in response to the modification of the light environment (sunlight versus shade) of the plants. This result was found in different highly inbred lines that have fully or nearly fixed genomic backgrounds. It connects indirectly epigenetic variation to the ecology of natural populations in different genomic backgrounds. Equivalent examples of this biological link can be found in the literature [44,45]. Here we call for more studies in non-model organisms. This will be necessary before we can obtain a clear picture of the ecology and evolution of genetic and epigenetic variation at the level of populations [46].

Our results suggest that DNA methylation variation was spread across the genome because the different dimensions of the PCA that summarized the variation of DNA methylation patterns across the genome provided a balanced explanation of the variation. Although the coordinates of most PCA dimensions varied between the light versus shade environments, only a few of these dimensions underwent a strong significant effect: the others were only indicative of trends in a low statistical power context. We therefore cannot conclude to the presence of global epigenetic response to shade across the genome. Instead, our results suggest that a limited number of epigenomic regions were involved in a strong modification of DNA methylation patterns in response to light versus shade environments. One could speculate about the interest of precisely identifying these regions but since the epiGBS approach covers a small percentage of the genome, whole-genome approaches would be more suitable for this aim. This limitation is inherent to reduced representation sequencing methods. Our findings, therefore, imply some strong but regionally restricted epigenomic changes in snapdragon plants in response to light versus shade environments. Epigenetic variation in snapdragon plants therefore participates to the schoolbook example of ecologically and evolutionary relevant examples of phenotypic plasticity in plants.

To date, our study is one of the very few that investigates the potential link between the phenotypic plasticity of plants in response to shade and epigenetic variation. For example, clonal lines of longstalk starwort plants (*Stellaria longipes*) submitted to different light treatments showed that stem elongation correlated with reduced methylated cytosine content measured by High Performance Liquid Chromatography (HPLC) [9]. In *Arabidopsis thaliana*, histone acetylation of H3/H4 and H3K4me3/H3K36me3 promoted the expression of shade responsive genes in the Col-0 genotype [10]. Collectively, these findings and ours suggest that beyond the widely documented genetic mechanisms underlying the phenotypic plasticity of plants in response to shade [4,7], epigenetic variation might potentially be involved. Further work is necessary before any finding can be generalized to other plant species.

4.3. The Epigenetic Basis of Phenotypic Variation or Lack Thereof

Although we found that phenotypic variation was often associated with variation in DNA methylation patterns in different highly inbred lines, no clear relationship between trait phenotypic plasticity and epigenetic change emerged. We found an epigenetic basis for trait variation, and its absence, in many scenarios. For example, it was the case for traits that did not change in response to shade, for traits that changed although no epigenetic modification was found, when neither the trait nor DNA methylation patterns were modified by the light versus shade treatment, but also when both responded. Our analysis was inconclusive and neither confirmed nor denied the hypothesis that epigenetic modifications played a role in snapdragon trait phenotypic plasticity in response to shade.

A clear pattern of DNA methylation variation among snapdragon lines emerged from the analysis of trait epigenetic variation. This finding confirms at the epigenomic level the results found previously by using chemical analyses (8); there are differences in the epigenetic variation of traits between genomic backgrounds. This is highly expected because DNA methylation variation is linked to the DNA sequence. DNA sequence polymorphism at potentially methylated cytosine sites results in methylation variation [47]. Other mechanisms link DNA sequence polymorphism to methylation patterns, e.g., the mobility of Transposable Elements (TEs) enabled by changes in DNA methylation [48] and epigenetically facilitated DNA mutation [17]. Our study illustrates that methodological developments are still necessary in non-model species to overcome limits in the study of the ecological and evolutionary significance of epigenetic variation.

5. Conclusions

Our findings and others suggest that epigenetic variation might be associated with the phenotypic plasticity of plants in response to shade. This plasticity is likely influencing the ability of most plant populations to adapt. Beyond its general ecological relevance in nature, it has implications for the ongoing challenges linked to climate change. This is because contemporary changes of the vegetation cover can be observed in many ecosystems worldwide because of fragmentation [49] and land-use changes [50]. Plant vegetative architecture and photosynthetic related traits also play a key role in the evolution, adaptation, and plasticity of crop plants (crop breeding and domestication [51–54]). We therefore call for more work on the potential epigenetic variation associated with the phenotypic plasticity of plants in response to shade because it would improve our understanding of the potential ecological and evolutionary significance of epigenetic variation in natural populations.

Supplementary Materials: The following are available online at https://www.mdpi.com/2073-4425/12/2/227/s1, Photo: Experimental garden at ENSFEA at Castanet-Tolosan (France), epiGBS protocol, Power analyses Figure S1: Power as a function of the effect size (r) in the case of our Mann–Whitney tests for light treatment effect ($n = 6$ in each treatment group). Dotted lines represent $r = 0.3$, $r = 0.5$, and $r = 0.7$, Figure S2: Power as a function of the Spearman correlation coefficient (r_s) in the case of our correlations between phenotypic traits and methylation PCA coordinates ($n = 12$ in each treatment group). Dotted lines represent $r_s = 0.3$, $r_s = 0.5$, and $r_s = 0.7$, Table S1: Phenotypic measurements (median [IQR]) in the shade and light treatments for each line, effect sizes of the treatment on phenotypic traits and their 95% confidence interval, Table S2: Relative contribution of each dimension to the explanation of DNA methylation data variation presented for each PCA, Table S3: Effect sizes of methylation differences between light treatments for each line, each PCA dimension and each methylation protocol (CHG, CHH, CpG) applied on apex tissue, Table S4: Effect sizes of methylation differences between light treatments for each line, each PCA dimension and each methylation protocol (CHG, CHH, CpG) applied on leaf tissue, Table S5: Spearman correlation coefficients (r_s) between each phenotypic trait and methylation PCA coordinates presented for cases showing methylation differences in apical tissue, Table S6: Spearman correlation coefficients (r_s) between each phenotypic trait and methylation PCA coordinates presented for cases showing methylation differences in leaf tissue, R SCRIPT: PCA_analysis CG_context.R, R SCRIPT: PCA _analysis CHH_CHG_ contexts.R, R SCRIPT: Data analysis of phenotypic traits, PCA coordinates and correlation of phenotypic and methylation data.

Author Contributions: B.P. conceived the study; B.P. and D.G. designed the plant experiment and molecular biology methodology; B.P., P.M. and N.L.A. designed the data analysis; J.-L.H. contributed to the experimental design; C.G. participated to the bioinformatics analytic design; D.G., M.L., and S.M. collected the data; P.M. and N.L.A. analyzed the data; B.P., D.G., N.L.A. and P.M. wrote the manuscript. All authors have read and agreed to the published version of the manuscript.

Funding: This project has received funding from the European Research Council (ERC) under the European Union's horizon 2020 research and innovation program (grant agreement No ERC-CoG-2015-681484-ANGI) awarded to BP. This work was supported by funding from the French "Agence Nationale de la Recherche" (ANR-13-JSV7-0002 "CAPA") to BP.

Data Availability Statement: Data will be made available upon request.

Acknowledgments: This study is set within the framework of the "Laboratoires d'Excellences (LABEX)" TULIP (ANR-10-LABX-41). With the support of LabEx CeMEB, an ANR "Investissements d'avenir" program (ANR-10-LABX-04-01) through the Environmental Epigenomics Platform. We thank the editors and anonymous reviewers for their comments and suggestions, which significantly contributed to improving the quality of this article.

Conflicts of Interest: The authors declare no conflict of interest. The funders had no role in the design of the study; in the collection, analyses, or interpretation of data; in the writing of the manuscript, or in the decision to publish the results.

References

1. Mousset, M.; Marin, S.; Archambeau, J.; Blot, C.; Bonhomme, V.; Garaud, L.; Pujol, B. Genetic variation underlies the plastic response to shade of snapdragon plants (*Antirrhinum majus* L.). *Bot. Lett.* **2020**. [CrossRef]
2. Ballare, C.L.; Pierik, R. The shade-avoidance syndrome: Multiple signals and ecological consequences. *Plant Cell Environ.* **2017**, *40*, 2530–2543. [CrossRef] [PubMed]
3. Pierik, R.; Testerink, C. The art of being flexible: How to escape from shade, salt, and drought. *Plant Physiol.* **2014**, *166*, 5–22. [CrossRef] [PubMed]
4. Schmitt, J.; Stinchcombe, J.R.; Heschel, M.S.; Huber, H. The adaptive evolution of plasticity: Phytochrome-mediated shade avoidance responses. *Integr. Comp. Biol.* **2003**, *43*, 459–469. [CrossRef] [PubMed]
5. Bell, D.L.; Galloway, L.F. Plasticity to neighbour shade: Fitness consequences and allometry. *Funct. Ecol.* **2007**, *21*, 1146–1153. [CrossRef]
6. Sessa, G.; Carabelli, M.; Possenti, M.; Morelli, G.; Ruberti, I. Multiple pathways in the control of the shade avoidance response. *Plants Basel Switz.* **2018**, *7*, 102. [CrossRef] [PubMed]
7. Donohue, K.; Schmitt, J. The genetic architecture of plasticity to density in *Impatiens capensis*. *Evolution* **1999**, *53*, 1377–1386. [CrossRef]
8. Gourcilleau, D.; Mousset, M.; Latutrie, M.; Marin, S.; Delaunay, A.; Maury, S.; Pujol, B. Assessing Global DNA Methylation Changes Associated with Plasticity in Seven Highly Inbred Lines of Snapdragon Plants (*Antirrhinum majus*). *Genes* **2019**, *10*, 256. [CrossRef]
9. Tatra, G.S.; Miranda, J.; Chinnappa, C.C.; Reid, D.M. Effect of light quality and 5-azacytidine on genomic methylation and stem elongation in two ecotypes of *Stellaria longipes*. *Physiol. Plant.* **2000**, *109*, 313–321. [CrossRef]
10. Peng, M.; Li, Z.; Zhou, N.; Ma, M.; Jiang, Y.; Dong, A.; Shen, W.-H.; Li, L. Linking PHYTOCHROME-INTERACTING FACTOR to histone modification in plant shade avoidance. *Plant Physiol.* **2018**, *176*, 1341–1351. [CrossRef]
11. Baulcombe, D.C.; Dean, C. Epigenetic regulation in plant responses to the environment. *Cold Spring Harb. Perspect. Biol.* **2014**, *6*, a019471. [CrossRef]
12. Nicotra, A.B.; Atkin, O.K.; Bonser, S.P.; Davidson, A.M.; Finnegan, E.J.; Mathesius, U.; Poot, P.; Purugganan, M.D.; Richards, C.L.; Valladares, F.; et al. Plant phenotypic plasticity in a changing climate. *Trends Plant Sci.* **2010**, *15*, 684–692. [CrossRef]
13. Mirouze, M.; Paszkowski, J. Epigenetic contribution to stress adaptation in plants. *Curr. Opin. Plant Biol.* **2011**, *14*, 267–274. [CrossRef]
14. Richards, C.L.; Alonso, C.; Becker, C.; Bossdorf, O.; Bucher, E.; Colomé-Tatché, M.; Durka, W.; Engelhardt, J.; Gaspar, B.; Gogol-Döring, A.; et al. Ecological plant epigenetics: Evidence from model and non-model species, and the way forward. *Ecol. Lett.* **2017**, *20*, 1576–1590. [CrossRef]
15. Liu, J.; Feng, L.; Li, J.; He, Z. Genetic and epigenetic control of plant heat responses. *Front. Plant Sci.* **2015**, *6*, 267. [CrossRef] [PubMed]
16. Banerjee, A.; Wani, S.H.; Roychoudhury, A. Epigenetic control of plant cold responses. *Front. Plant Sci.* **2017**, *8*, 1643. [CrossRef] [PubMed]
17. Danchin, E.; Pocheville, A.; Rey, O.; Pujol, B.; Blanchet, S. Epigenetically facilitated mutational assimilation: Epigenetics as a hub within the inclusive evolutionary synthesis. *Biol. Rev.* **2019**, *94*, 259–282. [CrossRef]
18. Danchin, E.; Pujol, B.; Wagner, R.H. The double pedigree: A method for studying culturally and genetically inherited behavior in tandem. *PLoS ONE* **2013**, *8*, e61254. [CrossRef] [PubMed]
19. Thomson, C.E.; Winney, I.S.; Salles, O.C.; Pujol, B. A guide to using a multiple-matrix animal model to disentangle genetic and nongenetic causes of phenotypic variance. *PLoS ONE* **2018**, *13*, e0197720. [CrossRef]
20. van Gurp, T.P.; Wagemaker, N.C.A.M.; Wouters, B.; Vergeer, P.; Ouborg, J.N.J.; Verhoeven, K.J.F. epiGBS: Reference-free reduced representation bisulfite sequencing. *Nat. Methods* **2016**, *13*, 322–324. [CrossRef]
21. Andalo, C.; Cruzan, M.B.; Cazettes, C.; Pujol, B.; Burrus, M.; Thébaud, C. Post-pollination barriers do not explain the persistence of two distinct Antirrhinum subspecies with parapatric distribution. *Plant Syst. Evol.* **2010**, *286*, 223–234. [CrossRef]
22. Tavares, H.; Whibley, A.; Field, D.L.; Bradley, D.; Couchman, M.; Copsey, L.; Elleouet, J.; Burrus, M.; Andalo, C.; Li, M.; et al. Selection and gene flow shape genomic islands that control floral guides. *Proc. Natl. Acad. Sci. USA* **2018**, *115*, 11006–11011. [CrossRef]
23. Khimoun, A.; Burrus, M.; Andalo, C.; Liu, Z.-L.; Vicédo-Cazettes, C.; Thébaud, C.; Pujol, B. Locally asymmetric introgressions between subspecies suggest circular range expansion at the *Antirrhinum majus* global scale. *J. Evol. Biol.* **2011**, *24*, 1433–1441. [CrossRef]
24. Pujol, B.; Archambeau, J.; Bontemps, A.; Lascoste, M.; Marin, S.; Meunier, A. Mountain landscape connectivity and subspecies appurtenance shape genetic differentiation in natural plant populations of the snapdragon (*Antirrhinum majus* L.). *Bot. Lett.* **2017**, *164*, 111–119. [CrossRef]
25. Khimoun, A.; Cornuault, J.; Burrus, M.; Pujol, B.; Thebaud, C.; Andalo, C. Ecology predicts parapatric distributions in two closely related *Antirrhinum majus* subspecies. *Evolut. Ecol.* **2013**, *27*, 51–64. [CrossRef]
26. Marin, S.; Gibert, A.; Archambeau, J.; Bonhomme, V.; Lascoste, M.; Pujol, B. Potential adaptive divergence between subspecies and populations of snapdragon plants inferred from Q_{ST}–F_{ST} comparisons. *Mol Ecol.* **2020**, *29*, 3010–3021. [CrossRef] [PubMed]

27. Khattak, A.M.; Pearson, S. Light quality and temperature effects on *Antirrhinum* growth and development. *J. Zhejiang Univ. Sci. B* **2005**, *6*, 119–124. [CrossRef]

28. Munir, M.; Jamil, M.; Baloch, J.; Khattak, K.R. Impact of light intensity on flowering time and plant quality of *Antirrhinum majus* L. cultivar Chimes White. *J. Zhejiang Univ. Sci. A* **2004**, *5*, 400–405. [CrossRef] [PubMed]

29. Schneider, C.A.; Rasband, W.S.; Eliceiri, K.W. NIH Image to ImageJ: 25 years of image analysis. *Nat. Methods* **2012**, *9*, 671. [CrossRef] [PubMed]

30. Kumar, S.; Beena, A.S.; Awana, M.; Singh, A. Salt-induced tissue-specific Cytosine methylation downregulates expression of HKT genes in contrasting wheat (*Triticum aestivum* L.) genotypes. *DNA Cell Biol.* **2017**, *36*, 283–294. [CrossRef]

31. Zhang, M.; Xu, C.; von Wettstein, D.; Liu, B. Tissue-specific differences in Cytosine methylation and their association with differential gene expression in Sorghum. *Plant Physiol.* **2011**, *156*, 1955–1966. [CrossRef] [PubMed]

32. Gawehns, F.; Postuma, M.; van Gurp, T.P.; Wagemaker, N.C.; Fatma, S.; Van Antro, M.; Mateman, C.; Milanovic-Ivanovic, S.; van Oers, K.; Vergeer, P.; et al. epiGBS2: An improved protocol and automated snakemake workflow for highly multiplexed reduced representation bisulfite sequencing. *bioRxiv* **2020**. [CrossRef]

33. Meröndun, J.; Murray, D.L.; Shafer, A.B.A. Genome-scale sampling suggests cryptic epigenetic structuring and insular divergence in Canada lynx. *Mol. Ecol.* **2019**, *28*, 3186–3196. [CrossRef] [PubMed]

34. Luviano, N.; Lopez, M.; Gawehns, F.; Chaparro, C.; Arimondo, P.; Ivanovic, S.; David, P.; Verhoeven, K.; Cosseau, V.; Grunau, C. The methylome of Biomphalaria glabrata and other mollusks: Enduring modification of epigenetic landscape and phenotypic traits by a new DNA methylation inhibitor. *Authorea Prepr.* **2020**. [CrossRef]

35. Rueger, F. Trim Galore: A Wrapper Tool around Cutadapt and FastQC to Consistently Apply Quality and Adapter Trimming to FastQ Files, with Some Extra Functionality for MspI-Digested RRBS-Type (Reduced Representation Bisufite-Seq) Libraries. 2012. Available online: http://www.bioinformatics.babraham.ac.uk/projects/trim_galore/ (accessed on 30 December 2020).

36. Li, M.; Zhang, D.; Gao, Q.; Luo, Y.; Zhang, H.; Ma, B.; Chen, C.; Whibley, A.; Zhang, Y.; Cao, Y.; et al. Genome structure and evolution of *Antirrhinum majus* L. *Nat. Plants* **2019**, *5*, 174. [CrossRef]

37. Xi, Y.; Li, W. BSMAP: Whole genome bisulfite sequence MAPping program. *BMC Bioinformatics* **2009**, *10*, 232. [CrossRef]

38. Akalin, A.; Kormaksson, M.; Li, S.; Garrett-Bakelman, F.E.; Figueroa, M.E.; Melnick, A.; Mason, C.E. methylKit: A comprehensive R package for the analysis of genome-wide DNA methylation profiles. *Genome Biol.* **2012**, *13*, R87. [CrossRef]

39. Mangiafico, S. rcompanion: Functions to Support Extension Education Program Evaluation. Version 2.3. 21. 2020. Available online: https://CRAN.R-project.org/package=rcompanion (accessed on 30 December 2020).

40. Hothorn, T.; Hornik, K.; van de Wiel, M.A.; Zeileis, A. Implementing a class of permutation tests: The coin package. *J. Stat. Soft.* **2008**, *28*, 1–23. [CrossRef]

41. R Core Team. R: A Language and Environment for Statistical Computing. R Foundation for Statistical Computing. 2020. Available online: https://www.R-project.org/ (accessed on 30 December 2020).

42. Ada, N.-L.; Farkash, L.; Hamburger, D.; Ovadia, R.; Forrer, I.; Kagan, S.; Michal, O.-S. Light-scattering shade net increases branching and flowering in ornamental pot plants. *J. Hortic. Sci. Biotechnol.* **2008**, *83*, 9–14. [CrossRef]

43. Alonso, C.; Perez, R.; Bazaga, P.; Herrera, C.M. Global DNA cytosine methylation as an evolving trait: Phylogenetic signal and correlated evolution with genome size in angiosperms. *Front. Genet.* **2015**, *6*, 4. [CrossRef]

44. Trap-Gentil, M.-V.; Hébrard, C.; Lafon-Placette, C.; Delaunay, A.; Hagège, D.; Joseph, C.; Brignolas, F.; Lefebvre, M.; Barnes, S.; Maury, S. Time course and amplitude of DNA methylation in the shoot apical meristem are critical points for bolting induction in sugar beet and bolting tolerance between genotypes. *J. Exp. Bot.* **2011**, *62*, 2585–2597. [CrossRef] [PubMed]

45. Gourcilleau, D.; Bogeat-Triboulot, M.-B.; Le Thiec, D.; Lafon-Placette, C.; Delaunay, A.; El-Soud, W.A.; Brignolas, F.; Maury, S. DNA methylation and histone acetylation: Genotypic variations in hybrid poplars, impact of water deficit and relationships with productivity. *Ann. For. Sci.* **2010**, *67*, 208. [CrossRef]

46. Richards, E.J. Population epigenetics. *Curr. Opin. Genet. Dev.* **2008**, *18*, 221–226. [CrossRef]

47. Greally, J.M. Population epigenetics. *Curr. Opin. Syst. Biol.* **2017**, *1*, 84–89. [CrossRef]

48. Lanciano, S.; Mirouze, M. Transposable elements: All mobile, all different, some stress responsive, some adaptive? *Curr. Opin. Genet. Dev.* **2018**, *49*, 106–114. [CrossRef]

49. Haddad, N.M.; Brudvig, L.A.; Clobert, J.; Davies, K.F.; Gonzalez, A.; Holt, R.D.; Lovejoy, T.E.; Sexton, J.O.; Austin, M.P.; Collins, C.D.; et al. Habitat fragmentation and its lasting impact on Earth's ecosystems. *Sci. Adv.* **2015**, *1*, e1500052. [CrossRef] [PubMed]

50. Poyatos, R.; Latron, J.; Llorens, P. Land use and land cover change after agricultural abandonment. *Mt. Res. Dev.* **2003**, *23*, 362–368. [CrossRef]

51. Pujol, B.; Salager, J.-L.; Beltran, M.; Bousquet, S.; McKey, D. Photosynthesis and leaf structure in domesticated cassava (*Euphorbiaceae*) and a close wild relative: Have leaf photosynthetic parameters evolved under domestication? *Biotropica* **2008**, *40*, 305–312. [CrossRef]

52. Pujol, B.; Mühlen, G.; Garwood, N.; Horoszowski, Y.; Douzery, E.J.P.; McKey, D. Evolution under domestication: Contrasting functional morphology of seedlings in domesticated cassava and its closest wild relatives. *New Phytol.* **2005**, *166*, 305–318. [CrossRef] [PubMed]

53. Richards, R.A. Selectable traits to increase crop photosynthesis and yield of grain crops. *J. Exp. Bot.* **2000**, *51*, 447–458. [CrossRef]
54. Hlaváčová, M.; Klem, K.; Rapantová, B.; Novotná, K.; Urban, O.; Hlavinka, P.; Smutná, P.; Horáková, V.; Škarpa, P.; Pohanková, E.; et al. Interactive effects of high temperature and drought stress during stem elongation, anthesis and early grain filling on the yield formation and photosynthesis of winter wheat. *Field Crops Res.* **2018**, *221*, 182–195.

genes

Article

The Role of Microbiome and Genotype in *Daphnia magna* upon Parasite Re-Exposure

Lore Bulteel [1,*], Shira Houwenhuyse [1], Steven A. J. Declerck [2,3] and Ellen Decaestecker [1]

[1] Laboratory of Aquatic Biology, Department of Biology, University of Leuven-Campus Kulak, E. Sabbelaan 53, 8500 Kortrijk, Belgium; shira.houwenhuyse@kuleuven.be (S.H.); ellen.decaestecker@kuleuven.be (E.D.)
[2] Netherlands Institute of Ecology (NIOO-KNAW), Droevendaalsesteeg 10, 6700 AB Wageningen, The Netherlands; S.Declerck@nioo.knaw.nl
[3] Laboratory of Aquatic Ecology, Evolution and Conservation, Department of Biology, KULeuven, 3000 Leuven, Belgium
* Correspondence: lore.bulteel@kuleuven.be; Tel.: +32-4-9772-1255

Abstract: Recently, it has been shown that the community of gut microorganisms plays a crucial role in host performance with respect to parasite tolerance. Knowledge, however, is lacking on the role of the gut microbiome in mediating host tolerance after parasite re-exposure, especially considering multiple parasite infections. We here aimed to fill this knowledge gap by studying the role of the gut microbiome on tolerance in *Daphnia magna* upon multiple parasite species re-exposure. Additionally, we investigated the role of the host genotype in the interaction between the gut microbiome and the host phenotypic performance. A microbiome transplant experiment was performed in which three germ-free *D. magna* genotypes were exposed to a gut microbial inoculum and a parasite community treatment. The gut microbiome inocula were pre-exposed to the same parasite communities or a control treatment. *Daphnia* performance was monitored, and amplicon sequencing was performed to characterize the gut microbial community. Our experimental results showed that the gut microbiome plays no role in *Daphnia* tolerance upon parasite re-exposure. We did, however, find a main effect of the gut microbiome on *Daphnia* body size reflecting parasite specific responses. Our results also showed that it is rather the *Daphnia* genotype, and not the gut microbiome, that affected parasite-induced host mortality. Additionally, we found a role of the genotype in structuring the gut microbial community, both in alpha diversity as in the microbial composition.

Keywords: *Daphnia magna*; diversity; dysbiosis; genotype; gut microbiome; parasite re-exposure; tolerance

check for **updates**

Citation: Bulteel, L.; Houwenhuyse, S.; Declerck, S.A.J.; Decaestecker, E. The Role of Microbiome and Genotype in *Daphnia magna* upon Parasite Re-Exposure. *Genes* **2021**, *12*, 70. https://doi.org/10.3390/genes12010070

Received: 30 November 2020
Accepted: 5 January 2021
Published: 7 January 2021

Publisher's Note: MDPI stays neutral with regard to jurisdictional claims in published maps and institutional affiliations.

1. Introduction

Recently, it has been shown that the microbial community is involved in multiple processes in host biology, such as food digestion, metabolic regulation, developmental signaling, behavior, and social interactions [1–3]. A part of these microbiota resides in the gut, where they are in direct and continuous contact with host tissues [4,5]. These gut symbionts, composed of bacteria, archaea, anaerobic fungi, protozoa, and viruses, provide nutrients, detoxify toxins, and contribute to the host's development and growth [2,6–8]. The gut microbiome can also provide protection against parasites [9–11]. There is growing recognition that the effects of the gut microbiota and parasites on the host are intertwined. Infection with parasites, e.g., intestinal helminths, can significantly disrupt or restructure the host's microbial community, both in invertebrates [12–14] and in vertebrates [15,16].

The gut microbiome can shape and enhance the host's immune system by up-regulation of mucosal activity and induction of antimicrobial peptides [17–19]. Additionally, a stable and diverse gut microbiome can prevent colonization and limit detrimental effects of invading parasites, which is generally associated with interactions between the gut microbiome and the immune system [20,21]. An unbalanced or maladapted microbiome, i.e., dysbiosis of the microbial community, may result in a microbial community which can be both in low-diversity and modified metabolic state [22]. This unbalanced microbiome can increase susceptibility to parasites [23], and is set in motion by a number of direct or indirect mechanisms in the gut [24], e.g., by subverting the immune system, leading to further negative effects [25]. A lower microbial diversity has also been correlated with either a higher abundance of, or an increased susceptibility to, low abundant, opportunistic parasites [26–28]. Even though studies mostly focus on the beneficial effects of the gut microbiota, particular commensal gut communities can also increase host susceptibility to disease [29], and even turn harmful under certain conditions (i.e., pathobiont) [22,30].

The reciprocal role between parasites and the host's microbiome has been studied in multiple invertebrates (beetles [14], bumblebees [9,11,31], *C. elegans* [32], fruit flies [33], oysters [34]). The honey bees are an especially well-studied group in that perspective; a protective function of the gut microbiome against parasites has been shown through controlled microbial inoculations [35,36]. However, only one study so far has been undertaken on the experimental model system, the water flea *Daphnia* and its parasites [37]. Sison-Mangus et al. [38] found no evidence that a host's microbiota regulated resistance against the bacterial parasite *Pasteuria ramosa*, but did find significant differences between microbial communities depending on the host's genotype. *Daphnia* studies also showed clear genotype-specific responses of the gut microbiome, mainly upon toxic cyanobacterial exposure in *Daphnia* tolerance [39–41] or exposure to environmental microbial pools [42,43].

Studies have been undertaken on the reciprocal role between the host microbiome and parasites. No studies, however, have looked into the mediating capacity of the microbiome in parasite tolerance after re-exposure to that same parasite community. Filling this knowledge gap must be one of the next priorities in studying the complex interplay between the gut microbiome and parasite infection, especially because organisms are often infected by multiple parasites. To address this knowledge gap, we set up an experiment to investigate the reciprocal role of the gut microbial community and parasite exposure in host–multiple parasite interactions in *Daphnia magna* after parasite re-exposure. We first performed a microbiome adaptation experiment in which we exposed *Daphnia* populations to two parasite communities and one control treatment, assuming the host gut microbiomes will adapt to the parasite communities. Additionally, we included three *Daphnia* genotypes in our experimental design to reveal possible intraspecific responses in *Daphnia*-microbiome-multiple parasite interactions, which could induce microbiome mediated evolutionary responses linked to particular genotypes [2]. The gut microbial community can drive eco-evolutionary dynamics through its host by impacting life history traits. Combined with the important role of parasites and the genotype in regulating and shaping individual responses and host populations, such population effects can mediate changes up to community level and the whole ecosystem.

2. Materials and Methods

2.1. Microbiome Adaptation Experiment

We first performed a microbiome adaptation experiment in which we pre-exposed *Daphnia* and their microbiomes to three types of parasite treatments. By doing so, we obtain gut microbiomes which communities have been altered by exposure to a parasite community. The gut microbiomes are then to be inoculated in fixed recipient genotypes (see further below). The experimental design consisted of *Daphnia* populations crossed with three parasite treatments (Figure 1), with three replicates per multifactorial combination. The parasite treatments consisted out of two different parasite communities and one control treatment. Parasite community 1 (further referred to as P1) consisted of a pool of the iridovirus causing White Fat Cell Disease (further referred to as WFCD, Figure 2a, [44]) and *Binucleata daphniae* (Figure 2b, [45]). WFCD is an iridovirus which infects the adipose tissue of *Daphnia* and is a highly virulent parasite, as it induces mortality in its host. Infection by WFCD is visible as a greenish, iridescent shine from the fat cells in reflected light. *Binucleata daphniae* is a microsporidian parasite known to infect the integument cells lining the hemocoel of its *D. magna* host. Infection with *B. daphniae* results in a reduced reproduction and survival. Parasite community 2 (further referred to as P2) consisted of a pool of *Pasteuria ramosa* (Figure 2c, [46,47]), *Ordospora colligata* (Figure 2d, [48]), and *Mitosporidium daphniae* (Figure 2e, [49]). *P. ramosa* is an obligate endospore-forming bacterial parasite in *Daphnia*, and it infects the hemocoel. Infection with *P. ramosa* often results in castration of the host (partial or complete stop of reproduction), indirectly resulting in an increase in *Daphnia* body size (i.e., gigantism). Infection with *P. ramosa* is also known to be highly genotype-specific and has little to no impact on survival. *O. colligata* and *M. daphniae* are both microsporidian parasites infecting the *D. magna* gut cells, with *O. colligata* infecting the foregut and *M. daphniae* the hindgut. Both endo-parasites are avirulent, as they induce small reductions in survival and reproduction success. Prevalence of these avirulent endoparasites can reach up to 100% in natural populations. All parasites used in this experiment (P1 and P2) are known to transmit horizontally, i.e., infection is not passed down from mother to offspring, but originates from the environment. The P1 pool was sampled from infected *D. magna* individuals from Blauwe hoeve in Kortrijk (50°48′57.8″ N 3°16′19.6″ E; WFCD and *B. daphniae*) and Muinkpark in Gent (51°02′33.1″ N 3°43′54.4″ E; WFCD). The P2 pool was sampled from infected *D. magna* individuals from Pottelberg pond in Aalbeke (50°47′01.6″ N 3°14′13.7″ E). The control treatment (PC) was not exposed to any parasite community. *Daphnia* populations receiving P1 and P2 originated from two natural haphazardly chosen ponds (the Kennedy Pond in Kortrijk, 50°48′05.9″ N 3°16′33.3″ E and the Morinne Pond in Kortrijk, 50°48′20.8″ N 3°18′45.2″ E). *Daphnia* populations receiving PC were pooled lab genotypes to avoid possible parasite influx from natural populations. All populations, the natural populations as well as the non-infected cultures in the laboratory, were exposed to a mixture of pond water from two *Daphnia*-free ponds to attempt a similar bacterioplankton community for all populations. Pond water was obtained from a mixture from the Kulak Pond (50°48′30.2″ N 3°17′38.3″ E) and the Libel Pond (50°47′44.3″ N 3°15′22.8″ E), both located in Kortrijk, Belgium. Pond water was subsequently filtered over 140 µm and 10 µm and pooled before exposing to the *Daphnia* populations from the microbiome adaptation experiment. Guts from the infected individuals in the microbiome adaptation experiment were dissected after infection reached its peak (after two weeks). This time period was based on research by Macke et al. [39] and Houwenhuyse et al. [50], which indicates a stable microbial gut community after perturbation by the biotic antagonist *Microcystis aeruginosa* after 7 days. To additionally ensure a stable community, we prolonged this period to the timepoint (2 weeks in this experiment) where infection rates started to decrease after reaching its peak. Dissected guts were utilized as microbial donor inocula for recipient *Daphnia* individuals in the microbiome transplant experiment. Microbial inocula were pooled per parasite treatment and per replicate vial. Each microbial pool inoculum was filtered over a sterile glass microfiber filter (grade GF/C; mesh size 1.1 µm) utilizing vacuum filtration to remove parasite spores from the microbial community.

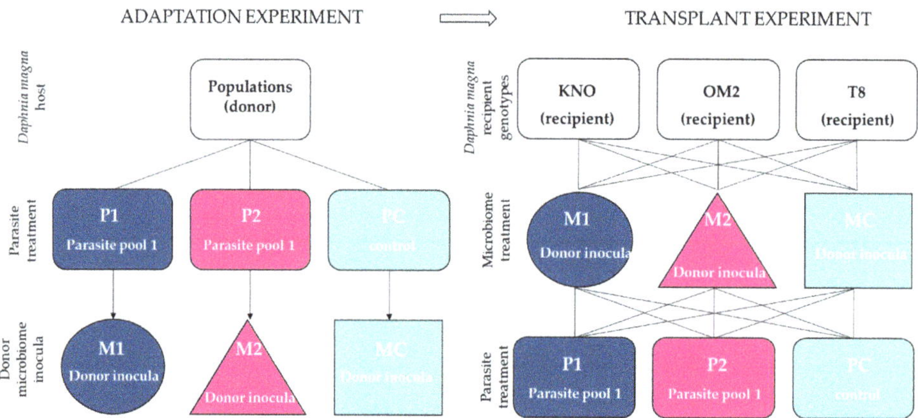

Figure 1. Schematic representation of the experimental design. In the microbial adaptation experiment (**left panel**), all donor populations received a parasite treatment (P1, P2, PC), after which the donor microbial inocula were obtained (M1, M2, MC) for the transplant experiment (**right panel**). P1 consisted of a pool of White Fat Cell Disease (WFCD) and *Binucleata daphniae*. P2 consisted of a pool of *Pasteuria ramosa, Ordospora colligata* and *Mitosporidium daphniae*. PC is the control treatment and consisted of a pool of healthy squashed *Daphnia* individuals. In the transplant experiment (**right panel**), sterile recipient individuals of three *Daphnia* genotypes (KNO, OM2, T8) were exposed to the donor microbiomes obtained from the adaptation experiment (M1, M2, or MC). Each of these microbiome treatments were crossed with the three parasite treatments (P1, P2, or PC; similar as in the microbiome adaptation experiment).

2.2. Microbiome Transplant Experiment

After obtaining the microbiome inocula, we performed the microbiome transplant experiment to examine the role of the gut microbiome on host tolerance upon parasite re-exposure. The experimental design consisted of three microbiome treatments × three parasite treatments × three recipient genotypes (Figure 2). Each multifactorial combination of microbiome treatment, parasite treatment, and genotype was replicated independently three times (*Daphnia* individuals were isolated from independently cultured maternal lines). Ultimately, six individuals per replicate were set up, which totals 486 *Daphnia* individuals. All individuals were made germ-free (0 days old; axenity performed via an adapted protocol of Callens et al. [8]; see Appendix A) and were individually placed in 20 mL sterile filtered tap water in closed off sterile vials. Each germ-free *Daphnia* individual (i.e., recipient) was inoculated with one of the microbial inocula (i.e., microbiome treatment: M1, M2, or MC), receiving the equivalent of 0.75 gut per *Daphnia* individual. As the experimental individuals displayed some mortality, we decided to further boost survival of all individuals. At day 3, all individuals were given a broader microbiome pool derived from non-infected whole-organism *Daphnia* individuals with the equivalent of 1 squashed *Daphnia* per 6 *Daphnia* individuals. By doing so, we expected to increase general performance of our *Daphnia* as prior administration of bacterioplankton with a highly diverse community to our stock *Daphnia* in the lab resulted in an increase in survival and reproduction (personal observation). Adding additional microbial strains to our vials could possibly interfere with our experimentally manipulated donor microbiome inocula. We do, however, assume little impact by adding these additional microbial communities as the community present in the donor microbial inocula already colonized the gut. Additionally, the boosting communities were pooled and given in equal quantities to all experimental individuals. In this manner, limited effects on the host microbiome are equal across all treatments. Additionally, all recipient individuals received the same amount and same composition of pooled microbiota.

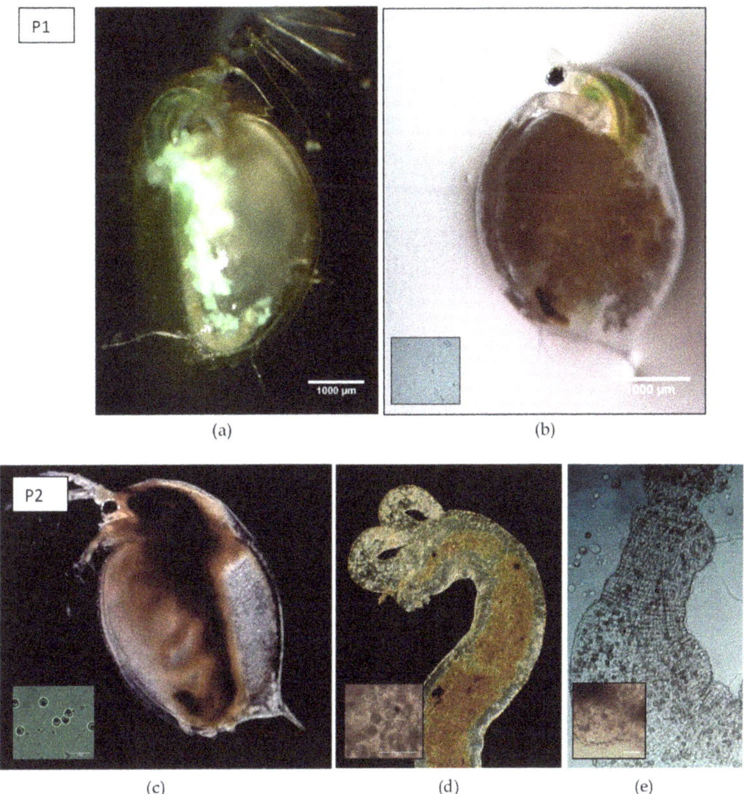

Figure 2. Overview of parasite species in the different parasite communities. (**a**) *Daphnia magna* individual infected with the causative agent of WFCD: Infected fat cells display a greenish, iridescent shine in reflected light, (**b**) heavily infected individual with *Binucleata daphniae* showing accumulated spores in the hemocoele of the carapace with detail of spore cluster, (**c**) individual infected with *Pasteuria ramosa* 30 days after exposure displaying a reddish appearance and larger body size (adapted photograph by Nina Schlotz, distributed under a Creative Commons Attribution 4.0 International license [51]), with detail of the final spore stages (400× magnification), (**d**) infected foregut with *Ordospora colligata*, (photograph by Dieter Ebert, distributed under a Creative Commons Attribution-Share Alike 4.0 International license) with detail of spore cluster (400× magnification), (**e**) infected hindgut with *Mitosporidium daphniae* with detail of spore cluster (100× magnification). Parasite treatment P1 consisted of a mixture of a and b, whereas parasite treatment P2 was composed of c, d, and e.

At day 5 and 6, all individuals were exposed to their respective parasite treatment (i.e., re-exposure with P1, P2, or PC). 5.8×10^3 spores of *B. daphniae* for P1 and 1.4×10^3 spores of *P. ramosa* for P2 were added per vial on the two consequent days. Spore solutions were obtained by squashing infected *Daphnia*. Spore counts for WFCD were not possible, as it is caused by an iridovirus [44], which cannot be routinely quantified under the microscope. Spore count for P2 was based on *P. ramosa* concentrations as this ensured sufficient exposure for infection, as spore counts for *O. colligata* and *M. daphniae* are generally higher compared to *P. ramosa* [52].

Prior to the exposure in the microbial transplant experiment, each parasitic inoculum was filtered over a sterile glass microfiber filter (grade GF/C according to the protocol as described for the microbiome filtration) to remove dominant contaminating microbiota from the parasite suspensions. In this manner, we want to examine the effect of the parasite community without possible interfering effects of the associated microbial strains. Measurement of the

P1 suspension through qPCR showed that the microbial load in the parasite treatments was reduced to 19% due to the filtration. Although our parasite spore suspension were not axenic, we expect little impact on the recipient gut community as administered volume is low and prior to colonization from our microbiome inocula. Parasite spores were brought in resuspension after filtration by placing the filter in 15 mL of sterile distilled water and gently shaking the vial to optimize detachment of the spores from the filter. The obtained spore-solution was then administered to the respective recipient *Daphnia* individuals. For PC, a similar volume as the parasite treatments (P1 and P2) of filtered squashed non-infected individuals was added per vial. The parasite inocula utilized in this experiment were derived from the same parasite communities utilized in the microbiome adaptation experiment to infect *Daphnia*. In this manner, we compared the same parasite communities in the microbiome adaptation and microbiome transplant experiment.

Daphnia were given a relatively low daily amount of 0.5 mg C/L of axenic *Chlorella vulgaris* between day 0 to day 6 to ensure high uptake of microbiota and parasite spores for the microbiome exposure and parasite exposure, respectively. From day 7 onwards, *Daphnia* individuals received 1 mg C/L of axenic *C. vulgaris*. *C. vulgaris* (strain SAG 211-11 B) cultures were started from an axenic slent and cultured in sterile WC medium enriched with $NaNO_3$ (425.05 mg/L) and $K_2HPO_4.3H_2O$ (43.55 mg/L) (adapted from [53]). The algae was cultured under sterile conditions in a climate chamber at 22 °C ($\pm$2 °C) and under fluorescent light (120 $\mu mol \cdot m^{-2} \cdot s^{-1}$) at a 16:8 h light:dark cycle in 2 L glass bottles. Algae cultures were maintained in batch cultures and were constantly stirred and aerated. Filters (0.22 μm) were placed at the input and output of the aeration system to avoid contamination. Algae were weekly harvested in stationary phase and checked for axenity via DAPI staining and LB and R2A medium agar plates.

The amount of sterile filtered tap water in each vial was increased from day three onwards on a daily base until 45 mL was reached. *Daphnia* individuals were monitored for survival and reproduction on a daily base. On day 11, all individuals were measured for their body size and their guts were dissected to analyse the microbial communities. Body size was defined as the distance between the head and the base of the tail. Visual screening for parasites or spores is generally possible two weeks after infection and not incorporated in this experiment. We, nonetheless, opted to dissect at day 11 to obtain sufficient gut microbial material as individuals were dying, and otherwise too little genetic material would be available for amplicon sequencing.

2.3. Statistical Analysis of Body Size and Survival

To examine *Daphnia* performance, we analyzed differences in body size. Body size was squared transformed. Normality of body size was tested for using a Shapiro–Wilk and Bartlett test. Differences in body size were analyzed using a nested linear mixed-effects model (LMER) with microbiome treatment, parasite treatment, and genotype as a fixed effect and maternal line as a random effect (Satterthwaite's method). A Tukey HSD test was used to make post hoc pairwise comparisons. All statistical tests on body size were performed in R 4.0.2 [54]. Analyses on reproduction were not possible due to few data points, as the experiment ended at the age of maturation.

To examine *Daphnia* tolerance and performance, we analyzed differences in survival. Survival was analyzed with a Cox proportional-hazards model regression using the SAS 9.4 software (PHREG procedure). Genotype, microbiome treatment, and parasite treatment were specified as fixed factors. The survival times of individuals that were still alive at the end of the experiment were coded as censored. As ties in survival were numerous, the Efron approximate likelihood was applied. Pairwise comparisons were performed using the CONTRAST statement, which provided both the hazard ratios (HR) between groups for the variable of interest, and the associated p-values. Survival curves were obtained with the ggsurvplot() function [Survminer package] in R 4.0.2 [54].

2.4. MiSeq Library Preparation

To identify the bacterial composition present in the gut, the guts of the surviving *Daphnia* per replicate were dissected under a stereo-microscope with sterile dissecting needles at the end of the experiment and pooled per replicate (mean = 4.309 guts/sample; sd = 1.357 guts/sample; Table S1). Pearson correlations were executed between the number of sequenced guts and the alpha diversity-diversity variables to check for interdependence. Genotype, microbiome treatment, parasite treatment, all two-way interactions, and the three-way interaction, all showed no significant correlation, dismissing the issue of interdependence (Table S2).

Guts were transferred to 10 µl of sterile MilliQ water. Samples were stored under $-20\,^\circ$C until further processing. DNA was extracted using a PowerSoil DNA isolation kit (MO BIO laboratories, Carlsbad, CA, USA) and dissolved in 20 µL MilliQ water. The total DNA yield was determined using a Qubit dsDNA HS assay (Invitrogen, Merelbeke, Belgium) on 1 µL of sample. Because of initially low bacterial DNA concentrations, a nested PCR was applied to increase specificity and amplicon. For the external amplification, a PCR reaction was run using primers 27F and 1492R on all of the template ($98\,^\circ$C-30 s; $98\,^\circ$C-10 s, $50\,^\circ$C-45 s, and $72\,^\circ$C-30 s; 30 cycles; $72\,^\circ$C-5 min; $4\,^\circ$C-hold) using the Platinum SuperFi DNA polymerase (Thermofisher, Merelbeke, Belgium). PCR products were subsequently purified using the QIAquick PCR purification kit (Qiagen, Antwerp, Belgium). To obtain dual-index amplicons of the V4 region, an internal PCR was performed on 5 µL of PCR product using a unique combination of a forward and revers primer per sample (Table S3; $98\,^\circ$C-30 s; $98\,^\circ$C-10 s, $55\,^\circ$C-45 s and $72\,^\circ$C-30 s; 30 cycles; $72\,^\circ$C-5 min; $4\,^\circ$C-hold). Both primers contained an Illumina adapter and an 8-nucleotide (nt) barcode at the $5'$-end. For each sample, PCRs were performed in triplicate, pooled, and gel-purified using the QIAquick gel extraction kit (Qiagen). An equimolar library was prepared by normalizing amplicon concentrations with a SequalPrep Normalization Plate (Applied Biosystems, Geel, Belgium) and subsequent pooling to standardize DNA concentrations. Amplicons were sequenced using a v2 PE500 kit with custom primers on the Illumina Miseq platform (KU Leuven Genomics Core, Leuven, Belgium), producing 2×250-nt paired-end reads.

2.5. Analysis of Microbial Communities

Sequence reads, statistical analyses, and plots were performed using R 4.0.2 [54] following [55]. Sequences were trimmed (the first 10 nucleotides and all nucleotides from position 190 onward were removed) and filtered (maximum two expected errors per read) on paired ends jointly. Sequence variants were inferred using the high-resolution DADA2 method [56,57], and chimeras were removed. Taxonomy was assigned with a naive Bayesian classifier using the Silva v132 training set. Amplicon sequence variants (ASVs, hereafter called OTUs) which had no taxonomic assignment at phylum level or were assigned as "Chloroplast" or "Cyanobacteria" were removed from the dataset. The final recipient dataset contained, after trimming, 2,295,594 reads with an average of 31,883,25 reads per sample (minimum = 1512 reads, maximum = 102,065 reads).

To examine differences in community composition within the different recipient samples and variables, alpha diversity was determined by calculating OTU richness and Shannon diversity number using the vegan package in R [58] following [59]. OTU richness was calculated as the sum of the present OTUs. Shannon diversity number was calculated as the exponential function of Shannon entropy and will be further referred to in this text as 'Shannon entropy'. First, all samples were rarified to a depth of 10,000 reads. The effect of genotype, microbiome treatment, and parasite treatment on OTU richness and Shannon entropy' was examined using a generalized linear model assuming a Poisson distribution of the data and accounted for overdispersion, as the observed residual deviance was higher than the degrees of freedom. Pairwise comparisons among significant variables and their interactions were performed by contrasting least-squares means with Tukey adjustment. To examine differences in gut microbial community composition among samples, a Bray–Curtis dissimilarity matrix was calculated. Differences between main effects and interaction were examined through a permutation MANOVA (Adonis function, vegan package). Multivariate community responses to genotype and treatments were investigated by means of Principal Coordinates Analysis. Obtained *p*-values were adjusted for multiple comparisons through the control of the false discovery rate (FDR). To identify which bacterial classes differed significantly between main and interaction effects, differential abundance analyses were performed (DESeq2 function) on the raw sequencing data from which low data counts were removed (less than 1000). Additionally, donor samples were analyzed to compare alpha diversity and overlap between donor and recipients and within the donors. Due to loss of samples, only three samples from the donor P1 treatment and one sample from the donor P2 treatment were recovered.

3. Results

3.1. Effect on Recipient Daphnia Tolerance and Performance in Terms of Body Size and Survival

The differences in body size were best explained by the microbiome inocula treatment (Table 1). Within the microbiome inocula treatments, *Daphnia* exposed to microbiome treatment 1 (M1, pre-exposure to WFCD + *B. daphniae*) were significantly smaller (Figure 3; Table S4) than animals exposed to microbiome treatment 2 (M2, pre-exposure to *P. ramosa* + *O. colligata* + *M. daphniae*). No significant differences between M1 and MC, and M2 and MC treatments were observed (Table S4, Figure 3). There was no role of the microbiome upon parasite re-exposure as no significant microbiome x parasite interaction was observed for body size and survival (Table 1).

Table 1. Significant results of the statistical analysis on the effect of genotype, microbiome treatment, parasite community treatment, and their interactions on body size, survival, and alpha diversity variables (OTU richness, Shannon entropy'). Degree of freedom (DF) is indicated per main and interaction effect. Obtained *p*-values were adjusted for multiple comparisons through the control of the false discovery rate (FDR).

		Adjusted *p*-Value				
	DF	Body Size	Survival	OTU Richness	Shannon Entropy'	Microbial Community Composition
Genotype	2		<0.001	<0.001	<0.001	0.014
Microbiome treatment	2	0.034				
Parasite treatment	2		0.002			
Genotype x Microbiome treatment	4			0.015	0.004	
Genotype x Parasite treatment	4		0.003			
Microbiome treatment x Parasite treatment	4				0.047	
Genotype x Microbiome treatment x Parasite treatment	8					

The differences in *Daphnia* survival were best explained by the genotype, parasite community treatment, and the *Daphnia* genotype × parasite community treatment interaction (Table 1; Table S4). Within the KNO genotype, *Daphnia* receiving the P1 or P2 treatment survived significantly shorter than *Daphnia* receiving the PC treatment (Figure 4; Table S4). Within the OM2 and T8 genotypes, none of the parasite treatments induced a significant reduction in *Daphnia* survival (Figure 4; Table S4). Within the P1 and P2 treatment, OM2 individuals survived significantly longer than the KNO and T8 individuals (Figure 4; Table S4). Within the PC treatment, KNO individuals survived significantly longer than T8 individuals (Figure 4; Table S4). Within the main genotype effect, OM2 individuals had a significantly higher survival compared with both T8 and KNO individuals (Table S4; Figure 4). Within the main parasite effect, individuals receiving the control treatment had a significantly higher survival compared with individuals receiving a parasite treatment (Table S4). There was no significant microbiome x parasite interaction (Table 1).

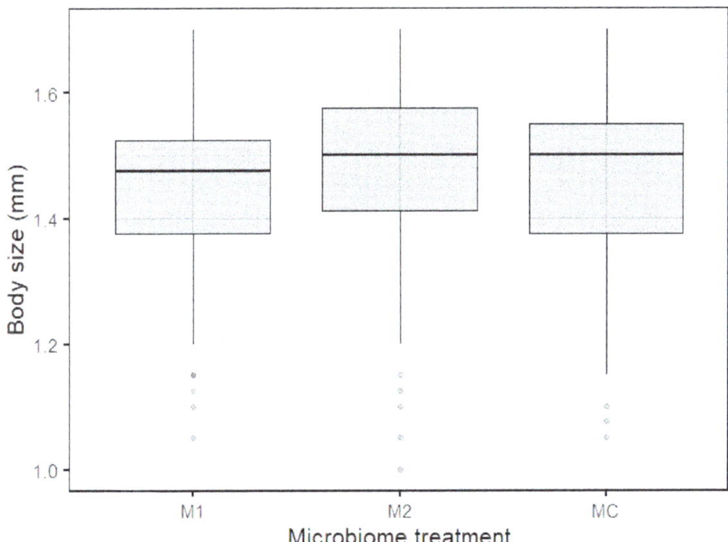

Figure 3. Boxplots of *Daphnia* body size exposed to the different microbiome treatments (M1, M2, and MC).

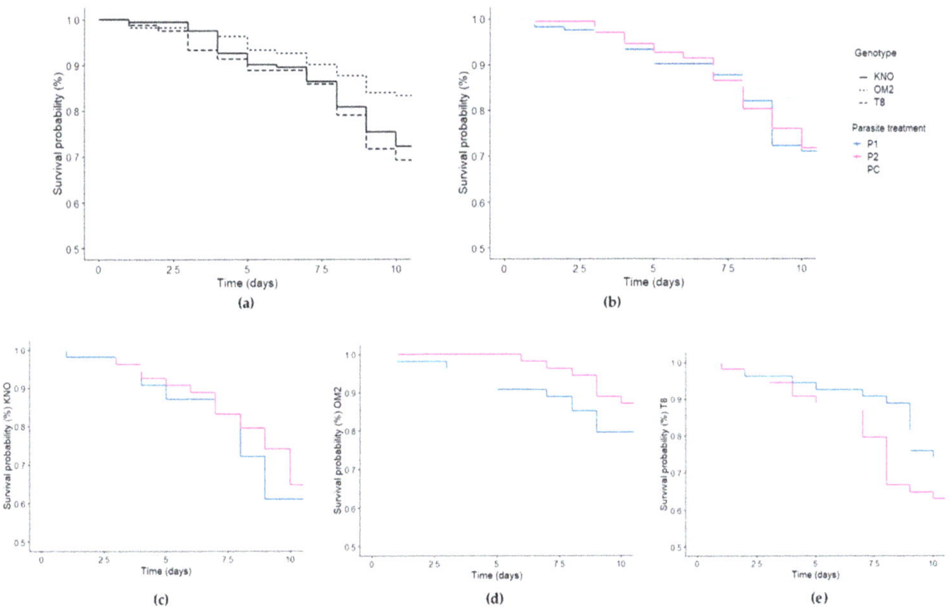

Figure 4. Survival curves for the *Daphnia* individuals for the different (**a**) genotypes, (**b**) parasite community treatment, (**c**) parasite community treatment for the KNO genotype, (**d**) parasite community treatment for the OM2 genotype, and (**e**) parasite community treatment for the T8 genotype. Line type indicates the genotype (solid = KNO, dotted = OM2, dashed = T8). Line color indicates the parasite treatment (dark blue = P1, pink = P2, light blue = PC).

3.2. Characterisation and Experimental Treatment Effects on Recipient Daphnia Gut Microbial Communities

Composition of the gut microbiomes: Eleven days after the microbial inoculation in the recipient *Daphnia*, the gut microbiomes were mainly dominated by Gammaproteobacteria (mean = 61.91%; sd = 29.00%), followed by Alphaproteobacteria (mean = 15.13%; sd = 21.87%), Bacteroidia (mean = 9.22%; sd = 12.55%), and Bacilli (mean = 8.40%; sd = 17.50%; Figure 5). The most dominant OTUs in all samples were Burkholderiaceae sp. (mean = 29.94%; sd = 28.01%; Gammaproteobacteria), *Methylobacterium* sp. (mean = 6.59%; sd = 15.64%; alphaproteobacteria), and *Streptococcus* sp. (mean = 6.09%; sd = 13.60%; Bacilli; Table S5). Gammaproteobacteria, more in particular, the Burkholderiaceae sp., were the most abundant taxa across all main treatments and two-way interactions. Interestingly, samples derived from M2 exposed individuals contained all sequenced classes, whereas all other treatments lacked one or more classes. Samples derived from OM2 individuals and P1 exposed individuals contained all sequenced classes, except Babeliae, Fimbriimonadia, and Gemmatimonadetes. Results of the Log2Fold test on the main effects on OTU level can be found in Table S6.

3.2.1. Alpha Diversity of Recipient Daphnia Gut Microbiomes

The differences in recipient OTU richness were best explained by *Daphnia* genotype and the *Daphnia* genotype × microbiome inoculum interaction (Figure 6; Table 1). Within the OM2 genotype, the microbial alpha diversity in the M1 treatment was significantly higher than the MC treatment (Table S4; Figure 6) before applying FDR correction. After applying FDR correction, no significant differences within all three genotypes were observed (Table S4). When examining the main genotype effect, OTU richness of the recipient guts of the KNO genotype was significantly lower compared with the OM2 genotype. No significant differences within the T8 genotype were observed: T8 showed a non-significant intermediate OTU richness. The differences in Shannon entropy′ were best explained by *Daphnia* genotype, *Daphnia* genotype × microbiome inoculum interaction, and microbiome inoculum × parasite community interaction (Table 1). Within the OM2 genotype, the M1 treatment was significantly higher compared than the MC treatment before correction (Table S4). After applying FDR correction, no significant differences within all three genotypes were observed (Table S4). No significant differences were observed within the microbiome inoculum × parasite community interaction for the Shannon entropy′. When examining the main genotype effect, the Shannon entropy′ for the KNO genotype was significantly lower compared with the OM2 genotype (Table S4). Results of analyses including the donor samples can be found in Figure S2.

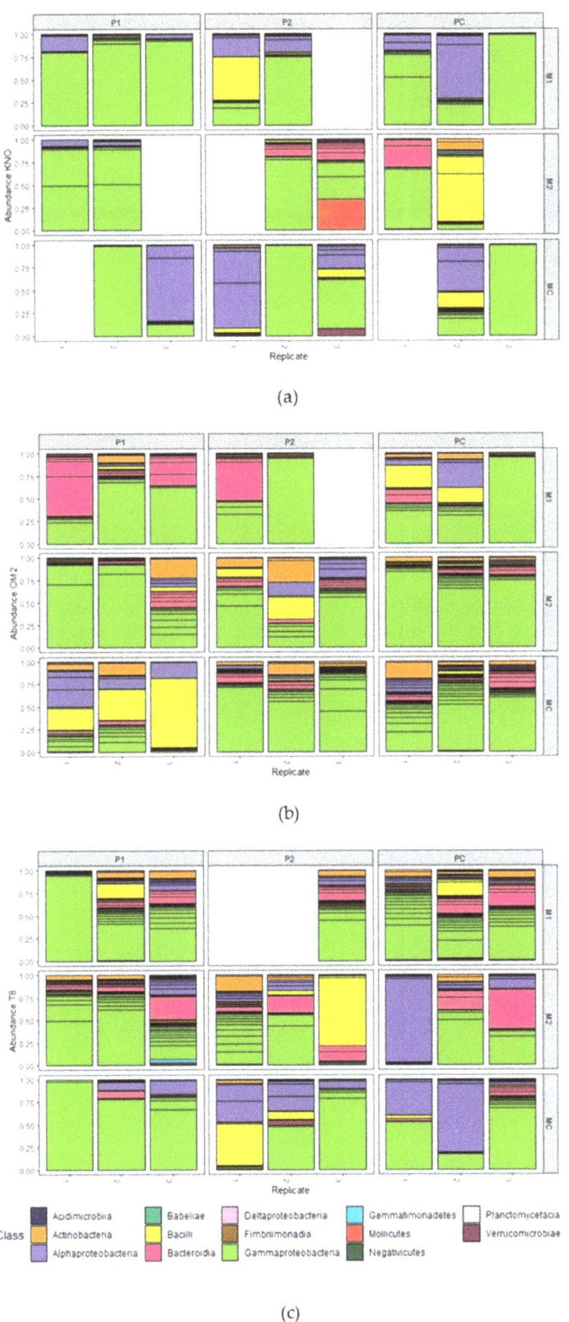

(a)

(b)

(c)

Figure 5. Gut microbial composition of *Daphnia* in the microbiome transplant experiment. OTU relative abundances are portrayed for (**a**) the KNO genotype, (**b**) the OM2 genotype, and (**c**) the T8 genotype. Recipient populations are grouped per multifactorial combination of microbiome and parasite community. Colors indicate the bacterial class. OTUs with a relative abundance lower than 1% are not included. Each bar represents one sample containing gut material of one to six recipients which was pooled for analysis.

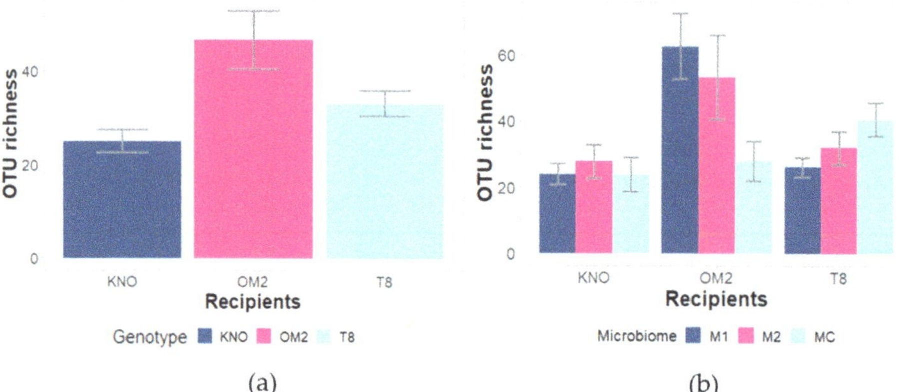

Figure 6. Effect of (**a**) genotype and (**b**) genotype × microbiome interaction for recipient samples on OTU richness. Error bars indicate standard error.

3.2.2. Treatment Effects on Gut Microbial Community Composition

Variation in gut microbial community composition was mainly explained by *Daphnia* genotype (Figure 7; Table 1). KNO differed significantly from OM2 and T8 (Table S4). Bray–Curtis ordinations demonstrated a complete overlap between OM2 and T8 individuals, indicating that the bacterial community of the OM2 and T8 genotype was similarly structured. KNO, on the other hand, showed complete overlap with OM2 and T8 individuals, but grouped closer together, suggesting a more homogeneous community between the different individuals in KNO than in OM2 and T8. Results of analyses including the donor samples can be found in Figure S3.

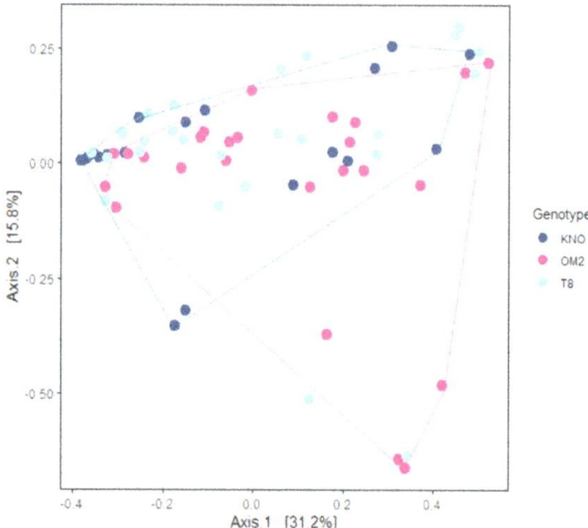

Figure 7. PCA of the gut microbial communities of recipients using weighted UniFrac distance for genotype. Colors indicate genotype.

4. Discussion

We here investigated the role of the gut microbiome on host tolerance upon parasite re-exposure through a microbiome adaptation and microbiome transplant experiment in germ-free individuals. We compared a pool of an iridovirus and a microsporidian parasite, and a pool of an endobacterium (*P. ramosa*, cfr. [38]) and two microsporidian parasites with a control. These gut microbiomes were obtained and used as microbial inocula (donors) in the microbiome transplant experiment. Germ-free *Daphnia* in our microbiome transplant experiment (recipients) received the gut microbiome inocula and were then exposed to the same three parasite treatments. Additionally, we examined intraspecific responses by including three *Daphnia* genotypes in our study. We here, thus, focus on a broader pool of parasites and on parasite-induced virulence effects than in Sison-Mangus et al. [38], who focused on parasite resistance for *P. ramosa*. Additionally, we included parasite re-exposure, whereas Sison-Mangus et al. [38] focused on initial parasite exposure. In conclusion, we aimed to examine the role of the gut microbiome on host tolerance upon parasite re-exposure in our recipients. We expected that the gut microbiome would affect *Daphnia* performance (survival and body size) in our recipients upon re-exposure to these parasite communities and that parasite specific responses could be detected. Three outcomes were thus possible upon re-exposure to the same parasite community: (1) No role of the microbiome, (2) a negative role of the microbiome, reflected in a reduced tolerance in case of parasite-mediated dysbiosis through the microbiome, or (3) a positive role of the microbiome, reflected in an improved tolerance of the gut microbiomes in case certain beneficial bacterial strains were selected for. Our results showed that (i) the gut microbiome plays no role in mediating *Daphnia* tolerance towards certain parasite communities upon re-exposure, (ii) the gut microbiome community affected *Daphnia* body size in a parasite specific way, (iii) it is the genotype rather than the microbiome affecting *Daphnia* survival, and (iv) *Daphnia* genotype plays an important role in shaping the gut microbiome community, both in alpha diversity and in the composition.

No microbiome × parasite effect was found on *Daphnia* body size, which implies that the gut microbiome has no substantial impact on body size upon parasite re-exposure. We did find that variation in *Daphnia* body size was dependent on the main gut microbiome inoculum treatment reflecting parasite specific responses (of the donors). Individuals exposed to gut microbiomes extracted from individuals exposed to the WFCD-*B. daphniae* pool (M1) were significantly smaller compared with individuals exposed to the gut microbiomes extracted from individuals exposed to the *P. ramosa*-*O. colligata*-*M. daphniae* pool (M2). This was surprising at first, as we assumed body size to be different in the recipients receiving the parasite pre-exposed gut microbiomes compared with the individuals receiving the control gut microbiomes, which would reflect parasite induced dysbiosis. This difference in body size between M1- and M2-exposed individuals could possibly be due to the presence of particular microbial communities (more or less diverse) or of particular bacterial strains associated with these parasite communities. Interestingly, *P. ramosa* (present in M2) is known to induce gigantism, i.e., an increased body size in infected *Daphnia* [37,60]. Alternatively, the presence of the less virulent (sometimes even mutualistic) microsporidian gut parasites may have induced protection in the P2 and M2 treatments, an effect which has been suggested in [61], especially in low food quality conditions. The amplicon sequencing results should reveal possible links between *Daphnia* body size and gut microbial communities. Studies have shown that an increased diverse bacterial community in the gut can be associated with a larger body size in *Daphnia*, e.g., [50,62]. Similarly, reduced growth and metabolic capacities due to parasite-induced reduced gut bacterial diversity has been observed in other species, such as mice [63]. The amplicon sequencing analysis reflected a lower alpha diversity in the inocula, but did not show a significantly reduced alpha diversity in the M1 treatment compared with the M2 and MC treatment in the recipients. Our amplicon analysis did reveal that M1 exposed individuals had a significantly lower proportion of, e.g., *Methylobacterium* sp. (Gammaproteobacteria) in their gut compared with M2 and MC exposed individuals. These body size related results imply that the gut microbiome plays an additional role in the already complex food web *Daphnia* is a part of in freshwater populations, especially because multiple infections often occur [52]. Body size is a critical trait in shaping consumer–prey interactions. A priori parasite exposure will alter the gut microbial community, impacting *Daphnia* body size, and as such impact, e.g., grazing and predation by zooplanktivorous fish.

No microbiome × parasite effect was found on *Daphnia* survival, which implies the gut microbiome has no substantial impact on survival size upon parasite re-exposure. Additionally, no main microbiome effect was detected. Our results on *Daphnia* performance thus suggest that the microbiome plays no role on *Daphnia* tolerance upon parasite re-exposure. This is in contrast with studies on, e.g., bumblebees which show a reciprocal role between the gut microbiome and parasite exposure [9,11,31]. The study by Sison-Mangus et al. [38], however, also found no evidence on a gut microbial-mediated resistance against the parasite *P. ramosa*. It appears that host regulated defense against parasites in *D. magna* is mediated in lesser extent or even completely independent from the gut microbial community. Interestingly, these results also suggest that exposure can alter gut microbial communities, but not in a selective manner resulting in, e.g., an improved tolerance towards certain parasites. Even though an increasing amount of studies hints at a protective role of the gut microbiome when encountering parasite infection, our results attenuate its assumed protective role. In our study, host tolerance is dependent on host genotype [64], independent of the gut microbial community. We were also interested in intraspecific differences, which is why we incorporated three genotypes in this study. We found a strong *Daphnia* genotype effect on survival as the genotype × parasite community interaction, as well as the main genotype effect significantly impacted *Daphnia* survival. These results are in line with previous studies which report strong genotype × genotype host–parasite [47,65,66] or *Daphnia* genotype × parasite species interactions [64]. It appears that, in our study, the KNO genotype thrives best under control conditions, whereas the OM2 genotype suffered the least from the exploitation by the parasite communities compared with the KNO and T8 genotypes.

Host genotype does not only mediate host tolerance in terms of survival, but also appears to be a strong determinant of the *Daphnia* recipient gut microbiome. Firstly, we found a significant effect on alpha diversity (OTU richness in our study), both from the genotype × microbiome interaction as the main genotype effect. Secondly, we found a significant main genotype effect on the microbial community composition of the *Daphnia* gut. These effects can be attributed to different immune system pathway expressions of the *Daphnia* genotypes, e.g., innate immune system genes expression or the production of antimicrobial peptides structuring the gut microbiome [67], or due to differences in molting capacities affecting bacterial establishment [68], amongst others. The importance of the host genotype in shaping the gut microbial composition has also been suggested by correlational studies on humans and mice in which correlations between the gut microbiome and genes associated with diet, innate immunity, vitamin D receptors, and metabolism were revealed [69,70]. Host genotype can also reciprocally influence the microbial community composition. Genotype-specific gut microbiomes can also be found in, e.g., sponges [71], corals [72], and mice [73]. These studies also revealed a stronger role of host genotype compared with the environment [72] and sex [73] in driving gut microbial variation. Further investigations are necessary to get insights into the mechanisms behind these genotype-specific gut microbiomes. Interestingly with this respect is that *Daphnia* genotypes can display different grazing behavior, resulting in different feeding patterns and consumed bacteria [42,43,74,75]. Our results suggest that the *Daphnia* genotype is a stronger determinant of gut microbial alpha diversity than the pre-exposure of the microbial inocula. The main genotype effect reveals that the gut microbial composition was structured by the presence of particular OTUs. OM2 individuals differed significantly from both the KNO and T8 individuals in relative abundance of the OTU *Methylobacterium* sp., which has the highest abundance in the OM2 genotype. This increased proportion of the *Methylobacterium* genus is also observed in M2 individuals, which had a higher body size compared with M1 individuals. Mono-association experiments with *Methylobacterium* sp. could give us more insights in its role of this OTU in *Daphnia* functioning. The amplicon sequencing analyses revealed an effect of the genotype on the gut microbial communities using Bray–Curtis dissimilarity and weighted UniFrac distance.

5. Conclusions

In conclusion, we can state that the microbiome plays no role in mediating *Daphnia* tolerance upon parasite re-exposure in our study. Our study does suggest an impact of the gut microbial community on body size, reflecting parasite specific responses. We found that it was rather the *Daphnia* genotype which mediated *Daphnia* tolerance, as survival upon re-exposure was mainly determined by the host genotype. Additionally, our study suggests a host genotype-specific gut bacterial community on alpha diversity, microbial community composition, and on the presence of specific strains.

Supplementary Materials: The following are available online at https://www.mdpi.com/2073-4425/12/1/70/s1, Table S1: Overview of the number of pooled recipient guts per microbial sample, Table S2: Overview results for Pearson correlation between the number of guts per sample and OTU richness or Shannon entropy'. Raw and adjusted (adjusted for multiple comparisons through the control of the false discovery rate (FDR)) *p*-values are given, Table S3: Overview forward and reverse primers used for the internal PCR, Table S4: Results of post-hoc analyses on body size, survival, OTU richness, Shannon entropy', and microbial community composition for the significant results of the statistical analysis (see Table 1). Raw and adjusted (adjusted for multiple comparisons through the control of the false discovery rate (FDR)) *p*-values are given for the results on *Daphnia* gut microbial communities, Table S5: Overview of relative abundances of the 40 most common OTUs in *Daphnia* guts from the microbiome transplant experiment. Abundances were calculated on rarefied data. Sd: Standard deviation, Table S6: Results Deseq analysis on class level and OTU level between main effects, Table S7: Significant results of the statistical analysis on the effect of genotype, microbiome treatment, parasite community treatment, and their interactions on body size, survival, and alpha-diversity variables (OTU richness, Shannon entropy'). Obtained P-values were adjusted for multiple comparisons through the control of the false discovery rate (FDR). Significant data ($p < 0.05$) is indicated in bold. Highly significant data ($p < 0.001$) are underlined. Raw and adjusted (adjusted for multiple comparisons through the control of the false discovery rate (FDR)) *p*-values are given for the results on *Daphnia* gut microbial communities. Figure S1: Effect of the genotype x parasite interaction on OTU richness. Colors indicate the different genotypes. Error bars indicate standard error. Figure S2: Effect of microbiome (M1 and M2) for donor and recipient samples on OTU richness. Error bars indicate standard error. Colors indicate the different microbiome treatments. OTU richness in the M1 inoculum (mean = 38.333, sd = 13.051) was, on average, lower compared with the M2 inoculum (OTU richness = 47.000). Figure S3: PCA of the gut microbial communities of recipients using weighted UniFrac distance for the donors (P1 and P2) and matching recipients (M1 x P1, M2 x P2) using weighted UniFrac distance for donor/recipient type and parasite treatment. Analyses on donor (P1 and P2) and matching recipient (M1 x P1, M2 x P2) bacterial communities showed a significant difference in structure for P1 ($p = 0.005$; R2 = 0.147), but not for P2 ($p = 0.783$; R2 = 0.330). Bray–Curtis ordinations revealed that both P1 and P2 showed complete segregation between donors and recipients (Figure 7b), indicating that the donors and recipients for both the P1 and P2 treatment were differently structured (Figure 7b), however non-significant for P2.

Author Contributions: Conceptualization, L.B. and E.D.; methodology, L.B., S.H., and E.D.; software, L.B.; validation, L.B., S.H., S.A.J.D., and E.D.; formal analysis, L.B. and S.A.J.D.; investigation, L.B.; resources, E.D.; data curation, L.B.; writing—original draft preparation, L.B.; writing—review and editing, L.B., S.H., S.A.J.D., and E.D.; visualization, L.B.; supervision, E.D.; project administration, E.D.; funding acquisition, E.D. All authors have read and agreed to the published version of the manuscript.

Funding: This research was funded by the KU Leuven research project C16/17/002 and the FWO projects G092619N and G06216N.

Institutional Review Board Statement: Not applicable.

Informed Consent Statement: Not applicable.

Data Availability Statement: The datasets generated for this study can be found in the NCBI, under accession number PRJNA688519.

Acknowledgments: The authors thank Isabel Vanoverberghe for the assistance with the MiSeq library preparation, and Emilie Macke for the assistance with the survival analysis and setting up the experimental design.

Conflicts of Interest: The authors declare no conflict of interest.

Appendix A

Appendix A.1. Daphnia Magna Genotypes

Three genotypes of *D. magna* were utilized in our transplant experiment: OM2 11.3; KNO 15.4; and T8 (further referred to as OM2, KNO and T8). The genotype OM2 was originally isolated from "Oude Meren, Abdij van't park" from Leuven in Belgium (50°51′47.82″ N, 04°43′05.16″ E). The genotype KNO was originally isolated from a small pond (350 m^2) from Knokke in Belgium (51°20′05.62″ N, 03°20′53.63″ E). The genotype T8 was originally isolated from a shallow manmade pond from Oud Heverlee in Belgium (50°50′ N, 4°39′ E). All genotypes were maintained in the lab under standard stock conditions for several years prior to the transplant experiment. All genotypes were cultured in filtered tap water at a temperature of 19 °C (±1 °C) and under a 16:8 h light:dark cycle in 2L jars. They were fed every other day with a saturating amount of *Chlorella vulgaris*. Medium was refreshed every week. As all genotypes were hatched in the laboratory from resting eggs, and maintained in the laboratory for several years, it is unlikely that the experimental genotypes contain parasites and bacteria from the pond of origin.

Appendix A.2. Preparation Maternal Lines

For each genotype, three maternal lines were set up. Five individuals per maternal line were transferred in 500 mL filtered tap water and fed every day with a saturating amount of *Chlorella vulgaris*. Medium was refreshed every week. Juveniles from the first brood were discarded. Twenty juveniles from the second brood per maternal line were transferred in 2l jars of filtered tap water. Maternal *Daphnia* were discarded after releasing their second brood. This process was repeated for the every new batch of 2nd brood juveniles up to seven generations.

Appendix A.3. Preparation Axenic Daphnia

Recipient *Daphnia* were obtained from the maternal lines of OM2, KNO, and T8. Females carrying parthenogenetic eggs (second brood) were dissected under a stereomicroscope. Eggs which were no longer than 24 h deposited in the brood chamber were collected in sterile filtered tap water (n = 100 eggs per maternal line). These eggs were then disinfected under a laminar flow hood following an adjusted protocol of Callens et al. [8]. Eggs were placed in 6 mL of a 0.10% gluteraldehyde solution and gently agitated for 10 min. After this disinfection step, two washing steps were performed in which the eggs were transferred to sterile filtered tap water for each 10 min. Eggs were then transferred to six-well (cell culture) sterile plates, each well containing 6 mL of sterile filtered tap water, and incubated at 20 ± 0.5 °C. Eggs were allowed to hatch during 72 h under sterile conditions, and the resulting germ-free juveniles were used as recipients in the transplant experiment.

References

1. Tremaroli, V.; Bäckhed, F. Functional interactions between the gut microbiota and host metabolism. *Nature* **2012**, *489*, 242–249. [CrossRef] [PubMed]
2. Macke, E.; Tasiemski, A.; Massol, F.; Callens, M.; Decaestecker, E. Life history and eco-evolutionary dynamics in light of the gut microbiota. *Oikos* **2017**, *88*, 508–531. [CrossRef]
3. Monnin, D.; Jackson, R.; Kiers, E.T.; Bunker, M.; Ellers, J.; Henry, L.M. Parallel evolution in the integration of a co-obligate aphid symbiosis. *Curr. Biol.* **2020**, *30*, 1949–1957. [CrossRef] [PubMed]
4. Bäckhed, F.; Ley, R.E.; Sonnenburg, J.L.; Peterson, D.A.; Gordon, J.I. Host-bacterial mutualism in the human intestine. *Science* **2005**, *307*, 1915–1920. [CrossRef] [PubMed]
5. Amato, K.R. Incorporating the gut microbiota into models of human and non-human primate ecology and evolution. *Am. J. Phys. Anthr.* **2016**, *159*, 196–215. [CrossRef]
6. Belkaid, Y.; Hand, T.W. Role of the microbiota in immunity and inflammation. *Cell* **2014**, *157*, 121–141. [CrossRef] [PubMed]
7. Sison-Mangus, M.P.; Mushegian, A.A.; Ebert, D. Water fleas require microbiota for survival, growth and reproduction. *ISME J.* **2014**, *9*, 59–67. [CrossRef]

8. Callens, M.; Macke, E.; Muylaert, K.; Bossier, P.; Lievens, B.; Waud, M.; Decaestecker, E. Food availability affects the strength of mutualistic host-microbiota interactions in *Daphnia magna*. *ISME J.* **2016**, *10*, 911–920. [CrossRef]

9. Koch, H.; Schmid-Hempel, P. Socially transmitted gut microbiota protects bumble bees against an intestinal parasite. *Proc. Natl. Acad. Sci. USA* **2011**, *108*, 19288–19292. [CrossRef]

10. Ford, S.A.; King, K.C. Harnessing the power of defensive microbes: Evolutionary implications in nature and disease control. *PLoS Pathog.* **2016**, *12*, e1005465. [CrossRef]

11. Mockler, B.K.; Kwong, W.K.; Moran, N.A.; Koch, H. Microbiome structure influences infection by the parasite *Crithidia bombi* in bumble bees. *Appl. Environ. Microbiol.* **2018**, *84*, e02335-17. [CrossRef] [PubMed]

12. Hayes, K.S.; Bancroft, A.J.; Goldrick, M.; Portsmouth, C.; Roberts, I.S.; Grencis, R.K. Exploitation of the intestinal microflora by the parasitic nematode *Trichuris muris*. *Science* **2010**, *328*, 1391–1394. [CrossRef] [PubMed]

13. Zaiss, M.M.; Harris, N.L. Interactions between the intestinal microbiome and helminth parasites. *Parasite Immunol.* **2015**, *38*, 5–11. [CrossRef] [PubMed]

14. Fredensborg, B.L.; Fossdal, I.; Kálvalíð, I.; Johannesen, T.B.; Stensvold, C.R.; Nielsen, H.V.; Kapel, C.M.O. Parasites modulate the gut-microbiome in insects: A proof-of-concept study. *PLoS ONE* **2020**, *15*, e0227561. [CrossRef] [PubMed]

15. Gaulke, C.A.; Martins, M.L.; Watral, V.G.; Humphreys, I.R.; Spagnoli, S.T.; Kent, M.L.; Sharpton, T.J. A longitudinal assessment of host-microbe-parasite interactions resolves the zebrafish gut microbiome's link to *Pseudocapillaria tomentosa* infection and pathology. *Microbiome* **2019**, *7*, 1–16. [CrossRef] [PubMed]

16. Charania, R.; Wade, B.E.; McNair, N.N.; Mead, J.R. Changes in the microbiome of *Cryptosporidium*-infected mice correlate to differences in susceptibility and infection levels. *Microorganisms* **2020**, *8*, 879. [CrossRef]

17. McDermott, A.J.; Huffnagle, G.B. The microbiome and regulation of mucosal immunity. *Immunology* **2014**, *142*, 24–31. [CrossRef]

18. Bruno, R.; Maresca, M.; Canaan, S.; Cavalier, J.; Mabrouk, K.; Boidin-Wichlacz, C.; Olleik, H.; Zeppilli, D.; Brodin, P.; Massol, F.; et al. Worms' Antimicrobial Peptides. *Mar. Drugs* **2019**, *17*, 512. [CrossRef]

19. Paller, A.S.; Kong, H.H.; Seed, P.; Naik, S.; Scharschmidt, T.C.; Gallo, R.L.; Luger, T.; Irvine, A.D. The microbiome in patients with atopic dermatitis. *J. Allergy Clin. Immunol.* **2019**, *143*, 26–35. [CrossRef]

20. Rolhion, N.; Chassaing, B. When pathogenic bacteria meet the intestinal microbiota. *Philos. Trans. R. Soc. B Biol. Sci.* **2016**, *371*, 20150504. [CrossRef]

21. Ignacio, A.; Terra, F.F.; Watanabe, I.K.M.; Basso, P.J.; Câmara, N.O.S. Chapter 13—Role of the Microbiome in Intestinal Barrier Function and Immune Defense. In *Microbiome and Metabolome in Diagnosis, Therapy, and Other Strategic Applications*; Academic Press: Cambridge, MA, USA, 2019; pp. 127–138. [CrossRef]

22. Bass, D.; Stentiford, G.D.; Wang, H.; Koskella, B.; Tyler, C.R. The Pathobiome in Animal and Plant Diseases. *Trends Ecol. Evol.* **2020**, *34*, 996–1008. [CrossRef]

23. Contijoch, E.J.; Britton, G.J.; Yang, C.; Mogno, I.; Li, Z.; Ng, R.; Llewellyn, S.R.; Hira, S.; Johnson, C.; Rabinowitz, K.M.; et al. Gut microbiota density influences host physiology and is shaped by host and microbial factors. *eLife* **2019**, *8*, e40553. [CrossRef] [PubMed]

24. Leung, J.M.; Graham, A.L.; Knowles, S.C.L. Parasite-microbiota interactions within the vertebrate gut: Synthesis through an ecological lens. *Front. Microbiol.* **2018**, *9*, 843. [CrossRef] [PubMed]

25. Stecher, B.; Maier, L.; Hardt, W.D. 'Blooming' in the gut: How dysbiosis might contribute to pathogen evolution. *Nat. Rev. Microbiol.* **2013**, *11*, 277–284. [CrossRef] [PubMed]

26. Guinane, C.M.; Cotter, P.D. Role of the gut microbiota in health and chronic gastrointestinal disease: Understanding a hidden metabolic organ. *Ther. Adv. Gastroenterol.* **2013**, *6*, 295–308. [CrossRef]

27. Stothart, M.R.; Bobbie, C.B.; Schulte-Hostedde, A.I.; Boonstra, R.; Palme, R.; Mykytczuk, N.C.S.; Newman, A.E.M. Stress and the microbiome: Linking glucocorticoids to bacterial community dynamics in wild red squirrels. *Biol. Lett.* **2016**, *12*. [CrossRef]

28. Flandroy, L.; Poutahidis, T.; Berg, G.; Clarke, G.; Dao, M.-C.; Decaestecker, E.; Furman, E.; Haahtela, T.; Massart, S.; Plovier, H.; et al. The impact of human activities and lifestyles on the interlinked microbiota and health of humans and of ecosystems. *Sci. Total Environ.* **2018**, *627*, 1018–1038. [CrossRef]

29. Ijay-Kumar, M.; Aitken, J.D.; Carvalho, F.A.; Cullender, T.C.; Mwangi, S.; Srinivasan, S.; Sitaraman, S.V.; Knight, R.; Ley, R.E.; Gewirtz, A.T. Metabolic syndrome and altered gut Microbiota in mice Lacking Toll-like receptor 5. *Science* **2010**, *328*, 228–231. [CrossRef]

30. Chow, J.; Tang, H.; Mazmanian, S.K. Pathobionts of the gastrointestinal microbiota and inflammatory disease. *Curr. Opin. Immunol.* **2011**, *23*, 473–480. [CrossRef]

31. Koch, H.; Schmid-Hempel, P. Gut microbiota instead of host genotype drive the specificity in the interaction of a natural host-parasite system. *Ecol. Lett.* **2012**, *15*, 1095–1103. [CrossRef]

32. Schulenburg, H.; Félix, M. The natural biotic environment of *Caenorhabditis elegans*. *Genetics* **2017**, *206*, 55–86. [CrossRef] [PubMed]

33. Jaenike, J.; Unckless, R.; Cockburn, S.N.; Boelio, L.M.; Perlman, S.J. Adaptation via symbiosis: Recent spread of a *Drosophila* defensive symbiont. *Science* **2010**, *329*, 212–215. [CrossRef] [PubMed]

34. King, W.L.; Jenkins, C.; Seymour, J.R.; Labbate, M. Oyster disease in a changing environment: Decrypting the link between pathogen, microbiome and environment. *Mar. Environ. Res.* **2019**, *143*, 124–140. [CrossRef]

35. Kwong, W.K.; Moran, N.A. Gut microbial communities of social bees. *Nat. Rev. Microbiol.* **2016**, *14*, 374–384. [CrossRef] [PubMed]

36. Raymann, K.; Moran, N.A. The role of the gut microbiome in health and disease of adult honey bee workers. *Curr. Opin. Insect Sci.* **2018**, *26*, 97–104. [CrossRef] [PubMed]

37. Ebert, D. *Ecology, Epidemiology, and Evolution of Parasitism in Daphnia [Internet]*; National Library of Medicine (US), National Center for Biotechnology Information: Bethesda, MD, USA, 2005. Available online: http://www.ncbi.nlm.nih.gov/entrez/query.fcgi?db=Books (accessed on 30 December 2020).

38. Sison-Mangus, M.; César, M.J.A.M.; Ebert, D. Host genotype-specific microbiota do not influence the susceptibility of *D. magna* to a bacterial pathogen. *Sci. Rep.* **2018**, *8*, 9407–9410. [CrossRef] [PubMed]

39. Macke, E.; Callens, M.; De Meester, L.; Decaestecker, E. Host-genotype dependent gut microbiota drives zooplankton tolerance to toxic cyanobacteria. *Nat. Commun.* **2017**, *8*, 1–13. [CrossRef]

40. Macke, E.; Callens, M.; Massol, F.; Vanoverberghe, I.; De Meester, L.; Decaestecker, E. Diet and genotype of an aquatic invertebrate affect the composition of free-living microbial communities. *Front. Microbiol.* **2020**, 11. [CrossRef]

41. Massol, F.; Macke, E.; Callens, M.; Decaestecker, E. A methodological framework to analyze determinants of host-microbiota networks, with an application to the relationships between *Daphnia magna*'s gut microbiota and bacterioplankton. *J. Anim. Ecol.* **2020**, 1–18. [CrossRef]

42. Mushegian, A.A.; Arbore, R.; Walser, J.C.; Ebert, D. Environmental Sources of Bacteria and Genetic Variation in Behavior Influence Host-Associated Microbiota. *Appl. Environ. Microbiol.* **2019**, *85*, e01547-18. [CrossRef]

43. Callens, M.; De Meester, L.; Muylaert, K.; Mukherjee, S.; Decaestecker, E. The bacterioplankton community composition and a host genotype dependent occurrence of taxa shape the *Daphnia magna* gut bacterial community. *FEMS Microbiol. Ecol.* **2020**, 96. [CrossRef]

44. Toenshoff, E.R.; Fields, P.D.; Bourgeouis, Y.X.; Ebert, D. The End of a 60-year Riddle: Identification and Genomic Characterization of an Iridovirus, the Causative Agent of White Fat Cell Disease in Zooplankton. *G3 Genes Genomes Genet.* **2018**, *8*, 1259–1272. [CrossRef] [PubMed]

45. Refardt, D.; Decaestecker, E.; Johnson, P.T.J.; Vávra, J. Morphology, Molecular Phylogeny, and Ecology of *Binucleata daphniae* n. g.; n. sp. (Fungi: Microsporidia), a Parasite of Daphnia magna Straus, 1820 (Crustacea: Branchiopoda). *J. Eukaryot. Biol.* **2008**, *55*, 393–408. [CrossRef] [PubMed]

46. Ebert, D.; Rainey, P.; Embley, M.T.; Scholz, D. Development, life cycle, ultrastructure and phylogenetic position of *Pasteuria ramosa* Metchnikoff 1888: Rediscovery of an obligate endoparasite of *Daphnia magna* Straus. *Philos. Trans. R. Soc. B Biol. Sci.* **1996**, *351*, 1689–1701. [CrossRef]

47. Decaestecker, E.; Gaba, S.; Raeymaekers, J.A.M.; Stoks, R.; Van Kerckhoven, L.; Ebert, D.; De Meester, L. Host–parasite 'Red Queen' dynamics archived in pond sediment. *Nature* **2007**, *450*, 870–873. [CrossRef] [PubMed]

48. Larsson, R.J.I.; Ebert, D.; Vávra, J. Ultrastructural Study and Description of *Ordopsora colligata* gen. et sp. nov. (Microspora, Ordosporidae fam. Nov.), a new microsporidian parasite of *Daphnia magna* (Crustacea, Cladocera). *Eur. J. Protistol.* **1997**, *33*, 432–443. [CrossRef]

49. Haag, K.L.; James, T.Y.; Pombert, J.-F.; Larsson, R.; Schaer, T.M.M.; Refardt, D.; Ebert, D. Evolution of a morphological novelty occurred before genome compaction in a lineage of extreme parasites. *Proc. Natl. Acad. Sci. USA* **2014**, *111*, 15480–15485. [CrossRef]

50. Houwenhuyse, S.; Stoks, R.; Mukherjee, S.; Decaestecker, E. Locally adapted gut microbiomes mediate host stress tolerance. *ISME J.* **2021**. under minor revision.

51. Harold, S.; Henderson, C.; Baguette, M.; Bonsall, M.B.; Hughes, D.P.; Settele, J. Ecologyimage competition 2014: The winning images. *BMC Ecol.* **2014**, *14*, 24. [CrossRef]

52. Decaestecker, E.; Declerck, S.; De Meester, L.; Ebert, D. Ecological Implications of Parasites in Natural *Daphnia* Populations. *Oecologia* **2005**, *144*, 382–390. [CrossRef] [PubMed]

53. Guillard, R.R.L.; Lorenzen, C.J. Yellow-Green Algae with Chlorophyllidec. *J. Phycol.* **1972**, *8*, 10–14.

54. R Core Team. *R: A Language and Environment for Statistical Computing*; R Foundation for Statistical Computing: Vienna, Austria, 2020; Available online: https://www.R-project.org/ (accessed on 30 December 2020).

55. Callahan, B.J.; Sankaran, K.; Fukuyama, J.A.; McMurdie, P.J.; Holmes, S.P. Bioconductor Workflow for Microbiome Data Analysis: From raw reads to community analyses [version 2; peer review: 3 approved]. *F1000Research* **2016**, 5. [CrossRef]

56. McMurdie, P.J.; Holmes, S. Waste Not, Want Not: Why rarefying microbiome data is inadmissible. *PLoS Comput. Biol.* **2014**, *10*, e1003531. [CrossRef] [PubMed]

57. Callahan, B.J.; McMurdie, P.J.; Rosen, M.J.; Han, A.W.; Johnson, A.J.A.; Holmes, S.P. DADA2: High-resolution sample inference from Illumina amplicon data. *Nat. Methods* **2016**, *13*, 581–583. [CrossRef] [PubMed]

58. Jari Oksanen, F.; Blanchet, G.; Friendly, M.; Kindt, R.; Legendre, P.; McGlinn, D.; Minchin, P.R.; O'Hara, R.B.; Simpson, G.L.; Solymos, P.; et al. Vegan: Community Ecology Package. 2016. Available online: https://cran.r-project.org/package=vegan (accessed on 30 December 2020).

59. Borcard, D.; Gillet, F.; Legendre, P. *Numerical Ecology with R*; Springer Science + Business Media: New York, NY, USA, 2011.

60. Cressler, C.E.; Nelson, W.A.; Day, T.; McCauley, E. Starvation reveals the cause of infection-induced castration and gigantism. *Proc. R. Soc. B Biol. Sci.* **2014**, 281. [CrossRef]

61. Lange, B.; Reuter, M.; Ebert, D.; Muylaert, K.; Decaestecker, E. Diet quality determines interspecific parasite interactions in host populations. *Ecol. Evol.* **2014**, *4*, 3093–3102. [CrossRef]

62. Callens, M.; Watanabe, H.; Kato, Y.; Miura, J.; Decaestecker, E. Microbiota inoculum composition affects holobiont assembly and host growth in *Daphnia*. *Microbiome* **2018**, *6*, 56. [CrossRef]
63. Houlden, A.; Hayes, K.S.; Bancroft, A.J.; Worthington, J.J.; Wang, P.; Grencis, R.K.; Roberts, I.S. Chronic *Trichuris muris* Infection in C57BL/6 mice causes significant changes in host microbiota and metabolome: Effects reversed by pathogen clearance. *PLoS ONE* **2015**, *10*, e0125945. [CrossRef]
64. Decaestecker, E.; Vergote, A.; Ebert, D.; De Meester, L. Evidence for strong host clone–parasite species interactions in the *Daphnia* microparasite system. *Evolution* **2003**, *57*, 784–792. [CrossRef]
65. Carius, H.J.; Little, T.J.; Ebert, D. Genetic variation in a host-parasite association potential for coevolution and frequency-dependent selection. *Evolution* **2001**, *55*, 1136–1145. [CrossRef]
66. Little, T.J.; Watt, K.; Ebert, D. Parasite-host specificity: Experimental studies on the basis of parasite adaptation. *Evolution* **2006**, *60*, 31–38. [CrossRef] [PubMed]
67. Decaestecker, E.; Labbé, P.; Ellegaard, K.; Allen, J.E.; Little, T.J. Candidate innate immune system gene expression in the ecological model Daphnia. *Dev. Comp. Immunol.* **2011**, *35*, 1068–1077. [CrossRef] [PubMed]
68. Duneau, D.; Luijckx, P.; Ben-Ami, F.; Laforsch, C.; Ebert, D. Resolving the infection process reveals striking differences in the contribution of environment, genetics and phylogeny to host-parasite interactions. *BMC Biol.* **2011**, *9*, 11. [CrossRef] [PubMed]
69. Bonder, M.J.; Kurilshikov, A.; Tigchelaar-Feenstra, E.; Mujagic, Z.; Imhann, F.; Vila, A.V.; Deelen, P.; Vatanen, T.; Schirmer, M.; Smeekens, S.P.; et al. The effect of host genetics on the gut microbiome. *Nat. Genet.* **2016**, *48*, 1407–1412. [CrossRef] [PubMed]
70. Kurilshikov, A.; Wijmenga, C.; Fu, J.; Zhernakova, A. Host Genetics and Gut Microbiome: Challenges and Perspectives. *Trends Immunol.* **2017**, *38*, 633–647. [CrossRef]
71. Griffiths, S.M.; Antwis, R.E.; Lenzi, L.; Lucaci, A.; Behringer, D.C.; Butler, M.J.; Preziosi, R.F. Host genetics and geography influence microbiome composition in the sponge *Ircinia campana*. *J. Anim. Ecol.* **2019**, *88*, 1684–1695. [CrossRef]
72. Glasl, B.; Smith, C.E.; Bourne, D.G.; Webster, N.S. Disentangling the effect of host-genotype and environment on the microbiome of the coral *Acropora tenuis*. *PeerJ* **2019**, *7*, e6377. [CrossRef] [PubMed]
73. Kovacs, A.; Ben-Jacob, N.; Tayem, H.; Halperin, E.; Iraqi, F.A.; Gophna, U. Genotype is a stronger determinant than sex of the mouse gut microbiota. *Microb. Ecol.* **2011**, *61*, 423–428. [CrossRef]
74. Decaestecker, E.; De Meester, L.; Ebert, D. In deep trouble: Habitat selection constrained by multiple enemies in zooplankton. *Proc. Natl. Acad. Sci. USA* **2002**, *99*, 5481–5485. [CrossRef] [PubMed]
75. Arbore, R.; Andras, J.P.; Routtu, J.; Ebert, D. Ecological genetics of sediment browsing behaviour in a planktonic crustacean. *J. Evol. Biol.* **2016**, *29*, 1999–2009. [CrossRef]

MDPI

St. Alban-Anlage 66

4052 Basel

Switzerland

Tel. +41 61 683 77 34

Fax +41 61 302 89 18

www.mdpi.com

Genes Editorial Office

E-mail: genes@mdpi.com

www.mdpi.com/journal/genes